T0318754

Neurosurgical Neuropsychology

Neurosurgical Neuropsychology

The Practical Application of Neuropsychology in the Neurosurgical Practice

Edited by

Caleb M. Pearson
Departments of Neurology, Neurosurgery, and Psychiatry, University of Kansas Medical Center, Kansas City, KS, United States

Eric Ecklund-Johnson
Departments of Neurology and Psychiatry, University of Kansas Medical Center, Kansas City, KS, United States

Shawn D. Gale
Department of Psychology and Neuroscience Center, Brigham Young University, Provo, UT, United States

ACADEMIC PRESS

An imprint of Elsevier

Academic Press is an imprint of Elsevier
125 London Wall, London EC2Y 5AS, United Kingdom
525 B Street, Suite 1650, San Diego, CA 92101, United States
50 Hampshire Street, 5th Floor, Cambridge, MA 02139, United States
The Boulevard, Langford Lane, Kidlington, Oxford OX5 1GB, United Kingdom

British Library Cataloguing-in-Publication Data
A catalogue record for this book is available from the British Library

Library of Congress Cataloging-in-Publication Data
A catalog record for this book is available from the Library of Congress

ISBN: 978-0-12-809961-2

For Information on all Academic Press publications
visit our website at https://www.elsevier.com/books-and-journals

Working together
to grow libraries in
developing countries

www.elsevier.com • www.bookaid.org

Publisher: Nikki Levy
Acquisition Editor: Melanie Tucker
Editorial Project Manager: Carlos Rodriguez
Production Project Manager: Anusha Sambamoorthy
Cover Designer: Victoria Pearson

Typeset by MPS Limited, Chennai, India

Dedication

To my son, Ashton, who reminds me daily that being curious about the world is a gift.

Caleb M. Pearson

To Jennifer, Violet, and Hazel for the love, support, and patience that make it possible for me to do what I do for a living.

Eric Ecklund-Johnson

To Lynne, the love of my life, and to my parents for making me go to college.

Shawn D. Gale

Contents

4.   A Primer on Neuropsychology for the Neurosurgeon   63

Ioan Stroescu and Brandon Baughman

Section II
Methods

5.   Functional Neuroimaging in the Presurgical Workup   77

Leslie C. Baxter

Section III
Applications

Section IV
Treatment & Management

List of Contributors

Mark Barisa Performance Neuropsychology, Dallas/Frisco, TX, United States

Brandon Baughman Department of Neuropsychology, Semmes Murphey Clinic, Memphis, TN, United States

Leslie C. Baxter Department of Neurosurgery, Barrow Neurological Institute, St. Joseph's Hospital and Medical Center, Phoenix, AZ, United States

Kelley Beck Baylor Rehabilitation, Frisco, TX, United States

Erin D. Bigler Department of Psychology, Brigham Young University, Provo, UT, United States; Neuroscience Center, Brigham Young University, Provo, UT, United States

Kier Bison Baylor Rehabilitation, Frisco, TX, United States

Paul J. Camarata Department of Neurosurgery, University of Kansas Medical Center, Kansas, United States

Tsinsue Chen Department of Neurosurgery, Barrow Neurological Institute, St. Joseph's Hospital and Medical Center, Phoenix, AZ, United States

Jeffrey Cole Department of Neurology, Columbia University Medical Center, New York, NY, United States

Hugues Duffau Department of Neurosurgery, Hôpital Gui de Chauliac, Montpellier University Medical Center, Montpellier, France; National Institute for Health and Medical Research (INSERM), U1051, Team "Plasticity of the central nervous system, human stem cells and glial tumors," Institute for Neurosciences of Montpellier, Montpellier University Medical Center, Montpellier, France; University of Montpellier, Montpellier, France

Eric Ecklund-Johnson Departments of Neurology and Psychiatry, University of Kansas Medical Center, Kansas City, KS, United States

Marla J. Hamberger Department of Neurology, Columbia University Medical Center, New York, NY, United States

Guillaume Herbet Department of Neurosurgery, Hôpital Gui de Chauliac, Montpellier University Medical Center, Montpellier, France; National Institute for Health and Medical Research (INSERM), U1051, Team "Plasticity of the central nervous system, human stem cells and glial tumors," Institute for Neurosciences of Montpellier, Montpellier University Medical Center, Montpellier, France; University of Montpellier, Montpellier, France

Stacy W. Hill Clearwater Neurosciences, Lewiston, ID, United States

Patrick Landazuri Department of Neurology, University of Kansas Health System, Kansas City, KS, United States

Kyle Noll Department of Neuro-Oncology, The University of Texas M.D. Anderson Cancer Center, Houston, TX, United States

Adam Parks Departments of Neurology and Psychiatry, University of Kansas Health System, Kansas City, KS, United States

Michael W. Parsons Massachusetts General Hospital/Harvard Medical School, Pappas Center for Neuro-Oncology & Psychological Assessment Center, Boston, MA, United States

Caleb M. Pearson Departments of Neurology, Neurosurgery, and Psychiatry, University of Kansas Medical Center, Kansas City, KS, United States

Francisco A. Ponce Department of Neurosurgery, Barrow Neurological Institute, St. Joseph's Hospital and Medical Center, Phoenix, AZ, United States

Sujit Prabhu Department of Neurosurgery, The University of Texas M.D. Anderson Cancer Center, Houston, TX, United States

Caitlin Reese UT Southwestern, Dallas, TX, United States

David Sabsevitz Department of Neurology, Medical College of Wisconsin, Milwaukee, WI, United States

Mairaj T. Sami Department of Neurosurgery, University of Kansas Medical Center, Kansas, United States

Ioan Stroescu NeuroCog Trials & Triangle Neuropsychology, Durham, NC, United States

Alexander I. Tröster Department of Clinical Neuropsychology and Center for Neuromodulation, Barrow Neurological Institute, Phoenix, AZ, United States

Sally J. Vogel Advanced Behavioral Medicine, Tacoma, WA, United States

Jeffrey Wefel Department of Neuro-Oncology, The University of Texas M.D. Anderson Cancer Center, Houston, TX, United States

Foreword

The discipline of neurosurgery dates back millennia with evidence of rudimentary surgical wounds found in the skulls of early paleolithic man. Neuropsychology by contrast is a relatively new discipline, which has garnered rich contributions from many fields including anatomy, physiology, biology, psychology, and philosophy. Although relatively young by comparison, neuropsychology's relationship to the field of neurosurgery is unquestionable with collaborations between neurosurgeons and neuropsychologists dating to the 1940s. Neurosurgery's role in techniques such as cortical stimulation mapping has helped advance our understanding of the language system from a 200-year-old model based on lesion studies, to a contemporary and dynamic network, which is only now starting to be understood. More recently, as the science and practice of neurosurgery has advanced, there has been increased focus on minimizing the cognitive morbidities that are often associated with traditional neurosurgical interventions. To this end, neuropsychology has played a major role. With techniques such as functional neuroimaging and diffusion tensor tractography neuropsychologists and other neuroscientists have greatly advanced both our understanding of the brain, as well as our treatment of patients undergoing neurosurgical intervention for a variety of diseases. Indeed, brain lesions once thought to be inoperable are now being surgically treated with the help of these techniques and tools.

In this book, experts in the fields of both neurosurgery and neuropsychology discuss the various ways in which these two important disciplines interface with each other. We begin with the history of the relationship between neuropsychology and neurosurgery. We then provide an overview of neuropsychology for those new to the field, as well as a primer on the use of neuropsychology to measure cognitive functioning. Next, we aim to familiarize the reader with the various methods of cognitive assessment pertinent to neurosurgery. These chapters include discussions of functional magnetic resonance imaging, as well as more traditional methods of assessing function such as cortical stimulation and the Wada procedure. We then discuss the various applications of neuropsychology in a number of diseases for which neurosurgical treatment plays a role. Finally, we discuss the assessment of functional status and the role of neuropsychological rehabilitation.

Our aim with this text is to aid both neurosurgeons and neuropsychologists in understanding the role of neuropsychological techniques and procedures in the contemporary neurosurgery program. It is meant to be read as a practical manual and we are hopeful it will be helpful to not only neurosurgeons and neuropsychologists, but to neurologists, psychologists, speech and language pathologists, occupational and physical therapists, and all other disciplines interested in the functional properties of the brain. We are optimistic about the role of neuropsychology within the field of neurosurgery, and we hope that this optimism shines through in our work.

Caleb M. Pearson

Departments of Neurology, Neurosurgery, and Psychiatry,
University of Kansas Medical Center, Kansas City, KS, United States

Epigraph

"In spite of all these disquieting triumphs in the field of natural science, it's astonishing how little man has learned about himself, and how much there is to learn. How little we know about this brain which made social evolution possible, and of the mind. How little we know of the nature and spirit of man and God. We stand now before this inner frontier of ignorance. If we could pass it, we might well discover the meaning of life and understand man's destiny."

Wilder Penfield

Section I

Introduction

Chapter 1

The Historical Role of Neuropsychology in Neurosurgery

Mairaj T. Sami and Paul J. Camarata

Department of Neurosurgery, University of Kansas Medical Center, Kansas, United States

The more we discover about the brain, the more clearly do we distinguish between the brain events and the mental phenomena, and the more wonderful do both the brain events and the mental phenomena become.

John C. Eccles.

Trephined skulls and ancient papyri testify to the fact that practitioners of what we now know as neurosurgery have existed for millennia. Similarly, since the earliest times humans have puzzled over how human behavior is produced. It is only within the last century or so that the two fields have begun to cooperate extensively, and it is this cooperation that has led to profound advances in the respective sciences. In this chapter, we will describe the early evolution of each field, and the fruits of that ever-increasing cooperation.

ORIGINS OF NEUROSURGERY

Brain surgery is a terrible profession. If I did not feel it will become different in my lifetime, I should hate it.

Wilder Penfield.

Where written records are lacking, the telltale signs of manmade surgery on the human skull are evident in the paleological record of most every continent. It can be determined that many of these skull perforations were placed before death; some perhaps for medical reasons, and others perhaps in warfare or injury. In a 1954 monograph on trephination, Graña concluded that many of the skull perforations were made for medical reasons, with evidence

Neurosurgical Neuropsychology. DOI: https://doi.org/10.1016/B978-0-12-809961-2.00002-3

of brain tumors, fractured skulls, and cranial osteomyelitis (Grana & Rocca, 1954). However, the first written records regarding brain surgery date from about 3000 BC, where the writings from Imhotep, a physician priest of King Zoser, are detailed in what is now known as the Edwin Smith papyrus.

In this document which describes the wounds of 48 persons, 17 of them had injuries to the nervous system. He describes treating wounds of the head with underlying skull fractures by laying fresh meat upon the wound the first day, then afterward with grease, honey, and lint. However, if the dura were penetrated, it was "an ailment not to be treated" (Breasted, 1930).

In the time of Hippocrates, who lived from about 460 to 379 BC, physicians knew the brain to be the receptive organ for sensations and the source of all motor activity, but were unclear about how the brain communicated with the body. It was known, however, that in head wounds "if the wound is on the left side of the head, convulsion seizes the right side of the body, and if the wound is on the right side of the head, convulsion seizes the left side of the body. There are some patients who also become apoplectic, and in this condition they die within 7 days in summer or 14 days in winter" (Hanson, 1999). Aristotle, however, a near contemporary, believed that the heart was actually the seat of thinking and emotion, believing that the brain only served to cool the heart-generated heat (Benton & Sivan, 2007). Despite evidence to the contrary and perhaps somewhat surprisingly, this belief of a heart-centered mind had its proponents even through the 1600s. In the English language alone, we maintain vestiges of this belief in such aphorisms as "learning by heart," a loved one "has our heart," or someone with a "heart of gold."

Perhaps the most outstanding early Christian physician was the Greek Galen who was born in AD 130. As a physician to the gladiators for 4 years, Galen undoubtedly witnessed many brain injuries. Much of his neurological writing was unavailable to the western world, and indeed it was lost from the 9th century until the 19th century when two Arabic translations were found (Walker, 1998). Galen made numerous studies in animals observing the brain and spinal cord in vivo, recognizing sensory and motor nerves of the spinal cord, and seven pairs of cranial nerves. It was not until the 17th century when the brain had clearly displaced the heart as the seat of the mind and sentience, and with it came certain concepts of cerebral localization subserving various neurologic functions. Thomas Willis (1621–73), a physician in Oxford and London, attracted a number of talented associates who performed exquisite anatomical dissections which helped him postulate on the various functions of different areas of the brain. He placed memory in the cerebral cortex. Others had other theories of localization—Francois La Peyronie (1678–1747) sited intelligence in the corpus callosum and Emmanual Swedenborg (1688–1772) placed intellectual functions in the cerebral cortex (Benton & Sivan, 2007).

Another major advance in the concept of neurophysiology that would impact surgery of the brain took place when Franz Joseph Gall (1758–1828) postulated that the human brain was an "assemblage of organs" that subserved a specific cognitive ability or trait. This stimulated detailed anatomical studies of the brain throughout the scientific community, such that by 1860 illustrations of the human cortical surface are essentially the same as those found in today's texts. (Benton & Sivan, 2007, p. 143). Shortly thereafter, Paul Broca (1824−80) discovered that speech is primarily mediated by the left hemisphere. Though, as detailed earlier, Hippocrates had noted the contralateral control of extremities 2000 years earlier, the demonstration by Gustav Fritsch and Eduard Hitzig in 1870 that electrical stimulation of the precentral gyrus in a dog produced contralateral limb movement rocked the neuroscience community. This was the first time anyone had done any localized study regarding the brain and electric current.

Following shortly thereafter, in 1874, a Cincinnati physician, Roberts Bartholow, performed a similar "experiment" in one of his patients. A young woman dying of a malignant scalp lesion allowed him to insert insulated wires through the granulation tissue directly into the underlying cortex. When current was passed through the electrodes presumably placed in the left peri-Rolandic cortex, it elicited a pronounced contraction of the right arm, hand, and leg. With continued stimulation, he eventually produced a seizure. The woman died a few days later of overwhelming brain infection. His experiments were roundly criticized by scientists all over the world (Walker, 1998).

The work on cerebral localization and observations of eminent British Neurologist John Hughlings Jackson (1835−1911) set the stage for the nascent field of neurosurgery to remove brain tumors and treat focal causes of epilepsy. It was also occasioned by the introduction of antiseptic techniques and anesthesia. Sir Rickman John Godlee successfully removed a tumor from the right Rolandic cortex in 1884. When that was presented before the Medical and Chirurgical Society of London, Scottish physician William Macewen (Fig. 1.1A) related that he had removed a brain tumor in 1879. He is then largely regarded as the first surgeon to operate a brain tumor without any external clue that a tumor was present in the brain. It has since been learned that in Italy in 1883, Giacomo Novaro in Turin operated on a hyperostosing meningioma, and his colleague in Rome, Francesco Durante, removed an olfactory groove meningioma in a patient in 1885 (Guidetti, Giuffre, & Valente, 1985).

Pioneer neurosurgical procedures were fraught with an extraordinarily high rate of complications with a mortality at times reaching 65%! It took the wiles and skills of an American, Harvey Williams Cushing (1869−1939) to essentially rescue neurosurgery from its rapid waning at the turn of the last century (Fig. 1.1B). Cushing taught himself to operate on the brain, and through his meticulous surgical technique, learned at the hand of surgeon William Halstead at the Johns Hopkins Hospital, he reduced the mortality to only 10%. These pioneers and others ushered in the neurosurgical advances of the last century.

(A)   (B)

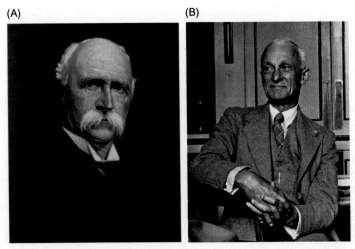

FIGURE 1.1 (A) Photo of William Macewan (first surgeon to operate brain tumor by intuiting its location) and (B) photo of Harvey Cushing.

ORIGINS OF NEUROPSYCHOLOGY

If a man has lost a leg or an eye, he knows he has lost a leg or an eye; but if he has lost a self—himself—he cannot know it, because he is no longer there to know it.

Oliver Sachs.

The discipline of neuropsychology comprises the study of brain function as it relates to human behavior and thinking. Over the course of history, neuropsychology has enjoyed rich contributions from many disciplines including anatomy, physiology, pharmacology, biology, and philosophy. Today, neuropsychology is a dynamic profession which serves many practical clinical applications—lateralization and localization of brain lesions, diagnosis of neurobehavioral conditions, detection of cognitive deficits or brain damage, and the establishment of functional baselines for assessment of neuropsychiatric improvement or deterioration (Parsons, 1991).

Although neuropsychology is a relatively young field-achieving fruition primarily during the 20th century, the series of events that led to its emergence as a distinct science, and the very nature of its origin are often debated.

It is uncertain as to when the term "neuropsychology" first entered common scientific lexicon. In 1913, Sir William Osler used the term while addressing the Phipps Psychiatric Clinic of Johns Hopkins Hospital in order to underscore the biological aspect of psychiatric disease (Bruce, 1985). Hebb (1949) was the first to capture the term in physical text, when he published his book, *The Organization of Behavior: A Neuropsychological*

Theory (Hebb, 1949). Karl Lashley used "neuropsychology" briefly in a 1936 presentation to the Boston Society of Psychiatry and Neurology (Bruce, 1985), as well as in later texts during the 1960s (Beach, Hebb, Morgan, & Nissen, 1960). Perhaps the most prominent use of the term, and the instance that gave "neuropsychology" its current meaning, was during a 1948 symposium presentation given by Hans-Lukas Teuber at the American Psychological Association (APA). During the symposium, which was aptly titled "Neuropsychology," Teuber described procedures that he developed with his colleague, Morris Bender, to study the behavioral consequences of penetrating brain injuries (Benton, 1987). Teuber's presentation at the APA is considered an influential work that led to widespread recognition of neuropsychology as an independent, separate field.

Numerous attempts have been made to attribute the creation of the field of Neuropsychology to a specific person or event. The history of neuropsychology is replete with such examples. The creation of Ward Halstead's laboratory for psychological evaluation of brain damaged patients in 1935 was a significant contribution. Karl Lashley, considered one of the founders of neuropsychology, performed much of his seminal work in the physical manifestation of memory and neuroplasticity during the 1920s. Fritsch and Hitzig's experiment detailing contralateral movement with electrical stimulation of the cortex heralded a "golden age" of cerebral localization where many subsequent studies aimed to define the functional purpose of every area of the cerebral cortex. The case could also be made that Broca's discovery of hemispheric cerebral dominance in the 1860s, or Gall's theory of the brain as an aggregate of specialized cortical areas rather than a single unitary organ, were also key in shaping the field of neuropsychology (Stringer & Cooley, 2002).

When these isolated events are viewed as a whole, it becomes apparent that the field of neuropsychology arose from evolution rather than revolution. There is no single inciting event or creator that is solely responsible for the precipitous existence of neuropsychology. It is a discipline that arose out of necessity, a tool developed to serve the clinical and intellectual curiosities of its many pioneers. Neuropsychology was born from vital contributions from seminal individuals, spanning many generations, avenues, and eras in time.

THE COLLABORATION OF NEUROSURGERY AND NEUROPSYCHOLOGY: A MUTUALLY BENEFICIAL PARTNERSHIP

But a man does not consist of memory alone. He has feeling, will, sensibilities, moral being—matters of which neuropsychology cannot speak.

A.R. Luria.

In an autobiographical work, Karl Pribram describes an early illustrative example of the symbiotic relationship between neurosurgery and neuropsychology (Fig. 1.2) (Pribram, 2002). Pribram was intrigued by the recently published classic text, *Physiology of the Nervous System*, by John Fulton. After training in neurosurgery at Northwestern with Percival Bailey and Paul Bucy, and in Memphis with Eustace Semmes, Pribram was working as a neurosurgeon in Jacksonville, Florida when he first met Karl Lashley in 1946. Pribram and Lashley devised a plan to perform experiments on spider monkeys in order to settle a dispute. Pribram recalls: "the issue between us was whether the brain was organized into systems, each of which served a separate psychological process." While Lashley agreed that localized systems could serve in sensory and motor tasks, he believed that higher order cognitive functions were distributed generally over the cerebral cortex. Drawing from clinical observations of his own patients with various agnosias, Pribram reasoned instead that higher order cognitive functions reside in specific areas of injured cerebral cortex.

The ensuing collaboration between Pribram and Lashley was emblematic of how the distinct characteristics of neurosurgery and neuropsychology are ideally suited and complementary to one another. In approaching Pribram, Lashley sought biological techniques to investigate psychological processes. Conversely, Pribram was in pursuit of behavioral techniques to investigate the organization of brain processes. Pribram jokingly asserted that Lashley also required his aid because Lashley's surgical skill and knowledge of aseptic technique left much to be desired. Pribram placed this sentiment in a larger context when he stated that "physiological psychologists fared no

FIGURE 1.2   Photo of Karl Pribram.

better; their lack of surgical skill and anatomical naïveté made it common knowledge that no two experimenters ever obtained the same results." However, Pribram was quick to note that contemporary studies within neurosurgery and neurology were flawed as well. He felt that human clinical studies endured difficulty in identifying the site and extent of a lesion, and that there was a prevalent lack of sophistication in experimental design. In a 1954 paper titled "Toward a Science of Neuropsychology," Pribram ultimately concluded that neuropsychology became a true science when the traditions and practices of clinical neurology and neurosurgery were mated with experimental psychology (Pribram, 1954).

Epilepsy

Pribram and Lashley serve as just one example of the many mutually beneficial partnerships between neurosurgery and neuropsychology found throughout history. Perhaps the earliest and most meaningful collaborations between neurosurgery and neuropsychology took place within the realm of epilepsy surgery. The collaborations we will mention are by no means the only such partnerships that flowered over the last 100 years.

The disciplines of neurosurgery and neuropsychology were each making independent progress toward treatment of epilepsy during the late 1800s. In 1870, British neurologist John Hughlings Jackson was the first to describe the origination of focal motor seizures from the contralateral cerebral hemisphere, rather than from the brainstem as previously believed (Eadie, 2007). Meanwhile, Sir Victor Horsely, was refining his surgical technique using animals as he would go on to perform the first epilepsy surgery in 1886. Horsely had been informed by Jackson's theory that all epilepsy was focal, deducing that removal of the focal abnormality should result in a cure from seizures. In this case, Horsely's 22-year-old patient, James B., suffered a skull fracture in a traffic accident and became seizure-free after a cortical scar had been surgically removed (Taylor, 1986).

While it had long been understood that uncontrolled epilepsy would lead to progressive neurological decline and dementia, the first formal studies of the effects of epilepsy on cognition were not performed until the late 19th and early 20th centuries. In a British psychology journal, Smith (1905) compared two groups of epilepsy patients to healthy controls using an array of tests that resemble those used in neuropsychology today. Word and picture tasks were used to assess memory. Smith found that epilepsy patients with dementia suffered from increased confusion with regard to "old and new" (Smith, 1905). The first study of cognition in epilepsy in the United States was published in 1912. Wallin used Binet–Simon intelligence scales to categorize patients with epilepsy into groups based on intelligence as reflected by mental age (MA). This study is remarkable for its use of a positive control group, as these epilepsy patients were then compared to "feeble-minded" institutionalized patients. Wallin found that patients with epilepsy had overall less cognitive impairment than "feeble-minded" patients based on intelligence according to MA (Wallin, 1912).

Collaboration between neurosurgery and neuropsychology at three major epilepsy centers gave rise to the modern era of epilepsy surgery. These centers include The University of Illinois, the University of London, and the Montreal Neurological Institute (MNI).

The epilepsy program at the University of Illinois was led by neurosurgeon, Percival Bailey. Bailey, whose former residents included Paul Bucy and Karl Pribram, avoided resection of the hippocampal apparatus during epilepsy surgery to prevent development of what has become known as the Kluver—Bucy syndrome. This is the likely explanation for poorer outcomes seen in the University of Illinois series of epilepsy patients in comparison to other similar institutions at the time. However, the University of Illinois epilepsy program enjoyed the advantage of being in close proximity to Chicago. Thus, Percival Bailey had all of his epilepsy patients evaluated pre- and postoperatively by Ward Halstead. Halstead employed a battery of neuropsychological tests and published his findings in 1958, noting that epilepsy surgery did not result in a generalized cognitive decline, and that most indicators actually demonstrated an improvement postoperatively. Patients showed no significant change in the Impairment Index. Halstead's work was critical in showing that surgical treatment of epilepsy did not have to result in significant cognitive impairment (Halstead, 1958; Hermann & Stone, 1989).

Murray Falconer was the neurosurgeon in charge of epilepsy treatment at University of London. Unlike his contemporary, Percival Bailey, Falconer was known for *en bloc* resections of the temporal lobe and was keenly aware of hippocampal involvement in the pathological process and treatment outcomes related to epilepsy. Falconer worked closely with psychologists Victor Meyer and Aubrey Yates, and their findings regarding the effects of epilepsy surgery and hippocampal/temporal lobe dysfunction remain relevant to this day. Meyer and Yates concluded that operation in the dominant temporal lobe commonly results in a severe deficit of auditory—verbal learning (Mayer & Yates, 1955).

Nonetheless, the greatest contributions to the fields of epilepsy surgery and neuropsychology occurred at McGill University and the MNI. The epilepsy program was led by the renowned neurosurgeon from Hudson, Wisconsin, Wilder Penfield (Fig. 1.3), who cofounded MNI in 1934. Penfield's early career was spent in a sort of neuroscience "Grand Tour" beginning with the aid of a Rhodes scholarship to study with the famous Oxford neurophysiologist Sir Charles Scott Sherrington. He also spent time in the Madrid laboratories of histopathologists Ramon y Cajal and Pio del rio Hortega. And finally, he spent time with the man who pioneered electroencephalography (EEG), Otfrid Foerster.

FIGURE 1.3   Neurosurgeon Wilder Penfield, Director of the Montreal Neurological Institute.

It is doubtless that Penfield's pioneering work in epilepsy surgery, operating on awake human brains, was occasioned by his time with Sherrington. Charles Sherrington won the Nobel Prize for his contributions to the study of the neuron and synapses, and later became a mentor to three Nobel laureates. Sherrington later wrote about the connection between the mind and the brain in his work, *Man: On His Nature* (Sherrington, 1940), where he waxed eloquently on the relationship of the mind to the myriad of neural impulses.

the brain is waking and with it the mind is returning. It is as if the Milky Way entered upon some cosmic dance. Swiftly the head-mass becomes an enchanted loom where millions of flashing shuttles weave a dissolving pattern, always a meaningful pattern though never an abiding one;

Sir Charles Scott Sherrington (Sherrington, 1940).

He would spend the later years of his career trying to solve the mind—brain conundrum, and hence the relationship between psychiatry and neurophysiology...

> *In the training and in the exercise of medicine a remoteness abides between the field of neurology and that of mental health, psychiatry. It is sometimes blamed to prejudice on the part of the one side or the other. It is both more grave and less grave than that. It has a reasonable basis. It is rooted in the energy-mind problem. Physiology has not enough to offer about the brain in relation to the mind to lend the psychiatrist much help.*
>
> Sherrington in The Brain Collaborates With Psyche (1940).

One can see Penfield's time with his mentor influencing his "second career," when a year before his death he would remark, "Throughout my own scientific career I, like other scientists, have struggled to prove that the brain accounts for the mind. But perhaps the time has come when we may profitably consider the evidence as it stands, and ask the question... Can the mind be explained by what is now known about the brain?" (Penfield, 1975).

Penfield's inspiration for surgical treatment of epilepsy was derived from the time he spent with Otfrid Foerster at the Brain Research Institute in Breslau, Germany. Penfield had received a grant to study with Foerster from the Rockefeller Foundation. Though not a surgeon by formal training, as a neurology consultant in World War I Foerster began operating on the wounded, especially those with peripheral nerve injuries. He described to Paul Bucy how he became a reluctant neurosurgeon...

> *I had to accompany the patient to the operating room. I had to tell Mickulz (a general surgeon) where to operate. When he got inside the skull, I had to tell him where the tumor was and what to do. And then the patients all died. I decided I could do no worse.*
>
> Bucy (1978) (Fig. 1.4).

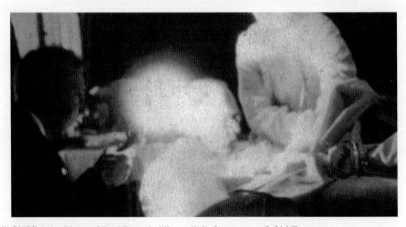

FIGURE 1.4   Photo of Paul Bucy holding a light for surgeon Otfrid Foerster.

In 1928, Penfield spent 6 months with Foerster in Breslau, learning awake cortical stimulation techniques while performing surgery on many of the World War I veterans who suffered from posttraumatic epilepsy (Sarikcioglu, 2007). Though Hans Berger is credited with recording the first human electroencephalogram (EEG), Foerster was the first to record an actual intraoperative electrocorticogram in 1934.

Penfield returned from Germany, and with William Cone, a surgeon who had also spent time in Breslau, founded the MNI at McGill, aided by another grant from the Rockefeller Foundation. At MNI, Penfield championed a multidisciplinary approach that was unique for its time. Penfield recruited Herbert Jasper, a neurophysiologist, who claimed that he could localize epileptic foci by detecting abnormal brain rhythms outside of the skull. Through multiple shared patients, Penfield was able to confirm the merits of EEG in localizing epileptogenic lesions (Feindel, 1999).

As a visionary, Penfield also established the first clinical psychology service at a major hospital in 1938 at the MNI. Though at the time, the sciences of psychology and medicine were largely separate entities, Penfield understood the value and insight that neuropsychologists could provide. In progressive fashion, Penfield hired Molly Harrower, who was the only female staff in the entire hospital, as his first clinical neuropsychologist (Loring, 2010).

Over the remainder of his career at MNI, Penfield continued a tradition of collaboration between neurosurgery and what would come to be known as neuropsychology. Penfield relied on another prominent neuropsychologist, Donald Hebb, to successfully map eloquent cortex through stimulation language mapping. During dominant temporal lobe resections for treatment of epilepsy, it was Hebb who would often converse with awake patients through a wide range of topics (Almeida, Martinez, & Feindel, 2005).

Hebb's most significant contributions to Penfield and MNI can be attributed to the innovative work he performed regarding the frontal lobes. Hebb showed that clinical frontal release signs were caused by frontal lesions and associated mass effect rather than large prefrontal lesions alone (Hebb, 1977). These findings afforded clinicians at MNI an opportunity to more rapidly diagnose and offer cognitive evaluation to these patients in comparison to their peers. In the course of his career, Hebb accrued a substantial series of patients with frontal lobe dysfunction.

Interestingly, Penfield's sister herself belonged to this series, after she underwent frontal lobe resection at his hand. His only sister Ruth had suffered from headaches for 20 years, and seizures since age 20. Penfield notes that she had developed "very definite mental blunting and slight confusion" over the previous 5 years. Radiographs revealed a calcified mass within the right frontal lobe, and after consultation with colleagues, Penfield decided to operate on his sister to remove the lesion. He proceeded to perform an awake craniotomy for resection of what proved to be an oligodendroglioma.

Postoperatively, though she returned to functioning in her family as the mother of six children, she was described as having difficulty with goal-directed behavior, such as organizing family meals. Penfield's later memoirs detail how this had a profound effect on him.

After the surgery, which was complicated by profound hemorrhage deep in the right frontal lobe, Penfield wrote about it asking advice from his former mentor, Dr. Harvey Cushing at the Peter Bent Brigham Hospital in Boston.

> *Dear Dr Cushing, I wish to ask you a word of advice. The day before yesterday I removed most of the right frontal lobe of a patient who had an Oligodendroglioma ... You will understand how much this case means to me when I tell you that it was my own sister ... I had to leave some tumour behind in the vicinity of the corpus callosum; I do not know whether it goes across to the other side or not. She had a close shave of it, but is doing very well. Would you advise my using x-ray therapy, or would it be better just to let her go and if the symptoms occur again re-operate? ... As a matter of fact I should have very much preferred bringing her to you, but the family, spurred on by a strange type of confidence, was anxious to have me do it. Will you please give me your advice about subsequent therapy?*

> *Cushing replied on 15 December 1928:*

> *Dear Penfield: I am distressed to learn of your sister's malady and that circumstances forced you to take the case on yourself, but I am glad to have been spared the anxiety of operating on a member of your family and rejoice to learn that she is doing well. I don't believe I would recommend radiation. We have not had a great deal of experience in radiation on these particular tumours, and they are as a matter of fact as benign as any form of glioma ... [Percival] Bailey is making just now a study of all our cases. They are curiously much alike, and I would suggest your writing him to ask some thing about the prognosis. As he has gone into the matter more recently than I have, his information will be more valuable than any I could give you. So for your own comfort do send him a couple of slides with an out line of the history, and ask him from our Brigham experience what you may expect; and then like a good man do let me know what he says — that is, whether he agrees with me or not about the radiation and with you as to the histogenesis of the lesion (Feindel & Leblanc, 2016).*

Her tumor returned 2 years later, and was then resected by Cushing. The tumor had returned in a more malignant form, and she died 6 months later. Autopsy revealed recurrent tumor in two different locations of the frontal lobe (Fig. 1.5). Penfield himself published his sister's clinical report, along with two other cases of right frontal lobe resection, concluding that this seemed to have produced a "lack of initiative and incapacity for planned action," while leaving intact insight, introspection, and the ability to follow instruction.

(A) (B)

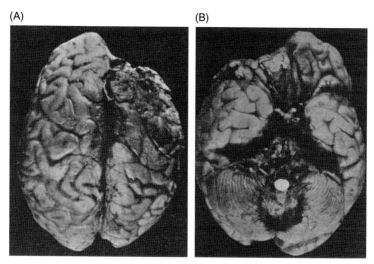

FIGURE 1.5 Ruth Penfield's brain at post mortem. *Reproduced from Compston, A. (2006). From the archives. Brain, 129(4), 827−829. https://doi.org/10.1093/brain/awl059.*

Regarding Ruth he remarked, "One day . . . she had planned to get a simple supper for one guest . . .. This was a thing she could have done with ease ten years before. When the appointed hour arrived, she was in the kitchen, the food was all there. . . but the salad was not ready, the meat had not been started and she was distressed and confused by her long continued effort alone." Penfield and Evans remarked that "maximum amputation of right or left frontal lobe has for its most important detectable sequel impairment of those mental processes which are prerequisite to planned initiative" (Penfield & Evans, 1935).

Penfield presented these three cases and the effect of frontal lobe resection at the 2nd International Congress of Neurology at University College in London in August of 1935. At the same conference, Richard Brickner reported on a patient in whom Walter Dandy had removed the frontal lobes, and Carlyle Jacobsen reported on the effect of bifrontal lobectomies in two chimpanzees (Feindel & Leblanc, 2016). In the audience at that meeting, was Portuguese diplomat, politician, neurologist and future Nobel laureate Egas Moniz, who within months persuaded a colleague neurosurgeon to perform the first "psychosurgery" by lesioning the frontal lobes of a psychotic patient. Hebb remarked that "Penfield changed the whole doctrine and theory of frontal lobe function and the basis of the so-called frontal lobe signs" (Penfield & Evans, 1935).

Hebb was also the first to administer IQ tests to patients following frontal lobe resection in 1940. Hebb's discovery that frontal lobe resection actually paradoxically resulted in increased Full-Scale IQ was controversial, as the frontal lobes had long been considered the center of human intelligence prior to his work (Loring, 2010).

These observations prompted Hebb to recommend to Penfield that one of his graduate students, Brenda Milner, evaluate Penfield's patients before and after surgery (Fig. 1.6). While Milner went on to become an incredibly influential figure in neuropsychology—her accomplishments within the field are vast—her greatest encounter with the field of neurosurgery occurred through her study of patient H.M. In 1953, William B. Scoville, a neurosurgeon at Hartford Hospital in Connecticut, performed a bilateral medial temporal lobectomy on H.M. in an effort to obtain better control of his epilepsy. The operation was successful in halting H.M.'s seizures, but H.M. was left with a peculiar new postoperative deficit. H.M. could easily recall events prior to surgery, but he was unable to form any new memories.

FIGURE 1.6   Photo of Brenda Milner.

Scoville had heard that Penfield presented two cases of profound short term memory loss following unilateral mesial temporal lobectomy at a meeting of the American Neurological Association in 1955. These cases had preexisting injury to the opposite temporal lobe. Scoville had designed the operation of bilateral mesial temporal lobectomy as a substitute for frontal lobotomy, thinking that severing the mesial temporal to orbitofrontal connections would have a psychiatric benefit, but avoid the undesirable effects of lobotomy (Scoville & Milner, 1957). He decided to use it in H.M. for epilepsy, but noted the profound effect on short term memory. He contacted Penfield and Brenda Milner and her students were invited to Hartford to further investigate H.M.'s condition (Loring, 2010; Milner, 1964).

In the early stages of her interactions with H.M, Milner sought to better understand the nature of his memory impairment. She assigned H.M. a perceptual motor task, asking him to draw a star while viewing it in a mirror and H.M. displayed progressive daily improvement. However, Milner noted that H.M. was still unable to recall any events that had taken place during the time that she spent with him. She observed that H.M.'s anterograde amnesia could be separated into episodic and procedural components, helping to change the way that memory is understood. Milner's work with H.M. was vital in demonstrating that separate components of memory can be simultaneously encoded in different locations within the brain, and that they function independently (Loring, 2010; Milner, 1964).

A substantial subset of epilepsy patients with "split-brain" emerged in the 1960s, permitting another opportunity for cooperation between neurosurgeons and neuropsychologists. During this time, Dr. Joseph Bogen and Dr. Phillip Vogel were performing corpus callosotomies in their epilepsy patients in Los Angeles, essentially severing the connection between both cerebral hemispheres to prevent contralateral spread of seizures. These patients usually benefited from greatly decreased seizure frequency. Roger W. Sperry, a neuropsychologist from Caltech working with Bogen and Vogel, studied these patients and, in the process, greatly challenged previous beliefs pertaining to the role of the right and left cerebral hemispheres (Kolb & Whishaw, 2009).

Paul Broca's theory of left cerebral hemispheric dominance had permeated throughout the scientific community since the 1860s. As a result, the left hemisphere was not only considered the dominant site for language, but also for skilled movements, and higher order cognitive functions. The right cerebral hemisphere was essentially considered nonfunctional by comparison. By taking advantage of the disjointed hemispheres in "split-brain" patients, Sperry was able to study the function of the right hemisphere in isolation. In his experiments, he found that the right hemisphere was able to comprehend spoken and written words, and allowed for the identification and matching of pictures or objects based on spoken and printed cues. Despite initial resistance to his findings, Sperry later received the 1981 Nobel Prize in

Physiology and Medicine for his efforts. Sperry's work was instrumental in defining the importance of the right cerebral hemisphere in spatial tasks (Benton & Sivan, 2007; Loring, 2010). Michael Gazzaniga is a neuropsychologist who was mentored by Sperry, and worked with Joseph Bogen. Gazzaniga became a pioneer in his own right, further developing numerous studies in "split-brain" patients, many of which evaluated visuospatial and laterality effects of commissurotomy (Gazzaniga, 2005).

Collaboration in the Surgical Treatment of Psychiatric Disorders

While shared treatment of epilepsy between neurosurgery and psychology grew to become more anatomical and physiological in nature, surgical treatment of psychiatric disorders, or psychosurgery, represented a true foray into the unknown. With respect to epilepsy, seizures were localized to an area of the brain using electrical recordings or clinical exam findings and were often ascribed to a lesion which could then be targeted for consequent removal. Morbidity and neurological deficit due to these operations could be expected as a function of the brain tissues affected.

In contrast, the roots of psychosurgery take hold within the most nebulous corners of medicine, formed at the junction of spirituality, socioeconomics, and ethics. Psychosurgery enjoyed a meteoric rise in the mid-20th century, but fell from grace just as quickly. What began as sincere desire to help the mentally ill eventually became fraught with missteps and complications. As with many scientific advances throughout history, progress within the field of psychosurgery unfortunately often reflected a human cost.

In 1888, a Swiss psychiatrist named Gottlieb Burckhardt performed the first series of psychosurgical procedures in Neuchâtel (Gross, 1998). He had carried out bilateral cortical resections in patients who demonstrated intractable aggressive behavior. Burckhardt was content with his results as five out of six patients survived and appeared more docile. While Burckhardt had done something truly novel, his findings were largely met with indifference or outright contempt (Mueller, 1960). Burckhardt discontinued these surgical interventions and the field of psychosurgery remained stagnant until the mid-1930s.

Neuroscientist and physician, John Farquhar Fulton, was influential in reviving psychosurgery. Fulton was a Rhodes Scholar who received his PhD in 1925 through work related to mapping the primate motor cortex under the tutelage of a renowned neurophysiologist that we have already profiled, Sir Charles Sherrington (Heller et al., 2006). Fulton completed medical school at Harvard and spent significant time on the neurosurgical service at the Peter Bent Brigham Hospital guided by Harvey Cushing. It was Fulton who wrote the definitive biography of his neurosurgical mentor (Fulton, 1946). Fulton ascended to a position as chairman of the Department of Physiology at Yale Medical School in 1930 (Davey, 1998). At Yale, Fulton established the first primate neurophysiology laboratory in the United States. His studies were primarily focused on ascertaining functions of the primate cortex via resection and subsequent observation of behaviors and reflexes (Horwitz, 1998).

In 1933, Fulton and his colleague, Carlyle Jacobsen, began training chimpanzees to perform certain tasks with positive reward reinforcement. They noted that the chimpanzees would exhibit behavior consistent with emotional frustration if they were not rewarded due to a failed task attempt. Fulton and Jacobsen performed bilateral frontal lobectomies in these chimpanzees and found that, while the chimpanzees were unhindered in their ability to perform these tasks, their emotional dispositions had changed. The primates were no longer distressed if a task was not completed or a reward had been withheld (Horwitz, 1998). Previously, one of the chimpanzees had to be repeatedly forced back into its cage. After the operation, it would enter its cage voluntarily (Crawford et al., 1948). Fulton and Jacobsen brought two chimpanzees to the 1935 International Neurological Congress in London to share their findings (Kopell & Rezai, 2003). Portuguese neurologist Egas Moniz was in the auditorium at the 1935 Neurologic Congress in London where Fulton and Jacobsen and Penfield described their work on frontal lobectomies and associated emotional/behavioral responses in primates. It is believed that Fulton and Jacobsen's work provided the impetus for Moniz to initially explore psychosurgery (Kotowicz, 2005).

Egas Moniz was born in Avanca, Portugal in 1874 and completed medical school at University of Coimbra in 1899. He became the chair of neurology in Lisbon in 1911, and also cultivated a career in Portuguese politics until the early 1920s under many posts including Minister for Foreign Affairs. Moniz earned a significant accolade in 1928, when he received a nomination for the Nobel Prize in Medicine for the first instance of cerebral angiography on a human patient (Gross & Schafer, 2011). However, his work on frontal leucotomy which predominated the 1930s would prove to be much more controversial.

Moniz believed that mental disorders occurred as a result of synaptic derangement and aberrant neuronal connections (Gross & Schafer, 2011). Drawing from Fulton's work in primates and his own experiences with war veterans who suffered personality changes as a result of frontal lobe injury, Moniz concluded that cutting neuronal fibers would allow patients' thoughts to change by offering alternate pathways of travel (Barahona, 1956). Moniz enlisted the help of Portuguese neurosurgeon, Almeida Lima, to perform frontal leucotomy on a series of 20 patients in 1936. They sought to sever the fiber tracts between the frontal lobes and the rest of the brain by administering alcohol. Moniz and Lima developed an instrument for the operation, called the "leucotome" and described the steps of a "standard leucotomy."

The first patient to undergo the operation was a female from an asylum with severe paranoia and agitation. Though her symptoms improved with the procedure, she became dull and indifferent afterward (Feldman & Goodrich, 2001).

Moniz asserted that, overall, 35% of patients in this series were cured, 35% were improved, 30% were unchanged, none of the patients became worse, and there were no deaths. Understandably, the procedure that Moniz had developed gained tremendous popularity. For example, there were more than 200 leucotomies performed in Brazil from 1936 to 1945 (Barahona, 1956). Praise for leucotomy reached its zenith when Moniz was awarded the Nobel Prize for Medicine 1949 based on the therapeutic value of the operation in treatment of psychoses (Lima, 1974).

The historical context surrounding leucotomy provides insight as to why it was so heavily lauded. World War I and widespread economic despair contributed to an increase of psychiatric disorders globally. Additionally, psychiatric patients, such as those with schizophrenia, had proven difficult to treat due to the severity of their symptoms and having failed multiple therapeutic modalities. At the time, lobotomy appeared to be the only viable treatment, as medications such as chlorpromazine did not appear until several years later.

As time passed, Moniz began receiving criticism for his work with lobotomy. American and British neurologists and psychologists questioned the logical and theoretical basis for leucotomy. They believed Moniz prematurely published his results to further secure a place in history during the waning years of his career, and they also questioned the sincerity of his intention to help patients (Gross & Schafer, 2011; Valenstein, 1990).

An American neurologist, Walter Freeman, was also in the audience during Fulton and Jacobsen's presentation at the 1935 International Neurological Congress. In America, Freeman was facing many of the same challenges as Egas Moniz across the Atlantic. World War I left a high proportion of veterans with psychiatric illness from mechanisms such as direct head trauma, posttraumatic stress disorder, and tertiary syphilis. There was tremendous strain on the nation's overcrowded asylums, and Freeman was also working in an era devoid of psychotropic medications.

In 1936, Freeman and his neurosurgeon colleague, Dr. James Watts, began their own series of frontal lobotomies. Freeman initially employed the surgical techniques described by Egas Moniz, but found that they were imprecise and dangerous, as placing lesions too deep into the frontal lobes led to death among other complications. Freeman began using x-ray and skeletal landmarks to target white matter tracts (Freeman, 1971). Freeman's first frontal lobotomy patient was a 63-year-old woman from Kansas who had anxiety and depression. After the procedure, her anxiety and agitation were alleviated. Freeman was so encouraged by the success of the procedure, he wished to devise a way that neurologists, psychiatrists, and general physicians could perform it without need of an operating theater (Freeman, 1948; Kopell & Rezai, 2003) (Fig. 1.7).

Based on work done by Italian psychiatrist Amarro Fiamberti, Freeman brought the transorbital leucotomy to the United States in 1946. A patient

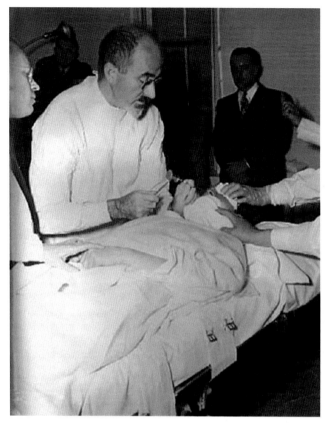

FIGURE 1.7 Photo of Freeman and leukotome.

undergoing this operation would be anesthetized with two electric shocks, then an ice pick would be driven by a mallet through the upper eyelid and roof of the orbit into the desired location. In his first 2 years of doing the procedure, Freeman claims to have performed 2000 transorbital leucotomies in his first 2 years, and a total of 4000 until the 1950s. Watts became dejected by Freeman's dubious practices regarding patient consent, patient selection, and ethics. Freeman continued to espouse the merits of transorbital leucotomy until his death in 1972 at age 77 (Feindel, 1999; Fulton, 1946; Kolb & Whishaw, 2009).

The impact that figures such as Egas Moniz and Walter Freeman had over the positive perception of psychosurgery from 1930 to 1950 is substantial. It is estimated that over 50,000 lobotomies were performed in the United States between 1945 and 1955. Freeman traveled across the United States in a Winnebago called the "Lobotomobile" to share the benefits of lobotomy with as many patients as possible (Davey, 1998; Feindel & Leblanc, 2016).

The Veterans Administration even recommended lobotomy for World War II veterans with psychiatric disability (Horwitz, 1998).

By the 1950s, stories of unscrupulous practices and outcomes regarding psychosurgery began to surface. One of Freeman's patients was Rosemary Kennedy, the younger sister of John F. Kennedy, Jr. Rosemary, who was 23 years old at the time of surgery, underwent transorbital lobotomy and was left incontinent, and profoundly cognitively impaired. She was institutionalized for 60 years until her death. There are also reports that Freeman operated on an agitated patient in a motel room while the patient was being held down by police officers. Freeman also operated on Frances Farmer, a famous Hollywood actress, who was relegated to multiple mental institutions after her procedure (Feldman & Goodrich, 2001; Heller et al., 2006; Lerner, 2005). Ultimately, lack of a scientific basis, poor ethical practices, severe complications, and superiority of antipsychotic medications led to an overwhelming negative public opinion of psychosurgery.

Recently, a resurgence in psychosurgery has manifested in the form of neuromodulation. Whereas traditional psychosurgery had focused on destruction of white matter tracts and creation of lesions in order to elicit changes in behavior, neuromodulation is characterized by regulation of brain function by means of a specifically placed stimulating electrode. At high stimulation frequencies, the electrode acts in the same manner as a lesion, but has the benefit of having its effect reversed and modified if necessary (Fins, 2003). Deep brain stimulation is being used to treat obsessive−compulsive disorder, Tourette syndrome, chronic pain, and major depression (Heller et al., 2006; Lapidus et al., 2013). Recent advancements such as diffusion-tensor imaging, fiber tractography, stem cell transplantation, and gene therapy may provide additional avenues for surgical treatment of psychiatric conditions. In its modern-day present iteration, the field of psychosurgery displays tremendous promise for growth in multiple future directions.

Collaboration between neurosurgeons and neuropsychologists has burgeoned over the last few decades with advances in the recognition and treatment of brain injury. The development of neuropsychometric tests such as the ImPACT test for athletes used at baseline and to assess cognitive changes after concussion has brought neurosurgeons and neuropsychologists together in the treatment of mild traumatic brain injury (Brett et al., 2016; Davis et al., 2017). It is now commonplace to have complex neuropsychological assessment batteries precede surgery for epilepsy, and for resection of tumors and vascular malformations of the frontal lobes. Neuropsychologists and neurosurgeons interact frequently in the operating room with the revival of awake craniotomy techniques to resect tumors in eloquent areas. As our knowledge of cognitive changes as a result of surgical interventions in the brain advances, it is clear that the alliance that has developed over the last century will only continue to grow and mature, resulting in better patient outcomes.

REFERENCES

Almeida, A. N., Martinez, V., & Feindel, W. (2005). The first case of invasive EEG monitoring for the surgical treatment of epilepsy: Historical significance and context. *Epilepsia, 46*(7), 1082−1085.

Barahona, F. (1956). Egas Moniz: Personality and work. *Journal do Medico (Oporto), 29*(692), 941−949.

Beach, F. A., Hebb, D. O., Morgan, C. T., & Nissen, H. W. (1960). *The neuropsychology of Lashley.* New York: McGraw-Hill.

Benton, A. (1987). *Evolution of a clinical specialty. The TCN guide to professional practice in clinical neuropsychology* (pp. 1−4). Amsterdam: Swets & Zeitlinger.

Benton, A. L., & Sivan, A. B. (2007). Clinical neuropsychology: A brief history. *Disease-a-Month: DM, 53*(3), 142−147.

Breasted, J. (1930). *The Edwin Smith surgical papyrus published in facsimile and hieroglyphic transliteration with translation and commentary in two volumes.* Chicago, IL: University of Chicago Press.

Brett, B. L., et al. (2016). Long-term stability and reliability of baseline cognitive assessments in high school athletes using ImPACT at 1-, 2-, and 3-year test−retest intervals. *Archives of Clinical Neuropsychology.*

Bruce, D. (1985). On the origin of the term "neuropsychology". *Neuropsychologia, 23*(6), 813−814.

Bucy, P. C. (1978). Neurosurgery in darkness. *Surgical Neurology, 9*(6), 360.

Crawford, M. P., Fulton, J. F., et al. (1948). Frontal lobe ablation in chimpanzee: A resume of Becky and Lucy. *Research Publications—Association for Research Nervous Mental Disease, 27*(1 vol.), 3−58.

Davey, L. M. (1998). John Farquhar Fulton. *Neurosurgery, 43*(1), 185−187.

Davis, G. A., et al. (2017). The Berlin International Consensus Meeting on Concussion in Sport. *Neurosurgery.*

Eadie, M. J. (2007). Cortical epileptogenesis—Hughlings Jackson and his predecessors. *Epilepsia, 48*(11), 2010−2015.

Feindel, W. (1999). Herbert Henri Jasper (1906−1999): An appreciation. *The Canadian Journal of Neurological Sciences, 26*(3), 224−229.

Feindel, W., & Leblanc, R. (2016). *The wounded brain healed: The golden age of the Montreal Neurological Institute 1934−1984.* Montreal: McGill Queen's University Press.

Feldman, R. P., & Goodrich, J. T. (2001). Psychosurgery: A historical overview. *Neurosurgery, 48*(3), 647−657. (discussion 657-9).

Fins, J. J. (2003). From psychosurgery to neuromodulation and palliation: History's lessons for the ethical conduct and regulation of neuropsychiatric research. *Neurosurgery Clinics of North America, 14*(2), 303−319, ix-x.

Freeman, W. (1948). Transorbital leucotomy. *Lancet, 2*(6523), 371−373.

Freeman, W. (1971). Frontal lobotomy in early schizophrenia. Long follow-up in 415 cases. *The British Journal of Psychiatry, 119*(553), 621−624.

Fulton, J. (1946). *Harvey Cushing: A biography.* Springfield, IL: Charles C. Thomas.

Gazzaniga, M. S. (2005). Forty-five years of split-brain research and still going strong. *Nature Reviews Neuroscience, 6*(8), 653−659.

Grana, F., & Rocca, E. (1954). *Las Trepanaciones Craneanas en el Peru en la Epoca Pre-Hispanica* (p. 340) Lima, Peru: Imprenta Santa Maria.

Gross, D. (1998). Gottlieb Burckhardt's (1836–1907) contribution to psychosurgery: Medicohistorical and ethical aspects. *Gesnerus*, *55*(3–4), 221–248.

Gross, D., & Schafer, G. (2011). Egas Moniz (1874–1955) and the "invention" of modern psychosurgery: A historical and ethical reanalysis under special consideration of Portuguese original sources. *Neurosurgical Focus*, *30*(2), E8.

Guidetti, B., Giuffre, R., & Valente, V. (1985). Italian contribution to the origin of neurosurgery. In P. Bucy (Ed.), *Neurosurgical giants: Feet of clay and iron* (p. 23). New York: Elsevier.

Halstead, W. C. (1958). Some behavioral aspects of partial temporal lobectomy in man. *Research Publication—Association for Research in Nervous and Mental Disease*, *36*, 478–490.

Hebb, D. (1977). Wilder Penfield: His legacy to neurology. The frontal lobe. *Canadian Medical Association Journal*, *116*(12), 1373–1374.

Hebb, D. O. (1949). *The organization of behavior: A neuropsychological theory*. New York: Wiley.

Heller, A. C., et al. (2006). Surgery of the mind and mood: A mosaic of issues in time and evolution. *Neurosurgery*, *59*(4), 720–733. (discussion 733-9).

Hermann, B. P., & Stone, J. L. (1989). A historical review of the epilepsy surgery program at the University of Illinois Medical Center: The contributions of Bailey, Gibbs, and collaborators to the refinement of anterior temporal lobectomy. *Journal of Epilepsy*, *2*(3), 155–163.

Hippocrates *on head wounds*. In M. Hanson *Corpus Medicorum Graecorum*. Berlin: CMG.

Horwitz, N. H. (1998). Library: Historical perspective. John Farquhar Fulton. *Neurosurgery*, *43*(1), 178–184.

Kolb, B., & Whishaw, I. Q. (2009). The development of neuropsychology. In B. Kolb, & I. Q. Whishaw (Eds.), *Fundamentals of human neuropsychology* (pp. 16–19). New York: Worth Publishers.

Kopell, B. H., & Rezai, A. R. (2003). Psychiatric neurosurgery: A historical perspective. *Neurosurgery Clinics of North America*, *14*(2), 181–197, vii.

Kotowicz, Z. (2005). Gottlieb Burckhardt and Egas Moniz—Two beginnings of psychosurgery. *Gesnerus*, *62*(1–2), 77–101.

Lapidus, K. A., et al. (2013). History of psychosurgery: A psychiatrist's perspective. *World Neurosurgery*, *80*(3–4), S27 e1–16, p.

Lerner, B. H. (2005). Last-ditch medical therapy—Revisiting lobotomy. *New England Journal of Medicine*, *353*(2), 119–121.

Lima, A. (1974). Egas Moniz, 1874–1955, Nobel-prize winner in medicine and physiology 1949. *Journal of Neurology*, *207*(3), 167–170.

Loring, D. W. (2010). History of neuropsychology through epilepsy eyes. *Archives of Clinical Neuropsychology*, *25*(4), 259–273.

Mayer, V., & Yates, A. J. (1955). Intellectual changes following temporal lobectomy for psychomotor epilepsy: Preliminary communication. *Journal of Neurology, Neurosurgery, and Psychiatry*, *18*(1), 44–52.

Milner, B. (1964). Some effects of frontal lobectomy in man. In J. M. Warren, K. Akert, & Pennsylvania State University (Eds.), *The frontal granular cortex and behavior*. New York: McGraw-Hill, p. x, 492 p.

Mueller, C. (1960). Gottlieb Burckhardt, the father of topectomy. *American Journal of Psychiatry*, *117*, 461–463.

Parsons, O. A. (1991). Clinical neuropsychology 1970–1990: A personal view. *Archives of Clinical Neuropsychology*, *6*(3), 105–111.

Penfield, W. (1975). *Mystery of the mind: A critical study of consciousness and the human brain.* Princeton, NJ: Princeton University Press.

Penfield, W., & Evans, J. (1935). The frontal lobe in man: A clinical study of maximum removals. *Brain, 58,* 115−133.

Pribram, K. H. (1954). *Toward a science of neuropsychology.* Pittsburgh, PA: University of Pittsburgh Press.

Pribram, K. H. (2002). Autobiography in anecdote: The founding of experimental neuropsychology. In A. Y. Stringer, E. L. Cooley, & A. Christensen (Eds.), *Pathways to prominence in neuropsychology: Reflections of twentieth century pioneers* (pp. 197−221). New York: Psychology Press.

Sarikcioglu, L. (2007). Otfrid Foerster (1873−1941): One of the distinguished neuroscientists of his time. *Journal of Neurology, Neurosurgery, and Psychiatry, 78*(6), 650.

Scoville, W. B., & Milner, B. (1957). Loss of recent memory after bilateral hippocampal lesions. *Journal of Neurology, Neurosurgery, and Psychiatry, 20*(1), 11−21.

Sherrington, C. (1940). *Man on his nature: The Gifford lectures,* Edinburgh. Cambridge, UK: Cambridge University Press.

Sherrington in The Brain Collaborates With Psyche (1940) *Man on his nature: The Gifford lectures, Edinburgh* (p. 283).

Smith, W. G. (1905). A comparison of some mental and physical tests in their application to epileptic and to normal subjects. *British Journal of Psychology, 2,* 240−260.

Stringer, A. Y., & Cooley, E. L. (2002). Neuropsychology: A twentieth century science. In A. Y. Stringer, E. L. Cooley, & A.-L. Christensen (Eds.), *Pathways to prominence in neuropsychology: Reflections of twentieth century pioneers* (pp. 3−24). New York: Psychology Press.

Taylor, D. C. (1986). One hundred years of epilepsy surgery: Sir Victor Horsley's contribution. *Journal of Neurology, Neurosurgery, and Psychiatry, 49*(5), 485−488.

Valenstein, E. S. (1990). The prefrontal area and psychosurgery. *Progress in Brain Research, 85,* 539−553. (discussion 553-4).

Walker, A. (1998). *The genesis of neuroscience.* Park Ridge, IL: AANS Publications.

Wallin, J. E. W. (1912). Eight months of psycho-clinical research at the New Jersey state village for epileptics, with some results from the Binet−Simon testing. *Epilepsia, A3,* 366−380.

Chapter 2

Foundations of Contemporary Neuropsychology

Eric Ecklund-Johnson[1] and Caleb M. Pearson[2]

[1]Departments of Neurology and Psychiatry, University of Kansas Medical Center, Kansas City, KS, United States, [2]Departments of Neurology, Neurosurgery, and Psychiatry, University of Kansas Medical Center, Kansas City, KS, United States

INTRODUCTION/EARLY DEVELOPMENTS

The science and clinical practice of neuropsychology arose mainly in the 20th century and particularly in the second half of the 20th century, with its most proximal roots in previously established traditions and scientific advances in several areas, primarily including behavioral neurology and assessment-oriented clinical psychology. There are other outstanding sources for readers interested in the historical development of neuropsychology (e.g., Barr, 2018; Stringer, Cooley, & Christensen, 2002) and Chapter 1, The Historical Role of Neuropsychology in Neurosurgery, provides an excellent review of the history of collaboration between the disciplines of neurosurgery and neuropsychology. In this chapter, we will briefly describe the history of neuropsychology highlighting some fruitful interactions between neuropsychology and neurosurgery that have played a critical role in this development and spend the remainder of the chapter discussing current approaches to neuropsychological practice and education/training. We focus heavily on developments in North America, which are relatively well documented in available historical accounts, recognizing that important contributions by such neuropsychology pioneers as Henry Hecaen, Andre Rey, Elizabeth Warrington, and many other contributors from Europe and elsewhere are neglected in an effort to keep the scope of this chapter manageable and limit redundancy with other chapters in this volume.

Applications of neuropsychological science/knowledge to neurosurgery began in very early stages of the development of neuropsychology as a specialty and have remained a key focus for many neuropsychologists, perhaps most notably for those who work in specialized epilepsy surgery centers.

Neurosurgical Neuropsychology. DOI: https://doi.org/10.1016/B978-0-12-809961-2.00003-5

In recent years, however, neuropsychologists' involvement with various other surgical populations (e.g., candidates for deep brain stimulation, neuro-oncology patients) seems to be on the increase and exciting new applications of neuropsychological knowledge in concert with advanced neuroimaging methods and sophisticated brain mapping techniques, as described in other chapters of this book, suggest that neuropsychology will continue to have much to offer to neurosurgical patients in the 21st century.

Descriptions of neuropsychology's roots often begin with discussion of early studies of localization of specific forms of aphasia that were first noted to be associated with certain lesion locations during the 19th century, including the work of Broca and Wernicke. This pioneering work was an important foundation for the field, by promoting an understanding that certain functions could be reliably correlated with specific brain areas, even though strict localizationism has given way to network-based theories that better account for the dynamic nature of many brain functions. Barr (2018) describes important contributions of other early forerunners in the more qualitative tradition, such as the theoretical and practical contributions of individuals such as Jean-Martin Charcot and Kurt Goldstein, as well as the achievements in measuring abilities that came out of the more quantitative tradition as pioneered by such individuals as James McKean Cattell and Shepard Ivory Franz. Out of these beginnings, a focus on the measurement of mental abilities in the service of understanding brain functions reached a more mature stage with the work of Luria and Halstead around the middle of the 20th century.

HISTORICAL UNDERPINNINGS OF MODERN NEUROPSYCHOLOGICAL METHODS

Aleksandr Luria overcame the constraints of Soviet-era restrictions on scientific endeavors (e.g., stipulations that all psychological research be consistent with a Pavlovian classical conditioning model for a significant period of his career) and managed to synthesize a unique combination of education and training (he was a professor and active researcher on a broad range of psychology topics prior to completing medical education/training) to develop an influential model of higher brain functions in the mid-20th century (Tranel, 2008). Rejecting both the strict localizationist approach and the opposing view that complex mental functions could not be localized, Luria recognized that higher cortical functions necessarily relied on cooperation of multiple functional "zones" within the brain, presaging modern network models. He emphasized qualitative analysis of behaviors with respect to their putative neural underpinnings and relied heavily on the method of double dissociation to determine zones that were critically involved in specific complex behaviors and were reliably associated with deficits in those behaviors when damaged. Luria's book *Higher Cortical Functions in Man* and other writings published primarily during the 1960s had an influence well beyond the

Soviet Union and stimulated subsequent theoretical and scientific developments in the West.

Meanwhile, beginning with the establishment of his laboratory at the University of Chicago during the mid-1930s, Ward Halstead began to research the construct that he called "biological intelligence" by studying the behavior of individuals with known brain damage in their daily lives to determine behaviors that could be sampled and measured by specific tests (Reitan, 2002; Reitan & Wolfson, 2009). Like his contemporary Luria, he focused a good deal on higher cortical functions including problem-solving skills, which might be less obvious unless elicited by appropriate tasks. Halstead recognized that intellectual tests of the day, having been developed to predict academic outcomes, were relatively insensitive to the broad range of deficits that could occur with various types of brain damage and that no single measure or scale was likely to be useful in identifying the many potential types of "organic" dysfunction. From early on, Halstead worked with neurosurgeons, which allowed him to focus on behavioral effects of known and clearly delineated brain lesions (rather than presumed brain damage). It is important to remember that this work occurred prior to the time when modern neuroimaging techniques would allow less invasive identification of brain pathology. Among the important applications of Haltead's approach was his collaboration with the epilepsy center at the University of Illinois, which was helpful in establishing that temporal lobectomy could be performed without a great deal of cognitive morbidity in appropriately selected patients. Through his studies of individuals with brain damage, Halstead developed a battery of tests that was subsequently revised and formalized by his student, Ralph Reitan, who also worked extensively with neurosurgeons to localize lesions while at the University of Indiana (Reitan, 2002). Reitan's approach and later variations on it became known as the primary example of a "fixed battery" approach to neuropsychological assessment in which an invariant grouping of tests is administered to every patient, traditionally interpreted prior to examining other "extra-test" information to localize and characterize the nature of brain lesions based on patterns of performance across a large battery of tests.

In contrast to the emphasis on psychometric measurement that came out of several Midwestern centers in the mid-20th century, an East Coast group centered around the Boston Veterans Association Medical Center championed an assessment method that Meier (2002) has characterized as a more "cognitive-experimental" approach, drawing on behavioral neurology principles similar to those espoused by Luria and informed by a model of "disconnection" syndromes elaborated by Norman Geschwind (Milberg, Hebben, & Kaplan, 2009). This approach has been known primarily as the "process" or "Boston process" approach to neuropsychological evaluation and draws heavily on the work of Edith Kaplan, Harold Goodglass, and others to characterize the complexities of deficits that can occur with different locations/types of brain lesions (e.g., subtypes of aphasia, apraxia). According to the process model, these sometimes subtle, but important differences in types of brain dysfunction are often best understood by observing qualitative aspects of an individual's performance of a task as opposed to his/her ultimate achievement of a correct or incorrect answer.

In Montreal, Donald Hebb, Brenda Milner, and colleagues at the Montreal Neurological Institute were among the first psychologists to be programmatically involved in the care of neurosurgical patients through their work with Wilder Penfield's epilepsy surgery program beginning by the late 1930s (Loring, 2010). Working with Penfield, Juhn Wada, and other collaborators, they helped to develop methods (including special procedures such as intracarotid amobarbital and cortical language mapping) with a focus not just on localization, but also on minimizing cognitive morbidity in epilepsy surgery. The work of Milner, Corkin, and others associated with the Montreal Neurological Institute and the contribution of the patient H.M., who was unfortunately rendered densely amnestic by a bilateral temporal procedure, provided many of the underpinnings of our modern understanding of memory including the critical contribution of medial temporal structures and separate systems for declarative and nondeclarative memory. Somewhat less well known is Milner's contribution to understanding frontal lobe functions through the study of patients who had undergone frontal lobectomy, particularly with respect to their performance on tasks used to elicit problem-solving and cognitive flexibility (Milner, 1964).

A middle ground between the psychometrically oriented "fixed battery" and behavioral neurology oriented "process" approaches came out of Arthur Benton's laboratory at the University of Iowa beginning in the 1950s (Meier, 2002; Tranel, 2009). Benton's approach to neuropsychological evaluation, which became known as the "hypothesis testing" or "flexible battery" approach, involved the use of a relatively brief core group of tests that were generally administered to every assessment patient and then supplemented by additional tests that were selected based on the specific referral question and performance of the patient on portions of the core battery. Tranel (2009) notes that Benton viewed neuropsychological assessment as a special case or extension of the neurological examination using quantitative methods to characterize higher brain functions that often do not lend themselves to identification through elicitation of pathognomonic signs in the typical neurological examination. The flexible battery approach that developed out of the Benton lab and similar approaches at a number of other centers has become the dominant approach to neuropsychological assessment by the early 21st century (Sweet, Benson, Nelson, & Moberg, 2016) and is now practiced by the large majority of clinical neuropsychologists, though a number of outstanding neuropsychologists continue to rely primarily on fixed battery or process approaches.

KEY PRINCIPLES OF NEUROPSYCHOLOGICAL ASSESSMENT

The currently widely used flexible battery approach melds the process and psychometric strains of the 20th century and largely follows the middle ground espoused by Benton and others who have practiced a flexible battery approach over the years. In the later decades of the 20th century and even the earliest part of the 21st century, before the number of neuropsychologists who use a flexible battery approach began to far exceed the number who use a fixed battery approach, there was at times animated debate about the relative merits of the two approaches in the scientific literature, on professional listserves, and perhaps at hotel bars during neuropsychology conferences. Those espousing the fixed battery approach (e.g., Russell, Russell, & Hill, 2005) have generally emphasized the importance of having an invariant group of tests that has been given to all individuals in the normative sample in order to make sound assumptions about the meanings of low test scores and other patterns of performance, an issue often referred to as "conorming" of tests from a battery. Those espousing a flexible battery approach have argued that, while conorming may be advantageous when possible, it has a downside of stifling the development of new methods (by requiring them to be normed on the same group of test-takers along with many other measures before being allowed into the canon of tests in the battery). Those arguing the flexible battery side also often note that the sensitivity of individual tests to various types of brain dysfunction is well established and that the administration of a full fixed battery, which may take more than 6 hours in some cases, is probably unnecessary to identify key clinical issues. Several studies (e.g., Larrabee, Millis, & Meyers, 2008; Miller, Fichtenberg, & Millis, 2010; Rohling, Meyers, & Millis, 2003) have supported equivalent utility of a flexible approach to that of a more extensive fixed battery. A number of neuropsychological test batteries have emerged that endeavor to combine some of the advantages of both approaches (e.g., Heaton, Miller, Taylor, & Grant, 2004; Meyers & Rohling, 2004; Schretlen, Testa, & Pearlson, 2010; White & Stern, 2003), allowing neuropsychologists to use groupings from a larger selection of conormed tests that are most relevant to the individual patient/referral question without necessitating the use of the entire potential battery of tests in every case.

Regardless of which approach to neuropsychological test batteries is favored, competent practice of neuropsychology requires in-depth understanding of psychometric techniques and principles of deficit measurement, theories of brain—behavior relationships, and effects of various brain diseases/injuries on cognitive and behavioral functioning (AACN Board of Directors, 2007). The rationale of deficit measurement is based on evaluation of deviations from expected patterns of performance on tests used to assess particular areas or "domains" of neuropsychological functioning (e.g., attention, memory, and executive functions). This may involve psychometric definitions of abnormal or "impaired" scores on specific tests relative to a demographically appropriate normative group (e.g., scores falling 1.0, 1.5, or 2.0 standard deviations below the mean score in the comparison group) and/ or differences in observed scores from a person's presumed previous (or "premorbid") level of functioning based on his/her performance on tasks that are relatively insensitive to most types of acquired brain dysfunction. However, there is ample evidence that abnormal scores are often obtained on one or more measures from an extensive battery of tests even by healthy individuals without any known brain dysfunction (Binder, Iverson, & Brooks, 2009; Schretlen, Testa, Winicki, Pearlson, & Gordon, 2008), indicating a need for caution in interpreting low scores on individual tests. One approach to this interpretive challenge involves comparing an individual's performance on several tests used to assess a particular area or "domain" of functioning to base rates of low scores in this domain in the normative population, which can provide psychometric guidance as to whether a grouping of low scores falls outside of expectations (Brooks, Iverson, & Holdnack, 2013).

Another important interpretive issue, given the potential for motivational factors (e.g., insufficient task engagement, pursuit of secondary gain) to influence performance on neuropsychological tests, is the objective assessment of performance validity through the use of specialized techniques that allow neuropsychologists to determine if test scores are a reasonable indication of an individual's abilities (Heilbronner et al., 2009). Neuropsychologists have extensive training in these and other important psychometric issues (e.g., assessment of clinically meaningful change over time), which sets them apart from other disciplines that use various forms of cognitive assessment. This technical knowledge can be extremely valuable in obtaining accurate interpretation of the meaning of test scores and there is a good deal of evidence from decades of research that actuarial data can outperform clinical judgment in making certain types of well-defined decisions such as weighing the likelihood of specific, well-defined diagnostic possibilities (Meehl, 1954). However, psychometric sophistication alone is not sufficient for good neuropsychological practice and extensive grounding in clinical neuroscience as it pertains to cognitive and behavioral effects of various brain disorders is also crucial for identifying recognizable neuropsychological syndromes by incorporating test scores with medical history and behavioral observations (AACN Board of Directors, 2007; Schretlen et al., 2008).

SPECIALIZATION AND TRAINING IN CLINICAL NEUROPSYCHOLOGY

As clinical neuropsychology has matured from a specialty practiced in a few centers to one that is practiced across North America and in many other places around the world, training of clinical neuropsychologists has evolved and become more formalized. Meier (2002) describes the development of neuropsychology training and specialization in the United States, beginning with a group of neuropsychologists from the Midwest and East Coast of the United States that was convened at the University of Minnesota in 1965, as well as with less formal collegial interactions following the formation of the International Neuropsychological Society (INS) in the late 1960s. A task force on specialization, initially within the INS and subsequently also under the auspices of the American Psychological Association's (APA) Division 40 (then the Division of Clinical Neuropsychology, since renamed the Society for Clinical *Neuropsychology*), grew out of these efforts and had Manfred Meier as its chair. This task force published a report in 1981 that provided the foundation for the establishment of neuropsychology as a distinct specialty (as opposed to an area of proficiency within the broader field of clinical psychology, as some advocated at the time). This vision later came to fruition with the recognition of neuropsychology as a specialty by the APA after formal petitioning of the Committee on the Recognition of Specialties and Proficiencies in Psychology during the 1990s. Another important development for the profession occurred in the early 1980s when a group of neuropsychologists that included Meier, Linas Bielauskas, Edith Kaplan, Muriel Lezak, Charles Matthews, Steven Mattis, Paul Satz, and Barbara Wilson met and formed the steering committee for a new specialty board, the American Board of Clinical Neuropsychology (ABCN), which reached agreement in 1982 with the umbrella board certifying organization the American Board of Professional Psychology (ABPP) to become the first new specialty board to be recognized since the four original specialties (Clinical, Counseling, Industrial/Organizational, and School Psychology) decades earlier. The new neuropsychology specialty board established a multistep examination process that has become a model for other specialty boards that have formed under ABPP in the years since. From the first clinical neuropsychology board examinations in 1983, the number of ABPP/ABCN board certified neuropsychologists reached the 500th diploma awarded in 2004 and 1000th diploma in 2014 (Lucas, Mahone, Westerveld, Bielauskas, & Baron, 2014; Yeates & Bielauskas, 2004). As of August 2017, there were 1068 currently active diplomates (Davis, 2017). An academy for ABCN certified neuropsychologists was established in 1996 in order to separate the advocacy mission of the academy from the examination/gatekeeping mission of the board itself (Yeates & Bielauskas, 2004) and subsequently added affiliate and student membership options for neuropsychologists and aspiring neuropsychologists who had not yet completed the board certification process. A subspecialization in pediatric neuropsychology awarded its first diplomas in 2014 and became the first subspecialty area recognized within any of the specialties subsumed under ABPP. It should be noted that there are also two other, smaller boards that have some type of formal examination process, the

American Board of Professional Neuropsychology and the American Board of Pediatric Neuropsychology. Aside from ABPP/ABCN and these organizations, other credentials claimed in neuropsychology are likely to be from vanity boards.

As the specialty of clinical neuropsychology matured and the scientific knowledge base on which the specialty was built expanded rapidly during the latter part of the 20th century, there was recognition of the relatively wide variation in approaches to education and training, making it difficult to define who was adequately prepared for specialty practice. In 1997, the Conference on Specialty Education and Training in Clinical Neuropsychology (Hannay et al., 1998) was held in Houston and has come to be known simply as the Houston Conference. This conference outlined a training model that emphasized the importance of both a strong base of general knowledge in psychology and focused specialty education/training. The conference participants developed a relatively flexible model that allowed for obtaining relatively greater amounts of generalist versus specialist knowledge/skills at different stages (doctoral education, internship, and postdoctoral residency), but affirmed postdoctoral training as a necessary part of the training sequence. Thus, neuropsychologists adhering to Houston Conference guidelines as of 2005 and later necessarily complete a 2-year postdoctoral training sequence in addition to internship training and a doctoral degree in clinical psychology, neuropsychology, or related area of psychology.

Reflecting the growing importance of a formal postdoctoral training sequence, the Association of Postdoctoral Programs in Clinical Neuropsychology (APPCN) was started in the 1990s, coming out of a model previously established by a smaller confederation of postdoctoral training programs centered in the Midwestern United States (Meier, 2002). The APPCN established procedures for self-study and program evaluation and eventually developed a match system for applicants to postdoctoral training programs. As of the time this chapter was prepared, there remained a number of well-respected training programs that do not participate in the match program and ongoing debate as to how best to manage the postdoctoral selection process (Belanger et al., 2013), though many of the North American formal postdoctoral programs (both members and nonmembers of APPCN) participate in the match.

NEUROPSYCHOLOGY AND NEUROSURGERY: PRESENT AND FUTURE

As described in the earlier sections, the science and clinical practice of neuropsychology have emerged in large part from a need to better understand cognitive and behavioral implications of brain diseases/injuries and have benefited greatly from collaboration with neurosurgery at key points in their development. At present, exciting developments are reinvigorating this collaborative relationship, several of which are summarized in other chapters of this book. While a grounding in the broader knowledge base of psychology including a sophisticated understanding of psychometric methods will continue to be important for the practice of clinical neuropsychology in the foreseeable future, the development of other tools such as computerized assessment, the clinical application of newer neuroimaging technologies (e.g., functional magnetic resonance imaging, tractography), and other increasingly sophisticated approaches to invasive and noninvasive mapping of brain functions underscore that the most crucial contribution of neuropsychology to the clinical management of neurosurgical patients is less likely to be tied to the specific technologies used, but rather to our ability to apply our specialized knowledge of brain–behavior relationships. With that said new methods such as functional neuroimaging techniques have much to contribute to our understanding of the brain and its function. Already, our understanding of language functions have evolved greatly with the widespread application of diffusion tensor imaging based tractography and functional language mapping combined with cortical stimulation mapping. New paradigms of cognitive function must be relied upon to develop new methods of assessment in order to improve our accuracy and predictive ability in the surgical field. As our knowledge increases and our field broadens to incorporate new technologies, our training model must also adapt. In the future it may be necessary to offer new trainees fellowship training in various imaging paradigms, or specific areas such as epilepsy and neuro-oncology.

During the still relatively brief course of the field's history, applications of neuropsychology to neurosurgical patients have moved from a primary focus on lateralization/localization of lesions based on psychometric test findings (a function that is now performed with more precision through the use of neuroimaging technologies) to a focus on understanding the functional implications of brain pathology and its surgical treatments. It is our contention that neuropsychologists have much more to offer than our tests and that the most important "tool" we have is the ability to apply our knowledge base to promote better understanding of brain–behavior relationships and improved quality of life for the patients with whom we work.

REFERENCES

American Academy of Clinical Neuropsychology Board of Directors (AACN Board of Directors). (2007). American Academy of Clinical Neuropsychology (AACN) practice guidelines for neuropsychological assessment and consultation. *The Clinical Neuropsychologist, 21*, 209–231.

Barr, W. B. (2018). Historical trends in neuropsychological assessment. In J. E. Morgan, & J. H. Ricker (Eds.), *Textbook of clinical neuropsychology* (2nd ed.). New York: Taylor & Francis.

Belanger, H. G., Vanderploeg, R. D., Silva, M. A., Cimino, C. R., Roper, B. L., & Bodin, D. (2013). Postdoctoral recruitment in clinical neuropsychology: A review and call for inter-organizational action. *The Clinical Neuropsychologist, 27*(2), 159–175.

Binder, L. M., Iverson, G. L., & Brooks, B. L. (2009). To err is human: "Abnormal" neuropsychological scores and variability are common in healthy adults. *Archives of Clinical Neuropsychology, 24,* 31–46.

Brooks, B. L., Iverson, G. L., & Holdnack, J. A. (2013). Understanding and using multivariate base rates with the WAIS-IV/WMS-IV. In J. A. Holdnack, L. W. Whipple, L. G. Weiss, & G. L. Iverson (Eds.), *WAIS-IV, WMS-IV, and ACS: Advanced clinical interpretation.* Cambridge, MA: Elsevier.

Davis, R. (2017). Personal communication.

Hannay, H. J., Bielauskas, L. A., Crosson, B. A., Hammeke, T. A., Hamsher, K. S., & Koffler, S. P. (1998). Proceedings: The Houston conference on specialty education and training in clinical neuropsychology. *Archives of Clinical Neuropsychology, 13*(2), 157–250.

Heaton, R. K., Miller, S. W., Taylor, M. J., & Grant, I. (2004). *Revised comprehensive norms for an expanded Halstead-Reitan battery.* Lutz, FL: Psychological Assessment Resources.

Heilbronner, R. L., Sweet, J. J., Morgan, J. E., Larrabee, G. J., Millis, S. R., & Conference Participants. (2009). American Academy of Clinical Neuropsychology consensus conference on the assessment of effort, response bias, and malingering. *The Clinical Neuropsychologist, 23,* 1093–1129.

Larrabee, G. J., Millis, S. R., & Meyers, J. E. (2008). Sensitivity to brain dysfunction of the Halstead-Reitan vs an ability-focused neuropsychological battery. *The Clinical Neuropsychologist, 22*(5), 813–825.

Loring, D. W. (2010). History of neuropsychology through epilepsy eyes. *Archives of Clinical Neuropsychology, 25,* 259–273.

Lucas, J. A., Mahone, E. M., Westerveld, M., Bielauskas, L., & Baron, I. S. (2014). The American Board of Clinical Neuropsychology and American Academy of Clinical Neuropsychology: Updated milestones 2005–2014. *The Clinical Neuropsychologist, 28*(6), 889–906.

Meehl, P. E. (1954). *Clinical versus statistical prediction: A theoretical analysis and a review of the evidence.* Minneapolis, MN: University of Minnesota.

Meier, M. (2002). In search of knowledge and competence. In A. Y. Stringer, E. L. Cooley, & A. L. Christensen (Eds.), *Pathways to prominence in neuropsychology.* New York: Psychology Press.

Meyers, J. E., & Rohling, M. L. (2004). Validation of the Meyers Short Battery on mild TBI patients. *Archives of Clinical Neuropsychology, 19,* 637–651.

Milberg, W. P., Hebben, N., & Kaplan, E. (2009). The Boston process approach to neuropsychological assessment. In I. Grant, & K. M. Adams (Eds.), *Neuropsychological assessment of neuropsychiatric and neuromedical disorders.* New York: Oxford Press.

Milner, B. (1964). Some effects of frontal lobectomy in man. In J. M. Warren, & K. Akert (Eds.), *The frontal granular cortex and behavior.* New York: McGraw-Hill.

Miller, J. B., Fichtenberg, N. L., & Millis, S. R. (2010). Diagnostic efficiency of an ability-focused battery. *The Clinical Neuropsychologist, 24,* 678–688.

Reitan, R. (2002). The best laid plans—and the vagaries of circumstantial events. In A. Y. Stringer, et al. (Eds.), *Pathways to prominence in neuropsychology.* New York: Psychology Press.

Reitan, R. M., & Wolfson, D. (2009). The Halstead-Reitan Neuropsychological Test Battery for adults—theoretical, methodological, and validational bases. In I. Grant, & K. M. Adams (Eds.), *Neuropsychological assessment of neuropsychiatric and neuromedical disorders.* New York: Oxford Press.

Rohling, M. L., Meyers, J. E., & Millis, S. R. (2003). Neuropsychological impairment following traumatic brain injury: A dose—response analysis. *The Clinical Neuropsychologist, 17*(3), 289—302.

Russell, E. W., Russell, S. L. K., & Hill, B. D. (2005). The fundamental psychometric status of neuropsychological batteries. *Archives of Clinical Neuropsychology, 20*(6), 785—794.

Schretlen, D. J., Testa, S. M., & Pearlson, G. D. (2010). *Calibrated neuropsychological normative system professional manual*. Lutz, FL: Psychological Assessment Resources.

Schretlen, D. J., Testa, S. M., Winicki, J. M., Pearlson, G. D., & Gordon, B. (2008). Frequency and bases of abnormal performance by healthy adults on neuropsychological testing. *Journal of the International Neuropsychological Society, 14*, 436—445.

Stringer, A. Y., Cooley, E. L., & Christensen, A. L. (2002). *Pathways to prominence in neuropsychology*. New York: Psychology Press.

Sweet, J. J., Benson, L. M., Nelson, N. W., & Moberg, P. J. (2016). The American Academy of Clinical Neuropsychology, National Academy of Neuropsychology, and Society for Clinical Neuropsychology (APA Division 40) 2015 TCN professional practice and 'salary survey': Professional practices, beliefs, and incomes of U.S. neuropsychologists. *The Clinical Neuropsychologist, 29*(8), 1069—1162.

Tranel, D. (2008). Theories of clinical neuropsychology and brain—behavior relationships: Luria and beyond. In J. E. Morgan, & J. H. Ricker (Eds.), *Textbook of clinical neuropsychology*. New York: Taylor & Francis.

Tranel, D. (2009). The Iowa-Benton school of neuropsychological assessment. In I. Grant, & K. M. Adams (Eds.), *Neuropsychological assessment of neuropsychiatric and neuromedical disorders*. New York: Oxford Press.

White, T., & Stern, R. A. (2003). *NAB psychometric and technical manual*. Lutz, FL: Psychological Assessment Resources.

Yeates, K. O., & Bielauskas, L. A. (2004). The American Board of Clinical Neuropsychology and American Academy of Clinical Neuropsychology: Milestones past and present. *The Clinical Neuropsychologist, 18*, 489—493.

Chapter 3

Components and Methods of Evaluating Reliable Change in Cognitive Function

Stacy W. Hill

Clearwater Neurosciences, Lewiston, ID, United States

INTRODUCTION: WHY EVALUATE FOR CHANGE?

The aim of the present chapter is to discuss the methods for evaluating reliable change (RC) in a practical way that can be applied to clinical practice and research. To that aim, discussion will review concepts relevant to RC evaluation, a number of the most commonly used RC models/equations, and provide a description of a practical approach to implementing RC in clinical practice or research. A discussion of advantages, challenges, and limitations of use of RC is included. The focus will not be on comparisons of findings across clinical conditions or mathematical discussion of differences of RC models.

Change in neurocognitive performance is an important indicator of clinical course and treatment effect which has been a topic of considerable research. Empirical studies have sought to develop and refine methods to quantify change and to evaluate if the degree of change represents significant or true change (Chelune, Naugle, Luders, Sedlak, & Awad, 1993; Crawford & Garthwaite, 2007; Hinton-Bayre, 2010; Iverson, 2001; Jacobson & Truax, 1991; Maassen, Bossema, & Brand, 2009; McSweeny, Naugle, Chelune, & Luders, 1993). However, methods to evaluate neurocognitive change have limitations which must be considered.

Neurocognitive change within a clinical context typically relates to monitoring clinical course or treatment response. The importance of serial neuropsychological evaluations and the ability to objectively measure change has become an increasing focus in the field (Chelune, 2010; Heilbronner et al., 2010). It is not possible to consistently and accurately consider within a given patient if the difference between two or more scores is large enough to indicate a real change beyond what could be explained by other factors such

Neurosurgical Neuropsychology. DOI: https://doi.org/10.1016/B978-0-12-809961-2.00004-7

as variability in performance across testing sessions, measurement error, and/ or having been previously tested (e.g., practice effects). The ability to accurately evaluate and monitor change has important implications for clinical populations in a number of ways. In sports related concussion, questions commonly include whether a patient has recovered sufficiently to return to play. In moderate to severe traumatic brain injury (TBI) questions may relate to whether there continues to be recovery and what functions continue to show deficits. In neurodegenerative conditions questions may relate to whether there is evidence of progressive worsening of cognition, whether the patient is responding to treatment, and whether recommendations related to supervision/restrictions need to be changed. In epilepsy, questions often relate to whether there are progressive changes in cognition prior to or following surgery.

In clinical research, neurocognitive change has obvious implications as common endpoints for evaluating clinical course and treatment outcome. The most common statistical method for evaluating clinical course and treatment effect has been group mean comparisons. However, this method does not account for individual change in performance but rather deals with aggregate change between groups in which actual change can be obscured due to averaging of extreme scores. Individual changes in performance may be lost by this method. In addition, this approach does not account for known psychometric factors that can affect interpretation of obtained scores and comparisons made. Specifically, neuropsychological test scores are not perfect measures of the construct of interest and they include measurement error. Tests are affected by patient/clinical factors and test factors. Consequently, use of RC equations to evaluate for significant change while accounting for potential sources of error has become increasingly important. RC models include reliable change indices (RCI) and standardized regression-based (SRB) equations which may be simple or multivariate (Chelune et al., 1993; Crawford & Garthwaite, 2007; Hinton-Bayre, 2010; Iverson, 2001; Jacobson & Truax, 1991; Maassen et al., 2009; McSweeny et al., 1993).

CLINICAL AND TEST/PSYCHOMETRIC FACTORS

The following will discuss factors relevant to use of RC models. The focus will primarily be related to test related factors and a few components used in calculation of RC models. An understanding of these concepts is important for discussion of RC models and the benefits/limits of their use.

Clinical Factors

Clinical factors may include the demographic variables and test setting variables (Duff, 2012). Most neuropsychological normative data attempt to take into account demographic variables to produce a standardized score from which comparisons or judgments can be made about a patient's performance. Important demographic variables include age, education, gender, and ethnicity. Clinical factors may include motivation/effort, fatigue, setting variables (e.g., noisy vs quiet environment), or other health factors (e.g., sleepiness, mood disturbance, and metabolic disturbance).

Test/Psychometric Factors

A number of factors are relevant in evaluating change in neurocognitive functioning including test−retest reliability of the measures, the standard error of measurement (SEM), standard error of the difference (SED), practice effects, regression to the mean, and retest interval. An understanding of these factors is important for use in RC models and when discussing how variance in these factors affect RC classification.

Test−retest reliability: Reliability of the measure used is an important factor that affects the accuracy of scores being compared. The uncorrected Pearson product moment correlation should be used in RC equations. Test−retest reliability refers to the consistency or correlation of obtained test scores on two separate occasions (Nunnally & Bernstein, 1994). An important concept regarding test−retest reliability, as discussed by Duff (2012) is that the strength of test reliability relates to the degree to which a case maintains their relative position in the distribution of case scores between testing sessions. The length of time between test and retest as well as the construct being evaluated affect test−retest reliability. Shorter time between testing tends to result in higher test−retest reliability. Most tests report test−retest intervals of days to weeks which will likely produce a stronger reliability coefficient than found in clinical settings where test−retest intervals are often 12 months or more. The Wechsler Memory Scale—Fourth Edition (WMS-IV) which consists of several subtests evaluating memory and attention, had a test−retest interval of 14−84 days (Wechsler, 2009). The Dementia Rating Scale—2 (DRS-2) reports a 1 week test−retest interval (Jurica, Leitten, & Mattis, 2001). It also appears that crystalized functions such as semantic word knowledge or fund of knowledge (e.g., Wechsler Adult Intelligence Scales—IV (WAIS-IV) Vocabulary or Information subtests; Wechsler, 2008) with which the information is either known or not, tends to have higher test−retest reliability when compared to more ability-based functions such as visuo-construction or visuo-spatial functions (e.g., WAIS-IV Block design or Matrix Reasoning; Dikmen, Heaton, Grant, & Temkin, 1999; Wechsler, 2008). Memory and executive functions may also produce comparatively lower reliability coefficients (Calamari, Markon, & Tranel, 2013). As discussed in Duff (2012), test−retest reliability may be affected by age (higher reliability at a younger age). Hinton-Bayre (2005) suggested that test−retest reliability should be greater than 0.70 for use in

RC calculations. Dikmen et al. (1999) suggested use of composite scores as opposed to single test scores may improve test—retest reliability. The improvement in reliability may also be seen when reviewing subtest versus composite scores for the WAIS-IV and WMS-IV (Wechsler, 2008; Wechsler, 2009).

Standard error of measurement: The SEM has been defined as "the standard deviation of obtained scores over...testing for a given individual..." (Nunnally & Bernstein, 1994, p. 211). The WAIS-IV technical and interpretive manual describes SEM as "an estimate of the amount of error in an individuals observed test score" (Wechsler, 2008, p. 45). Further, as reliability increases, SEM decreases, and accuracy of obtained scores increase (Wechsler, 2008). The SEM is calculated as the control group's baseline (time 1) standard deviation (SD_x) multiplied by the square root of 1 minus the test—retest reliability (r_{xy}; Pearson product moment correlation) of the measure (SEM $= SD_x\sqrt{1}-rxy$) as referenced by Chelune et al. (1993).

Standard error of the difference: The SED refers to the "spread of distribution of change scores that would be expected if no actual change had occurred" (Jacobson & Truax, 1991, p. 14). The SED can be calculated a number of ways. Chelune et al. (1993) calculated SED as the square root of 2 multiplied by the SEM to the second power (SED $= \sqrt{2(SEM)^2}$). Iverson in 2001, noted the SED should be calculated using the SEM calculated for the control baseline SD_x and control retest SD_y separately (SED $= \sqrt{SEMx + SEMy)^2}$) where $SEM_x = SD_x\sqrt{1}-r_{xy}$ and $SEM_y = SD_y\sqrt{1}-r_{xy}$. Thereby, SEM is not assumed to be the same at each testing interval.

Standard error of the estimate: The SEE is used in regression-based equations for comparing obtained and predicted posttest scores. The SEE (also known as residual SD) describes the variability of scores around the regression line in the sample for which the equation was created (Crawford & Garthwaite, 2006). The SEE can be calculated in a number of ways. Hinton-Bayre (2016) provided a calculation for SEE based on summary statistics for the SRB model of McSweeny et al. (1993), where, SEE $= SD_y \sqrt{(1 - r_{xy}^2)}$.

Practice effect: Having prior exposure to or experience with a specific test (even alternate forms) can result in higher scores on subsequent testing which is referred to as practice effect. The degree of practice effect has been found to be influenced by the length of time between testing and the construct (e.g., cognitive domain) being tested (Calamia, Markon, & Tranel, 2012). Greater length of time between testing tends to reduce, but not necessarily eliminate practice effect. Tests of crystalized intelligence or functions (e.g., word knowledge, general fund of knowledge, and reading), where the information is either known or not, tend to have a lower degree of practice effect, whereas tests of fluid intelligence or functions (e.g., problem-solving) tend to show a higher degree of practice effect. It was suggested by

McCaffrey and Westervelt (1995) that two or more baseline testings can stabilize practice effects. Prior exposure to a test's procedure may result in practice effect even if an alternate form is used (Wilson, Watson, Baddeley, Emslie, & Evans, 2000).

Differential practice: Differential practice (Maassen, 2004) refers to the difference or inequality on baseline and retest variance or SDs. Practice effect from repeat testing is unlikely to affect cases equally resulting in a different degree of variance or changes in SD between baseline and retest intervals. As discussed below, differential practice or inequality of variance affects obtained RC scores and classifications.

Regression to the mean: Regression to the mean is the tendency of test scores to move closer to the mean score on subsequent testing. This appears to be especially the case for extreme scores (either high or low). Therefore, changes in obtained scores on serial testing may in part be due to movement toward the mean as opposed to an actual change in function. Regression-based models account for regression to the mean (Hinton-Bayre, 2010).

METHODS TO EVALUATE FOR CHANGE

There have been a number of different methods employed to evaluate for significant change which generally include the simple difference method, RCI, and SRB methods. Some of these approaches require use of mean scores, SDs, and test—retest reliability data which can be obtained from test manuals or research articles. Obtaining statistics from samples similar to the one being studied is recommended (Duff, 2012; Hinton-Bayre, 2010). In addition, use of statistics (test—retest reliability, practice affect) from a sample with a similar interval between baseline and retest is recommended. Raw or standardized scores can be used to develop RCI or SRB equations but the same score metric (raw or standardized score) for each case and the equation should match. For example, in an RCI formula, if T-scores were used to identify the practice effect then the baseline and retest scores for each case should be in T-score units. If using SRB generated from a data set, then again the same score metric for the control data and actual cases should be used. The greatest debate regarding methods to evaluate RC methods centers around the method used to calculate the standard error (Hinton-Bayre, 2004; Maassen, 2004; Maassen et al., 2009; Temkin, 2004). Refer to Table 3.1 for common equations used in RCI and discussed SRB (for summary data).

Simple Difference

The simplest methods for evaluating change or score differences involve a simple difference between scores compared to base rates of the observed difference. A similar approach involves correction of a difference score for the test's SD to produce a standardized score for comparison. However, both of

TABLE 3.1 Equations for Reliable Change Models

RC Model	Equation	Standard Error	Predicted Score
Standard deviation index	$Z_y - Z_x/SD_x$		
RCI Jacobson and Truax (RCI$_{jt}$)	$Z_y - Z_x/SED$	$SED = \sqrt{2SD_x^2(1 - r_{xy})}$	
RCI Chelune (RCI$_c$)	$(Z_y - Z_x) - (M_y - M_x)/SED$	$SED = \sqrt{2SD_x^2(1 - r_{xy})}$	
RCI Iverson (RCI$_i$)	$(Z_y - Z_x) - (M_y - M_x)/SED$	$SED = \sqrt{(SD_x\sqrt{1 - r_{xy}})^2 + (SD_y\sqrt{1 - r_{xy}})^2}$	
SRB McSweeny (SRB$_{mc}$; estimated with summary data)	$(Z_y - Z')/SEE$	$SEE = SD_y\sqrt{(1 - r_{xy}^2)}$	$Z' = bZ_x + a$
			$b = r_{xy}(SD_y/SD_x)$
			$a = M_y - bM_x$
SRB Maassen (SRB$_{ma}$; estimated with summary data)	$(Z_y - Z')/SED$	$SED = \sqrt{(SD_x^2 + SD_y^2)(1 - r_{xy})}$	$Z' = bZ_x + a$
			$b = SD_y/SD_x$
			$a = M_y - bM_x$

Note: SRB formulas based on summary data as reported in Hinton-Bayre 2016. Z_x is the case baseline score. Z_y is the case retest score. Z' is the predicted case retest score. SD_x is the control group baseline SD. SD_y is the control group retest SD. M_x is the control group baseline mean. M_y is the control group retest mean. r_{xy} is mean group test–retest reliability (uncorrected Pearson product moment correlation). SED is the standard error of the difference. SEE is the standard error of the estimate. b is the slope of the regression line. a is the constant for the regression equation.

these methods do not account for practice effect, regression to the mean, or the reliability of the test. The simple difference method is calculated by subtracting cases' baseline score (Z_x) from their retest score (Z_y; e.g., simple difference = $Z_y - Z_x$) and then comparing the result to a table of base rates or percent of the tested population that had the same or similar score discrepancy. The WAIS-IV (Wechsler, 2008) uses a similar approach to determining the base rate for differences observed between indices (e.g., Verbal Comprehension Index and Perpetual Reasoning Index). Although this method is simple, it is reliant on base rate data being available for the tests of interest and it does not account for sources of error that may reduce the accuracy of findings. A variant of this approach adjusts the simple difference discrepancy score for the SD (($Z_y - Z_x$)/SD_x) resulting a z-score. Generally, a z-score ± 1.645 is considered statistically significant, indicating a true/RC which corresponds to the 90% confidence interval (CI) with a 5% chance of error for positive or negative change.

Reliable Change Indices

An early statistical approach to measure RCI was developed by Jacobson and Truax (1991) as a method to quantify if statistically significant change had occurred for psychotherapy clients. This method evaluates for significant change by correcting difference scores for the test's SD at time 1 as well as the test–retest reliability of the measure producing a z-score where significant change is indicated by score ± 1.645 at the 90% CI. Refer to Table 3.1 for this RCI model (RCI_{jt}).

Application of RCI in a neuropsychological context raises the potential need to not only account for the variability of scores around the mean and the reliability of the measure but also the possibility of practice effects. Chelune et al. (1993) modified the Jacobson and Truax (1991) RCI formula to account for practice effects by calculating the observed practice effects within the normative sample (e.g., control mean score at retest minus control test mean score at baseline). The reader is referred to Table 3.1 for details regarding this equation (RCI_c). Conceptually, the change in score from baseline to retest is adjusted for observed practice effect in the control data and then adjusted for the observed variability in the control sample and the test–retest reliability of the measure. However, this applies a uniform adjustment for practice effect across individuals. Iverson (2001) suggested the variability in scores (SDs) at baseline and retest should not be assumed to be the same and consequently, to more accurately identify significant change, the SD at baseline and retest in the control sample should be included in the calculation of SED as opposed to only including the SD at baseline (see Table 3.1, RCI_i). Both Chelune et al. (1993) and Iverson (2001) RCI scores are considered indicative of a significant change when scores are ± 1.645 at the 90% CI.

The Iverson (2001) and Chelune et al. (1993) are the most commonly cited approaches for calculating RCI in neuropsychological research/practice, however, there are other methods. Many empirical studies have addressed variation in the RCI equation typically with various methods to account for error. For a review of differing methods to calculate error in RCI, the reader is directed to Maassen et al. (2009), Hinton-Bayre (2004), and Temkin (2004). Sawrie (2002) noted a significant limitation of the RCI method is the assumption of uniform practice effects across patients and consequently reducing the accuracy of RCI scores and determination of change/no change. Sawrie (2002) also noted that RCI is further limited by not accounting for other relevant demographic variables such as age, education, and gender.

Standardized Regression-Based Models

Regression-based models of RC have also been created to identify presence of significant change in test scores. SRB equations can be based on actual data sets (e.g., McSweeny et al., 1993); however, it is possible to use summary statistics in various SRB equations (Crawford, Garthwaite, Denham, & Chelune, 2012; Hinton-Bayre, 2016). A regression-based program has been made available for use with summary statistics by Crawford et al. (2012). Regression-based formulas are very complicated and consequently have an increased risk for error. For developing regression-based formulas, the reader is referred to Crawford & Garthwaite (2006), Hinton-Bayre (2010), and Maassen et al. (2009). Standard RCI methods directly compare obtained scores (baseline and retest) as adjusted for practice affect and error to determine if there is a statistically significant change in the obtained score beyond expected variability. Regression-based RC compares an obtained case retest score to a predicted retest score to determine if the obtained retest score significantly deviates from the predicted retest score. The regression equation accounts for reliability, practice effects, regression the mean, and measurement error in the simple form but in a multivariate/complex approach other variables can be included such as other test scores (e.g., IQ) and other demographic data (e.g., age, education). As noted by Hinton-Bayre (2016), some SRB models include reliability in the SEE only (Maassen et al., 2009) while others account for reliability in the predicted score and SEE (McSweeny et al., 1993). See Table 3.1 for relevant formulas for SRB using summary data for the McSweeny et al. (1993) (SRB_{mc}) and the Maassen et al. (2009) (SRB_{ma}) models. In a SRB, the predicted retest score is based on the regression equation produced for the retest score's regression slope, intercept, and constant. A RC is evaluated for by subtracting the predicted retest score from obtained retest score ($M_y - M'$) and adjusting or dividing that difference by the regression equation's SEE. Thus, the equation for the time 2 predicted score is as follows: $M' = bZ_x + c$ where M' is the predicted retest score, b is the regression slope, Z_x is the case baseline score, and c is the intercept/constant (McSweeny et al., 1993). The equation for regression-based RC is as follows: $(M_y - M')/SEE$. As with other methods of RC, a change score of ± 1.645 is typically used to identify significant change. Hinton-Bayre (2010) provides a detailed description and comparison of methods to identify RC in RCI models and SRB models.

Comparison of Change Predictors

Methods to evaluate change have been used widely in control and clinical samples and have become an important method to evaluate changes in clinical groups, response to treatment, and effects of surgery. A challenging issue for researchers and clinicians is selecting the best method for evaluating change. There have been a number of studies (Frerichs & Tuokko, 2005; Heaton et al., 2001; Hinton-Bayre, 2016; Maassen et al., 2009; Temkin, Heaton, Grant, & Dikmen, 1999) that compared RCIs and SRBs within the same sample in an attempt to identify a preferred method. The following discusses a number of important studies comparing RCI and SRB models within the same sample.

Temkin et al. (1999) compared RCI without correction for practice effects, RCI with correction for practice effects, simple SRB, and complex SRB to evaluate for change on seven measures of neuropsychological functioning in neurologically stable cases. They found similar classification between RCI with practice effects, simple SRB, and complex SRB. The RCI without correction for practice effects had the largest error rate. The authors noted the greatest discrepancies in models occurred with extremes of initial test scores (especially lower scores) compared to the control mean at initial test. This would seem to be especially important in clinical cases where initial test performance may be below the normative mean if a healthy control group is used for data in the RC model.

Heaton et al. (2001) compared three RC models (RCI with practice effects, simple SRB, and complex SRB) in normal controls and a clinical sample. Results indicated similar classification rates when comparing RC models. However, it was stated that when evaluating for change, normative data used for RC models should be from neurologically stable cases (not necessarily neurologically normal) with similar initial test mean performance to the case or population of study.

Frerichs and Tuokko (2005) compared six RC models (SD, RCI without practice effects, RCI with practice effects, RCI corrected for regression to the mean (Hsu, 1989), simple SRB, and complex SRB) in classification of change in healthy controls across 5 years. Results showed similar classification using the RCI with practice, simple SRB, and complex SRB. The RCI without accounting for practice effects and SD methods performed least well. Complex/multivariate SRB did not show much advantage over simple SRB.

Maassen et al. (2009) compared RC models (RCI and SRB) in real and simulated data. They proposed adjustments to the error terms in the Chelune et al. (1993) RCI with practice model and the McSweeny et al. (1993) simple SRB model. Results essentially demonstrated that the simple SRB classified the most as changed and the adjusted simple SRB was the most conservative with prediction of change. The RCI with practice was between the other two in prediction.

Hinton-Bayre (2010) indicated that a consensus has not been reached regarding what RC model best evaluates for change and that there is no RC model that consistently produces more extreme or conservative predictions. It was stated that "agreement between RC models will depend on practice effect, reliability, variance inequality, and the individual cases' relative position to the control group at initial testing" (Hinton-Bayre, 2010, p. 251). It was suggested that as variances for the control sample on initial test and retest are equal or closer to equal, the agreement between RC models will increase. In SRB, differences in initial test—retest variance (differential practice) results in more extreme or conservative estimates depending on whether the case scored above or below the control group mean on initial testing.

In a more recent article, Hinton-Bayre (2016) used simulated data to examine the effect of inequality of variance (differential practice) and the case scores relative position when compared to the control mean. Data were manipulated to be able to investigate the effect of direction of differential practice (SD_x compared to SD_y) and the relative position (lower or higher) of the baseline test score (Z_x) compared to the control mean baseline score (M_x). Comparisons were made graphically of the "responsiveness" or degree of negative or positive change identified by each model. A number of important results were derived from this study (for results, see Tables 3.2 and 3.3). It was noted that classification of change will be similar between models when the baseline score is close to the control baseline mean and/or when the control baseline and retest variances are very similar (e.g., minimal differential practice). It was illustrated (see Table 3.2) that if cases are expected to have a decline in performance/negative change and the baseline score is expected to be below the control baseline mean (e.g., as may be seen in various clinical groups), the SRB model of McSweeny et al. (1993; SRBmc) is most responsive to further decline regardless of differential practice. However, the SRB_{mc} model was least responsive if the baseline score is above the control baseline mean. In the case of expected positive change/improvement (see Table 3.3) and when the baseline score is expected to be above the control baseline mean, then the SRB_{mc} is most responsive regardless of differential practice. This model was found to be least responsive when the baseline score was below the control baseline mean score regardless of differential practice. Two other consistent patterns of responsiveness were noted. For the RCI Iverson (2001) (RCIi) model (see Tables 3.2 and 3.3) with expected negative or positive change, moderate responsiveness

TABLE 3.2 Responsiveness to Differential Practice and Relative Position at Baseline (Negative Change)

Change in Variable	Order of Most (1) to Least (3) Responsive by RC Model		
	RCI_i	SRB_{mc}	SRB_{ma}
$SD_x < SD_y$			
$Z_x < M_x$	2	1	3
$Z_x > M_x$	2	3	1
$SD_x > SD_y$			
$Z_x < M_x$	3	1	2
$Z_x > M_x$	1	3	2

Note: RCI_i refers to Iverson (2001) RCI model. SRB_{mc} refers to standardized regression model of McSweeny et al. (1993). SRB_{ma} refers to Maassen et al. (2009) adjusted SRB model. SD_x is control data baseline SD. SD_y is control data retest SD. Z_x is cases baseline score. M_x is control data mean baseline score.

TABLE 3.3 Responsiveness to Differential Practice and Relative Position at Baseline (Positive Change)

Change in Variable	Order of Most (1) to Least (3) Responsive by RC Model		
	RCI_i	SRB_{mc}	SRB_{ma}
$SD_x < SD_y$			
$Z_x < M_x$	2	3	1
$Z_x > M_x$	2	1	3
$SD_x > SD_y$			
$Z_x < M_x$	1	3	2
$Z_x > M_x$	3	1	2

Note: RCI_i refers to Iverson (2001) RCI model. SRB_{mc} refers to standardized regression model of McSweeny et al. (1993). SRB_{ma} refers to Maassen et al. (2009) adjusted SRB model. SD_x is control data baseline SD. SD_y is control data retest SD. Z_x is cases baseline score. M_x is control data mean baseline score.

occurred when control retest variance exceeded control baseline variance regardless of the relative position of the baseline score compared to the control mean score. For the SRB Maassen et al. (2009) (SRBma) model with expected negative or positive change, moderate responsiveness occurred when control baseline variance exceeded control retest variance regardless of the relative position of the cases baseline score compared to the control mean score. Other patterns of responsiveness varied by differential practice and the relative baseline score position compared to the control baseline mean. Overall, results essentially illustrated that the SRB McSweeny et al. (1993) model was most responsive to decline when baseline case scores are below control mean baseline scores and was most responsive to improvement when the case baseline scores are higher than the control mean baseline scores. Similar classification between models will occur when the baseline score is near/equal to the control baseline mean and/or when differential practice is minimal/equal baseline and retest variance. It should be noted that this study used generated data to evaluate effects of the variables of classification of change and consequently replication is important with clinical cases.

In summary, there has been some recent evidence that the SRB model of McSweeny et al. (1993) may be more responsive to change depending on factors above (Hinton-Bayre, 2016). For example, in the event of a clinical condition where baseline case scores are expected to be lower than the control mean scores (e.g., mild cognitive impairment (MCI)) and there is expectation for possible further decline, then the SRB McSweeny model would likely be most responsive to further decline/change. If a case's baseline scores are near or at the control mean and/or when the baseline and retest variance are equal or near equal, then models are likely to have similar classification. If improvement in function is likely and the case's baseline score is at or above the control baseline mean, such as with some concussion cases, then the SRB McSweeny model may be most responsive. Therefore, the difference in classification of change appears to depend on how similar the cases initial test score is to the normative/control group's initial test score, the degree of inequality of variance/differential practice in the normative/control group, and the test–retest reliability. The RC model used should account for practice effects as models not accounting for practice effects have poorer classification.

PRACTICAL IMPLICATIONS FOR USE OF RELIABLE CHANGE MODELS

Advantages, Challenges, and Limitations in Reliable Change Models

When considering use of RC models in clinical practice or research, several potential advantages and challenges should be taken into account. Use of a well-defined, consistently applied, and replicable method for determining if change has occurred is clearly important when monitoring the course of conditions and evaluating for response to treatment. Use of RC models provides a common method that can be applied in a wide range of conditions and with a wide range of variables to objectively evaluate for measurable change beyond what could be attributed to clinical and test related variables (e.g., reliability of the measure, practice effects, regression to the mean, variance within the measure, and demographic variables). It also provides a common metric/method by which empirical studies can be compared.

There are a number of general challenges when using RC models. Summary data should be obtained from samples similar to the case being evaluated or studied. However, summary data is not always available for the variable selected within the selected clinical condition and at times it may be necessary to use summary data from the test's standardization sample until further data is available. In addition, for selected variables, it may be difficult to find a sample with initial mean test scores similar to the case of study. Often test−retest intervals used for test development normative data have shorter test−retest intervals (weeks) than compared to what is used in clinical cases (e.g., many months to years) which may affect summary data (e.g., test−retest reliability, practice effects, and equality of variance). Of notable importance, floor effects or restricted range of scores is particularly important when scores are being corrected for practice effects and sources of error. In clinical cases where scores may already be lower than the normative mean, this may result in an inability to obtain a significant decline in a score. This becomes even more of an issue if the measure used has lower reliability and/or greater SDs/variance in the normative group. Consequently, due to measurement issues, no change may be detected when real change may have occurred.

Use of RCI formulas (e.g., Chelune et al., 1993, Iverson, 2001) certainly has the advantage of calculations that are fairly simple with readily available data (e.g., means, SDs, and test−retest reliability) that can often be obtained through research articles, test technical manuals, or personal data sets. However, some SRB models can be used with summary data (e.g., Crawford & Garthwaite, 2006; Hinton-Bayre, 2016) and fairly simple equations similar to RCI equations (Hinton-Bayre, 2016). The RCI formulas corrected for practice effects typically produce similar classification of cases when compared to SRB models when the case's baseline score is near the controls' baseline mean score and/or when baseline and retest variance is equal or near equal. This method adjusts for important variables that affect obtained test scores including practice effects, test−retest reliability, and variability. Correction for practice effects is an important component of RCI formulas; however, a relevant criticism is that RCI formulas apply correction for practice effects uniformly across cases, which is likely an inaccurate method since other factors likely affect the degree to which practice effects occur (Sawrie, 2002). In addition, the Chelune et al. (1993) RCI formula treats variance as equal for both test and retest time points; however, the Iverson (2001) formula does not. Consequently, use of the Iverson (2001) RCI formula is preferred. Classification rates may be affected by case scores that are lower than initial mean test scores for the normative group which may be an issue for various clinical groups. Lastly, most RCI models do not account for regression to the mean which is a factor known to affect retest scores.

The SRB models have several advantages. These models have flexibility in variables (e.g., age, education, and test—retest interval) that can be used to predict retest scores. As with RCI, this approach accounts for test—retest reliability, practice effects, and variance but it also accounts for regression to the mean. Additional variables may be added such as demographic or clinical variables; however, multivariate approaches have not been consistently found to provide improved classification to a simple SRB model (Duff, 2012; Frerichs & Tuokko, 2005). Unlike RCI models, SRB models do not assume uniform practice effects. Formulas used in SRB are quite complicated and prone to clerical error; however, use of SRB models based on summary data can balance simpler equations and readily obtained input variables or use of established programs (e.g., Crawford & Garthwaite, 2007; Hinton-Bayre, 2016). As with RCI models, classification is affected by extremes in test scores. Classification of change in SRB is affected by the relative position of the case's baseline test score as compared to the control sample's baseline mean score, differential practice, and reliability.

Implications for Practice: Suggestions for Use

The following outlines a number of key factors for implementation of RC models to a clinical or research sample with the purpose of providing a practical template for use. This is not intended as a format for "best practice." Use of an objective measure of change in cognitive test scores across evaluations is important for monitoring change in clinical conditions, recovery from illness or injury, and response to treatment including neurosurgical intervention. This chapter was not aimed toward a review of findings using RC in specific neurosurgical or clinical samples. The reader is directed to Table 3.4 in the Appendix of this chapter for a list of relevant studies using RC in surgical related samples (epilepsy, deep brain stimulation, cardiac surgery, oncology, and related variables). Although the citations in Table 3.4 are limited to surgical and related variables, RC has been implemented in many other neurological conditions such as concussion, TBI, multiple sclerosis, MCI, and dementia.

RC methodology can be used in a practical way for clinical practice and research. Considering research to date, RC formulas have often been found to produce similar classification rates for RCI (that accounts for practice) and SRB models. Typically, limited improvement has been found with use of multivariate SRB models. Most recently, there has been some evidence that the SRB formula of McSweeny et al. (1993) used with summary data (Table 3.1) may have an advantage compared to the RCI of Iverson (2001) and the adjusted SRB of Maassen et al. (2009) with use of summary data. The SRB of McSweeny may especially have an advantage in clinical cases when the cases baseline scores are expected to be below the control baseline mean and when decline on retest score is expected. The SRB McSweeny also appears to be preferable when expecting improved performance and baseline case scores are expected to be above the controls' baseline score. If use of an RCI model (nonregression model) is considered, then the Iverson (2001) model is preferred since it accounts for practice effects and allows for inequality of variance.

If there is potential of serial testing in a clinical setting or for research purposes, several test related factors should be considered. Selecting tests with strong test–retest reliability is very important since this will influence the accuracy and responsiveness of RC models. Greater reliability results in less error which results in greater responsiveness (Hinton-Bayre, 2016). Use of measures with a Pearson product moment correlation (test–retest reliability) of >0. 70 is strongly recommended with higher correlations being preferable (Calamari et al., 2013). Use of tests with alternate forms does not necessarily preclude practice effects. Tests should have sufficient range of scores to allow for ability to measure change, especially if baseline scores are expected to be above or below the normative mean or if the test has a naturally skewed distribution of raw scores (e.g., Boston Naming Test). Obviously, test selection should also be based on appropriateness of use in the sample of study/clinical case and should have relevance for the reason for examination. For example, in a presurgical temporal lobe epilepsy case, specific tests should be selected that have been found useful in that group (e.g., list learning, paired associates, and confrontation naming) and tests that establish preoperative baseline functioning while maximizing evaluation of areas with potential for change (e.g., emphasize memory, naming/language, and executive functions). Likewise, other clinical conditions such as multiple sclerosis or concussion may require placing greater emphasis on attention, processing speed, and motor functions.

The normative/control group from which summary data is obtained should be considered carefully regardless of which RC model is selected. Generally, the control group's summary data (baseline and retest means and SDs, Pearson product moment correlation/reliability) should be matched as closely as possible to the group of study/case. In addition, the control sample and case(s) of study should approximate each other's clinical (e.g., condition, retest interval) and demographic (e.g., age, education) factors if possible. Typically, summary data for cases showing stable performance across the baseline to retest evaluation are used for control summary data as this allows for evaluation of decline or improvement in performance. If possible, a control sample with a similar test–retest interval should be used since this will likely most closely approximate the reliability, practice effects, and variance of the case/group of study. For example, summary data collected from patients with MCI that showed stable performance across the retest interval could be a good comparison group if evaluating for decline in an MCI patient's performance. An argument could also be made for use of similar aged healthy controls' summary data. In the case of presurgical epilepsy evaluation, a sample of cases with a similar epilepsy diagnosis without surgical intervention may be the most appropriate comparison group if evaluating for effects of surgery. It is important to the aware that the selected summary data will have a predictable effect on responsiveness of the RC model depending on the case's relative baseline position compared to the controls' baseline mean and presence of differential practice. Therefore, summary data will have an effect on classification.

The literature has not clearly identified whether there is a preferred metric of score (e.g., raw score or standardized score) to be used in RC models. As discussed by Duff (2012), case scores should be in the same metric as those used in the control data for the RC formula. Therefore, if raw scores were used to establish the mean, SD, and reliability in the control data, then raw scores from the case should be used.

Evaluation for significant change in baseline and retest scores depends on the degree of confidence desired in the comparison. Most typically, a RC score (RCI or SRB) of ± 1.645 is used to determine if the magnitude of change in score from baseline to retest was significant. This corresponds to 90% confidence that the difference in scores did not occur by chance (5% chance a decline and 5% chance an increase was erroneous). If the consequences are deleterious for misclassifying a case as unchanged when change has occurred (false negative), then a more lenient significance value could be used (80% or 85%). As discussed by Duff (2012), the use of 90% confidence as the mark for significance in RCI was not empirically derived. The resulting RC score can be used statistically as a continuous variable or to create categorical groups (e.g., significant decline, no change, and significant increase). It should be clear however, that statistically significant change does not necessarily equate to clinically significant change. Statistically significant change refers to whether the change occurred by chance not whether the change or degree of change is rare. Therefore, use of base rate data or effect size to clarify how common/uncommon the degree of change is may be useful; however, there is generally very limited availability of base rate information for a wide range of tests and within specific clinical conditions.

The effect of restricted range in scores or floor/ceiling effects must also be considered. There certainly are clinical conditions where a patient's score would preclude the ability to identify further change such as in the case of a patient with MCI or dementia with a delayed memory score of 0/15 on the Rey Auditory Verbal Learning Test (RAVLT). Obviously in this example, there would be no way to evaluate for a decline in function on this measure. However, this issue is not confined to the most extreme cases of floor or ceiling effect. To be classified as having a significant decline, a case would have to have sufficient raw score range below their baseline score to have a significant decline after accounting for other factors in the RC equation. Depending on summary data and the RCI equation used, a RAVLT delayed recall score would need to be able to change by 4 or more raw score points to be found significantly changed. This will vary depending on the reliability and variance in the measure. This would not necessarily be true for SRB models where each case may require a different amount of change to be significant depending on their baseline score and consequently the absolute floor or ceiling for determining change will vary. For RCI models, it would be possible to identify the absolute lowest score required for each test to measure significant change in order to screen for cases to exclude or to inform of limitations in the data/study. The best approach to handling this issue is not clear. In research, excluding cases due to low score (or very high scores) may result in a sample that does not well represent the condition of study (e.g., results in poor generalizability). Alternatively, if cases are included that achieve the floor or ceiling but are classified as unchanged this will affect categorical group analysis and will truncate the degree of change measured in the sample. This may in turn skew interpretation of clinical course or treatment effect.

CONCLUSION

Reliable change methods can be a useful part of objectively evaluating change in neuropsychological functioning across time. This can assist with evaluating for progression or recovery from a condition and response to treatment. Evaluating RC can improve research and clinical practice by providing an objective method to account for factors that affect serial test performance including reliability, variance, practice effects, regression to the mean, and the retest interval. Summary data (baseline and retest means and SDs, Pearson product moment correlation/test−retest reliability) can be easily obtained for most tests for use in RC equations. Further development of normative data with relevant clinical test−retest intervals is important. There has been some recent evidence that the SRB model of McSweeny et al. (1993) may provide greater responsiveness and consequently may be the preferred method for evaluating change under a variety of conditions; however, replication is required for the Hinton-Bayre (2016) study. Findings from the Hinton-Bayre (2016) study provide insight into the way differential practice and the case's relative position at baseline compared to the control groups baseline mean affect classification of change by model. Perhaps most relevant, is an understanding of the variables that will affect classification by the RC equations which will assist with selection of an RC model and will provide an understanding of potential differences in classification that occur when RC models are compared.

REFERENCES

Baxendale, S., & Thompson, P. (2018). Red flags in epilepsy surgery: Identifying the patients who pay a high cognitive price for an unsuccessful surgical outcome. *Epilepsy & Behavior*, *78*, 269–272.

Bogod, N. M., Sinden, M., Woo, C., Defreitas, V. G., Torres, I. J., Howard, A. K., ... Lam, R. W. (2014). Long-term neuropsychological safety of subgenual cingulate gyrus deep brain stimulation for treatment-resistant depression. *Journal of Neuropsychiatry and Clinical Neuroscience*, *26*, 126–133.

Busch, R. M., Floden, D. P., Prayson, B., Chapin, J. S., Kim, K. H., Ferguson, L., ... Najm, I. M. (2016). Estimating risk of word-finding problems in adults undergoing epilepsy surgery. *Neurology*, *87*, 2363–2369.

Busch, R. M., Lineweaver, T. T., Ferguson, L., & Haut, J. S. (2015). Reliable change indices and standardized regression-based change score norms for evaluating neuropsychological change in children with epilepsy. *Epilepsy & Behavior*, *47*, 45–54.

Busch, R. M., Love, T. E., Jehi, L. E., Ferguson, L., Yardi, R., Najm, I., ... Gonzalez-Martinez, J. (2015). Effect of invasive EEG monitoring on cognitive outcome after left temporal lobe epilepsy surgery. *Neurology*, *85*, 1475–1481.

Caine, C., Deshmukh, S., Gondi, V., Mehta, M., Tomé, W., Corn, B. W., ... Kachnic, L. (2016). CogState computerized memory tests in patients with brain metastases: Secondary endpoint results of NRG Oncology RTOG 0933. *Journal of Neurooncology*, *126*, 327–336.

Calamia, M., Markon, K., & Tranel, D. (2012). Scoring higher the second time around: Meta-analysis of practice effects in neuropsychological assessment. *The Clinical Neuropsychologist*, *26*, 543–570.

Calamari, M., Markon, K., & Tranel, D. (2013). The robust reliability of neuropsychological measures: Meta-analyses of test–retest correlations. *The Clinical Neuropsychologist*, *27*, 1077–1105.

Chelune, G. J. (2010). Evidence-based research and practice in clinical neuropsychology. *The Clinical Neuropsychologist*, *24*, 454–467.

Chelune, G. J., Naugle, R. I., Luders, H., Sedlak, J., & Awad, I. A. (1993). Individual change after epilepsy surgery: Practice effects and base-rate information. *Neuropsychology*, *7*, 41–52.

Correa, D. D., Root, J. C., Baser, R., Moore, D., Peck, K. K., Lis, E., ... Relkin, N. (2013). A prospective evaluation of changes in brain structure and cognitive functions in adult stem cell transplant recipients. *Brain Imaging and Behavior*, *7*, 478–490.

Crawford, J. R., & Garthwaite, P. H. (2006). Comparing patient's predicted test scores from a regression equation with their obtained scores: A significance test and point estimate of abnormality with accompanying confidence limits. *Neuropsychology*, *20*, 259–271.

Crawford, J. R., & Garthwaite, P. H. (2007). Using regression equations built from summary data in the neuropsychological assessment of the individual case. *Neuropsychology*, *21*, 611–620.

Crawford, J. R., Garthwaite, P. H., Denham, A. K., & Chelune, G. J. (2012). Using regression equations built from summary data in the psychological assessment of the individual case: Extension to multiple regression. *Psychological Assessment*, *24*, 801–814.

Dieleman, J., Sauër, A. M., Klijn, C., Nathoe, H., Moons, K., Kalkman, C., ... Van Dijk, D. (2009). Presence of coronary collaterals is associated with a decreased incidence of cognitive decline after coronary artery bypass surgery. *European Journal of Cardiothoracic Surgery*, *35*, 48–53.

Dikmen, S. S., Heaton, R. K., Grant, I., & Temkin, N. R. (1999). Test−retest reliability and practice effects of expanded Halstead-Reitan Neuropsychological Test Battery. *Journal of the International Neuropsychological Society, 5,* 346−356.

Duff, K. (2012). Evidence-based indicators of neuropsychological change in the individual patient: Relevant concepts and methods. *Archives of Clinical Neuropsychology, 27,* 248−261.

Evered, L., Scott, D. A., Silbert, B., & Maruff, P. (2011). Postoperative cognitive dysfunction is independent of type of surgery and anesthetic. *Anesthesia and Analgesia, 112,* 1179−1185.

Evered, L. A., Silbert, B. S., Scott, D. A., Maruff, P., & Ames, D. (2016). Prevalence of dementia 7.5 years after coronary artery bypass graft surgery. *Anesthesiology, 125,* 62−71.

Frerichs, R. J., & Tuokko, H. A. (2005). A comparison of methods for measuring cognitive change in older adults. *Archives of Clinical Neuropsychology, 20,* 321−333.

Gondi, V., Paulus, R., Bruner, D. W., Meyers, C. A., Gore, E. M., Wolfson, A., ... Movsas, B. (2013). Decline in tested and self-reported cognitive functioning after prophylactic cranial irradiation for lung cancer: Pooled secondary analysis of Radiation Therapy Oncology Group randomized trials 0212 and 0214. *International Journal of Radiation Oncology, Biology and Physics, 86,* 656−664.

Heaton, R. K., Temkin, N., Dikmen, S., Avitable, N., Taylor, M. J., Marcotte, T. D., & Grant, I. (2001). Detecting change: A comparison of three neuropsychological methods, using normal and clinical samples. *Archives of Clinical Neuropsychology, 16,* 75−91.

Heilbronner, R. L., Sweet, J. J., Attix, D. K., Krull, K. R., Henry, G. K., & Hart, R. P. (2010). Official position of the American Academy of Clinical Neuropsychology on serial neuropsychological assessments: The utility and challenges of repeat test administrations in clinical and forensic contexts. *The Clinical Neuropsychologist, 24,* 1267−1278.

Hermann, B. P., Seidenberg, M., Schoenfeld, J., Peterson, J., Leveroni, C., & Wyler, A. R. (1996). Empirical techniques for determining the reliability, magnitude, and pattern of neuropsychological change after epilepsy surgery. *Epilepsia, 37,* 942−950.

Higginson, C. I., Wheelock, V. L., Levine, D., King, D. S., Pappas, C. T., & Sigvardt, K. A. (2009). The clinical significance of neuropsychological changes following bilateral subthalamic nucleus deep brain stimulation for Parkinson's disease. *Journal of Clinical and Experimental Neuropsychology, 31,* 65−72.

Hill, S. W., Gale, S. D., Pearson, C., & Smith, K. (2012). Neuropsychological outcome following minimal access subtemporal selective amygdalohippocampectomy. *Seizure, 21,* 353−360.

Hinton-Bayre, A. D. (2004). Holding out for a reliable change from confusion to a solution: a comment on Maassen's "The standard error in the Jacobson and Truax Reliable Change Index". *Journal of the International Neuropsychological Society, 10,* 894−898.

Hinton-Bayre, A. D. (2005). Methodology is more important than statistics when determining reliable change. *Journal of the International Neuropsychological Society, 11,* 788−789.

Hinton-Bayre, A. D. (2010). Deriving reliable change statistics from test−retest normative data: Comparison of models and mathematical expressions. *Archives of Clinical Neuropsychology, 25,* 244−256.

Hinton-Bayre, A. D. (2016). Clarifying discrepancies in responsiveness between reliable change indices. *Archives of Clinical Neuropsychology, 31,* 754−768.

Hsu, L. M. (1989). Reliable changes in psychotherapy: Taking into account regression toward the mean. *Behavioral Assessment, 11,* 459−467.

Iverson, G. L. (2001). Interpreting change on the WAIS-III/WMS-III in clinical samples. *Archives of Clinical Neuropsychology, 16,* 183−191.

Jacobson, N. S., & Truax, P. (1991). Clinical significance: A statistical approach to defining meaningful change in psychotherapy research. *Journal of Consulting and Clinical Psychology, 59*, 12−19.

Jahanshahi, M., Torkamani, M., Beigi, M., Wilkinson, L., Page, D., Madeley, L., ... Tisch, S. (2014). Pallidal stimulation for primary generalised dystonia: Effect on cognition, mood and quality of life. *Journal of Neurology, 261*, 164−173.

Jones, D., Vichaya, E. G., Wang, X. S., Sailors, M. H., Cleeland, C. S., & Wefel, J. S. (2013). Acute cognitive impairment in patients with multiple myeloma undergoing autologous hematopoietic stem cell transplant. *Cancer, 119*, 4188−4195.

Jurica, P. J., Leitten, C. L., & Mattis, S. (2001). *DRS-2 dementia rating scale-2 professional manual*. Odessa, FL: PAR.

Maassen, G. H. (2004). The standard error in the Jacobson and Truax reliable change index: The classical approach to the assessment of reliable change. *Journal of the International Neuropsychological Society, 10*, 888−893.

Maassen, G. H., Bossema, E. R., & Brand, N. (2009). Reliable change and practice effects: Outcomes of various indices compared. *Journal of Clinical and Experimental Neuropsychology, 31*, 339−352.

Martin, R., Griffith, H. R., Sawrie, S., Knowlton, R., & Faught, E. (2006). Determining empirically based self-reported cognitive change: Development of reliable change indices and standardized regression-based change norms for the multiple abilities self-report questionnaire in an epilepsy sample. *Epilepsy & Behavior, 8*, 239−245.

Martin, R., Sawrie, S., Gilliam, F., Mackey, M., Faught, E., Knowlton, R., & Kuzniekcy, R. (2002). Determining reliable cognitive change after epilepsy surgery: Development of reliable change indices and standardized regression-based change norms for the WMS-III and WAIS-III. *Epilepsia, 43*, 1551−1558.

McCaffrey, R. J., & Westervelt, H. J. (1995). Issues associated with repeated neuropsychological assessments. *Neuropsychology Review, 5*, 203−221.

McSweeny, A. J., Naugle, R. I., Chelune, G. J., & Luders, H. (1993). "T-scores for change": An illustration of a regression approach to depicting change in clinical neuropsychology. *Clinical Neuropsychologist, 7*, 300−312.

Mikos, A., Zahodne, L., Okun, M. S., Foote, K., & Bowers, D. (2010). Cognitive declines after unilateral deep brain stimulation surgery in Parkinson's disease: A controlled study using Reliable Change, part II. *The Clinical Neuropsychologist, 24*, 235−245.

Meyers, C. A., & Hess, K. R. (2003). Multifaceted end points in brain tumor clinical trials: Cognitive deterioration precedes MRI progression. *Neuro-Oncology, 5*, 89−95.

Mohile, S. G., Lacy, M., Rodin, M., Bylow, K., Dale, W., Meager, M. R., & Stadler, W. M. (2010). Cognitive effects of androgen deprivation therapy in an older cohort of men with prostate cancer. *Critical Reviews in Oncology/Hematology, 75*, 152−159.

Nemeth, E., Vig, K., Racz, K., Koritsanszky, K. B., Ronkay, K. I., Hamvas, F. P., ... Gal, J. (2017). Influence of the postoperative inflammatory response on cognitive decline in elderly patients undergoing on-pump cardiac surgery: A controlled, prospective observational study. *BMC Anesthesiology, 17*, 113.

Nunnally, J. C., & Bernstein, I. H. (1994). (3rd ed.). *Psychometric theory*, (7). New York: McGraw-Hill. (Chapter 6).

Ouimet, L. A., Stewart, A., Collins, B., Schindler, D., & Bielajew, C. (2009). Measuring neuropsychological change following breast cancer treatment: An analysis of statistical models. *Journal of Clinical and Experimental Neuropsychology, 31*, 73−89.

Rinehardt, E., Duff, K., Schoenberg, M., Mattingly, M., Bharucha, K., & Scott, J. (2010). Cognitive change on the repeatable battery of neuropsychological status (RBANS) in Parkinson's disease with and without bilateral subthalamic nucleus deep brain stimulation surgery. *The Clinical Neuropsychologist, 24,* 1339−1354.

Sawrie, S. M. (2002). Analysis of cognitive change: A commentary on Keith et al. (2002). *Neuropsychology, 16,* 429−431.

Sawrie, S. M., Chelune, G. J., Naugle, R. I., & Luders, H. O. (1996). Empirical methods for assessing meaningful neuropsychological change following epilepsy surgery. *Journal of the International Neuropsychological Society, 2,* 556−564.

Schoenberg, M. R., Rinehardt, E., Duff, K., Mattingly, M., Bharucha, K. J., & Scott, J. G. (2012). Assessing reliable change using the repeatable battery for the assessment of neuro-psychological status (RBANS) for patients with Parkinson's disease undergoing deep brain stimulation (DBS) surgery. *The Clinical Neuropsychologist, 26,* 255−270.

Sweet, J. J., Finnin, E., Wolfe, P. L., Beaumont, J. L., Hahn, E., Marymont, J., . . . Rosengart, T. K. (2008). Absence of cognitive decline one year after coronary bypass surgery: Comparison to nonsurgical and healthy controls. *Annals of Thoracic Surgery, 85,* 1571−1578.

Temkin, N. R. (2004). Standard error in the Jacobson and Truax reliable change index: The "classical" approach" leads to poor estimates. *Journal of the International Neuropsychological Society, 10,* 899−901.

Temkin, N. R., Heaton, R. K., Grant, I., & Dikmen, S. S. (1999). Detecting significant change in neuropsychological test performance: A comparison of four models. *Journal of the International Neuropsychological Society, 5,* 357−369.

Tröster, A. I., Meador, K. J., Irwin, C. P., Fisher, R. S., & SANTE Study Group. (2017). Memory and mood outcomes after anterior thalamic stimulation for refractory partial epilepsy. *Seizure, 45,* 133−141.

Tröster, A. I., Woods, S. P., & Morgan, E. E. (2007). Assessing cognitive change in Parkinson's disease: Development of practice effect-corrected reliable change indices. *Archives of Clinical Neuropsychology, 22,* 711−718.

Tully, P. J., & Baker, R. A. (2013). The reliable change index for assessment of cognitive dysfunction after coronary artery bypass graft surgery. *Annals of Thoracic Surgery, 96,* 1529.

Vearncombe, K. J., Rolfe, M., Wright, M., Pachana, N. A., Andrew, B., & Beadle, G. (2009). Predictors of cognitive decline after chemotherapy in breast cancer patients. *Journal of the International Neuropsychological Society, 15,* 951−962.

Vogt, V. L., Delev, D., Grote, A., Schramm, J., Von Lehe, M., Elger, C. E., . . . Helmstaedter, C. (2017). Neuropsychological outcome after subtemporal versus transsylvian approach for selective amygdalohippocampectomy in patients with mesial temporal lobe epilepsy: A randomized prospective clinical trial. *Journal of Neurology, Neurosurgery, and Psychiatry, December, 22.* (jnnp-2017-316311).

Von Rhein, B., Nelles, M., Urbach, H., Von Lehe, M., Schramm, J., & Helmstaedter, C. (2012). Neuropsychological outcome after selective amygdalohippocampectomy: Subtemporal versus transsylvian approach. *Journal of Neurology, Neurosurgery, and Psychiatry, 83,* 887−893.

Wechsler, D. (2008). *WAIS-IV technical and interpretive manual.* San Antonio, TX: Pearson.

Wechsler, D. (2009). *WMS-IV technical and interpretive manual.* San Antonio, TX: Pearson.

Whelan, B. M., Murdoch, B. E., Theodoros, D. G., Hall, B., & Silburn, P. (2003). Defining a role for the subthalamic nucleus within operative theoretical models of subcortical participation in language. *Journal of Neurology, Neurosurgery and Psychiatry, 74,* 1543−1550.

Williams, A. E., Arzola, G. M., Strutt, A. M., Simpson, R., Jankovic, J., & York, M. K. (2011). Cognitive outcome and reliable change indices two years following bilateral subthalamic nucleus deep brain stimulation. *Parkinsonism Related Disorders, 17*, 321–327.

Wilson, B. A., Watson, P. C., Baddeley, A. D., Emslie, H., & Evans, J. J. (2000). Improvement or simply practice? The effects of twenty repeated assessments on people with and without brain injury. *Journal of the International Neuropsychological Society, 6*, 469.

York, M. K., Dulay, M., Macias, A., Levin, H. S., Grossman, R., Simpson, R., & Jankovic, J. (2008). Cognitive declines following bilateral subthalamic nucleus deep brain stimulation for the treatment of Parkinson's disease. *Journal of Neurology, Neurosurgery, and Psychiatry, 79*, 789–795.

Zahodne, L. B., Okun, M. S., Foote, K. D., Fernandez, H. H., Rodriguez, R. L., Kirsch-Darrow, L., & Bowers, D. (2009). Cognitive declines one year after unilateral deep brain stimulation surgery in Parkinson's disease: A controlled study using reliable change. *The Clinical Neuropsychologist, 23*, 385–405.

APPENDIX

TABLE 3.4 Examples of Reliable Change Models in Surgical and Related Clinical Research

Epilepsy	Deep Brain Stimulation
Baxendale and Thompson (2018)	Tröster, Meador, Irwin, Fisher, and SANTE Study Group (2017)
Vogt et al. (2017)	Bogod et al. (2014)
Busch et al. (2016)	Jahanshahi et al. (2014)
Busch, Love et al. (2015)	Schoenberg et al. (2012)
Busch, Lineweaver, Ferguson, and Haut (2015)	Williams et al. (2011)
Hill, Gale, Pearson, and Smith (2012)	Rinehardt et al. (2010)
Von Rhein et al. (2012)	Mikos, Zahodne, Okun, Foote, and Bowers (2010)
Martin, Griffith, Sawrie, Knowlton, and Faught (2006)	Zahodne et al. (2009)
Martin et al. (2002)	Higginson et al. (2009)
Hermann et al. (1996)	York et al. (2008)
Sawrie, Chelune, Naugle, and Luders (1996)	Tröster, Woods, and Morgan (2007)
McSweeny et al. (1993)	Whelan, Murdoch, Theodoros, Hall, and Silburn (2003)
Chelune et al. (1993)	

(Continued)

TABLE 3.4 (Continued)

Oncology, Radiation Therapy, Chemotherapy	Cardiac Surgery
Meyers and Hess (2003)	Nemeth et al. (2017)
Correa et al. (2013)	Evered, Silbert, Scott, Maruff, and Ames (2016)
Caine et al. (2016)	Tully and Baker (2013)
Gondi et al. (2013)	Evered, Scott, Silbert, and Maruff (2011)
Jones et al. (2013)	Dieleman et al. (2009)
Mohile et al. (2010)	Sweet et al. (2008)
Vearncombe et al. (2009)	
Ouimet, Stewart, Collins, Schindler, and Bielajew (2009)	

Chapter 4

A Primer on Neuropsychology for the Neurosurgeon

Ioan Stroescu[1] and Brandon Baughman[2]

[1]*NeuroCog Trials & Triangle Neuropsychology, Durham, NC, United States,* [2]*Department of Neuropsychology, Semmes Murphey Clinic, Memphis, TN, United States*

Clinical neuropsychology is an applied science concerned with the behavioral expression of brain dysfunction.

Muriel Lezak (1995)

INTRODUCTION

Neuropsychological assessment is based on principles of psychological measurement and psychometrics and behavioral neurology in the objective examination and characterization of normal and abnormal central nervous system (CNS) functioning (Lezak, Howieson, Bigler, & Tranel, 2012; Schoenberg & Scott, 2011; Scott, 2011; Urbina, 2004). In particular, clinical neuropsychological evaluation focuses on cognitive, emotional, and behavioral manifestations of brain disease, injury, and dysfunction (Scott, 2011). While early neuropsychology was chiefly concerned with the detection and localization of brain dysfunction, contemporary neuropsychology has increasingly evolved toward more precise evaluation, description and quantification of brain—behavior relationships, including the characterization and detection of even subtle cognitive deficits and treatment outcomes in a broad spectrum of CNS disorders and interventions (Grant & Adams, 2009; Scott, 2011). Especially in conjunction with modern neuroimaging modalities, neuropsychological assessment provides a powerful tool in clinical and neuroscience research and in diagnosis and treatment planning.

RELEVANCE OF COGNITIVE FUNCTION

Successful everyday functioning relies on intact cognitive processing (FDA, 2018; Nuechterlein et al., 2008). The relevance of cognitive function and its

Neurosurgical Neuropsychology. DOI: https://doi.org/10.1016/B978-0-12-809961-2.00005-9
63

assessment is reflected in accumulating evidence over the last several decades that: Cognitive domains are related to brain structure and function (including neural networks and neurophysiological processes), cognitive measures are sensitive to aging and age-related changes in brain structure and function, strong associations exist between objective cognitive measures and various neuropathological processes, and cognitive measures are markers of disease course, cognitive function can be a sensitive indicator of a therapeutic response and of treatment toxicity, and cognitive performance is related to everyday functioning abilities (Fields, Ferman, Boeve, & Smith, 2011; Gläscher et al., 2009; Goh et al., 2011; Marcotte & Grant, 2009; Millan et al., 2012; Morgan & Ricker, 2017).

Within a framework originally proposed by the World Health Organization (WHO), *impairment* is defined as a deficit of brain function caused by disease and is assessed via neurologic and neuropsychological evaluations/tests. The impact of the impairment on an individual's ability to perform activities is reflected in level of *disability*, which in turn impacts mood status, well-being, and quality of life (*handicap*). These variables are commonly (to varying degrees) assessed within the context of a comprehensive neuropsychological evaluation and play an important role in treatment and health-related outcomes.

NEUROSURGICAL CONSIDERATIONS

In some neurosurgical circumstances, the assessment of "higher" neurocognitive functions may seem unnecessary given the nature of the lesion and planned intervention. For example, a patient being considered for resection of a nondominant hemisphere tumor in the primary motor area would not typically be referred for neuropsychological consultation. While the lesion is in an eloquent cortical area, it does not obviously involve cortical territory that would result in cognitive morbidity postoperatively. However, even in cases as these, there may be a rationale to conduct neuropsychological consultation to assess the patient's cognitive function to understand and appreciate the proposed surgical intervention and associated risks.

It is rare for any patient, regardless of neuropathology, to have a firm understanding of the nuance associated with assessing their own cognitive impairment. Thus, during a routine outpatient neurologic examination, the ability to detect subtle, yet meaningful, cognitive or behavioral impairment may be limited when relying on the patient's subjective reporting (Ljunggren, Sonesson, Saveland, & Brandt, 1985; Sager et al., 1992). A good example of bias influencing subjective impressions comes from the traumatic brain injury literature. Iverson, Lange, Brooks, and Rennison (2010) describe the "good old days bias" in patients who have sustained mild traumatic brain injuries. In a subset of these patients, individuals tend to overestimate their premorbid strengths and underestimate their weaknesses, whereas, an inverse relationship emerges postinjury.

This is also true with global markers of functional outcome that may be used in clinical trials or in routine decision making. Indeed, these markers may over or underappreciate the degree of impairment compared to neuropsychological assessment. For example, in one of the largest outcomes studies to date in survivors of aneurysmal subarachnoid hemorrhage rupture (International Subarachnoid Aneurysm Trial: ISAT), the neuropsychological substudy (Scott et al., 2010) revealed meaningful neuropsychological impairments, even in individuals considered to be functionally intact and without significant disability on the Modified Rankin Scale (mRS scores = 0 or 1).

In the same vein, the Karnofsky Performance Status (KPS) has long been the standard clinician rating tool to assess global functioning in the field of neuro-oncology. Originally designed to assess functional outcome and performance status in chemotherapy patients, the measure has been criticized for its lack of sensitivity in individuals with cognitive impairment. In a recent paper, Martin, Gerstenecker, Nabors, Marson, and Triebel (2015) report on 71 adults with primary or metastatic brain cancer. They found that a substantial portion of patients failed a standardized measure of medical decision making capacity, even when they would be considered functionally intact and minimally disabled according to the KPS score.

STRUCTURE OF THE NEUROPSYCHOLOGICAL EVALUATION

The neuropsychological evaluation can be best conceptualized in a bottom-up framework in which sensory-perceptual and motor functioning are essential to the assessment of increasingly complex cognitive functions and domains, including attention, language, memory, and aspects of executive function (Keifer & Haut, 2014; Scott, 2011). Therefore, intact sensory processing and motor function are required for the examination of higher order processes. If there is any concern regarding the integrity of sensory-perceptual and motor abilities, these aspects should be examined first, or early during the neuropsychological evaluation. Characterizing any deficits in these areas is important in determining how the cognitive examination will be performed or adapted and how the results are interpreted. For example, on a test of visual memory in an individual with a unilateral visual field loss, stimulus materials may have to be placed in the intact visual field and the individual may have to be prompted during the exam to turn his/her head in order to compensate for the visual field defect. Otherwise, the memory results may be confounded by the perceptual deficit. For sensory-perceptual and gross motor examination, function is typically assessed on an ordinal scale (i.e., within normal limits/impaired), whereas increasingly complex cognitive domains are quantified on an integral scale (i.e., standardized value relative to normative group) (Lezak et al., 2012; Scott, 2011; Strauss, Sherman, & Spreen, 2006). Equally important is the early examination of attention and concentration, as impairments in aspects of attention can have downstream effects impacting performance in other cognitive domains, such as memory and problem solving.

RECORD REVIEW AND CLINICAL INTERVIEW

At the beginning of each neuropsychological evaluation is the review of available medical or referral records, including neuroimaging and lab results when available (Parsons & Hammeke, 2014; Scott, 2011). This will guide the initial case formulation and clinical interview with the patient or available informant. Neuropsychological assessment can range from brief bedside neurobehavioral examination and mental state testing to comprehensive formal neuropsychometric evaluation that covers multiple cognitive domains as well as mood and psychological status/functioning (Parsons & Hammeke, 2014; Scott, 2011). The referral question and patient disposition will in part dictate the type of evaluation the neuropsychologist will choose to conduct (Scott, 2011). For each evaluation, the clinical neuropsychologist will as necessary draw upon knowledge in functional neuroanatomy, neuropathology, cognitive psychology, test theory and psychometrics, psychopathology, and neurodevelopment in formulating and integrating all the relevant and obtained data into a case conceptualization and resultant neuropsychological report of the findings, impression, diagnosis, and recommendations (Scott, 2011).

NEUROBEHAVIORAL STATUS EXAMINATION AND SELECTION OF COGNITIVE TESTS

The neurobehavioral status examination at the conclusion of the clinical interview allows the neuropsychologist to conduct a brief screening of the patient's cognitive status that can inform the selection of the cognitive tests to be administered (Keifer & Haut, 2014; Scott, 2011). This can take the form of administering a formal mental state test such as the Montreal Cognitive Assessment (MoCA; Nasreddine et al., 2005) or a flexible neurobehavioral bedside exam tailored to the individual's neurocognitive status and particular pattern and degree of impairment (Strub & Black, 2000). In an individual with a history of recent large cerebrovascular infarction who obtains a MoCA score of <16, for example, the administration of a lengthy neuropsychological test battery consisting of measures with a broad range of difficulty would be unlikely to yield useful information due to potential floor effects (i.e., patients performing at or near the lowest possible score/level), patient fatigue, and reduced cognitive endurance. However, for that same individual who 6 months later obtains a MoCA score of 23, a more comprehensive cognitive evaluation and test battery with a wider psychometric range would be most appropriate. A flexible or tailored brief/bedside neurobehavioral status examination typically consists of some or all of the following components: Orientation and insight, motor functions, motor tasks assessing aspects of executive function, sensory functions, visual–spatial functions, attention and working memory, basic language, and memory (Keifer & Haut, 2014; Strub & Black, 2000).

The selection of the test battery will also depend on the referral question and information most critical to patient care and treatment planning in the particular state or stage of the disease (e.g., acute rehabilitation needs in a stroke patient with dense aphasia vs long-term care or return to work recommendations in an individual several months postacutely who has made cognitive and physical gains, or assessment of neurocognitive status and profile in a neurosurgery candidate). This flexible approach to neuropsychological assessment is the most commonly practiced in contemporary clinical neuropsychology. According to the most recent, large scale practice review of US neuropsychologists (Sweet, Benson, Nelson, & Moberg, 2015), approximately 82% of survey respondents reported using a flexible approach. Additional test and norms selection and interpretation factors include the patient's age, primary language, level of education, ethnicity and cultural factors, literacy or reading level, degree of and extent of cognitive impairment (e.g., ceiling and floor effects), and physical disability (Smith, Ivnik, & Lucas, 2008). For example, ethnicity may contribute to demographic variance through differences in cultural traditions, values, test-taking experience, and attitudes toward testing and some measures may require a minimum reading level (Smith et al., 2008).

COGNITIVE DOMAINS

Several different methods of comprehensive neuropsychological assessment exist (Grant & Adams, 2009). All approaches involve the formal assessment of several or all aspects of core neurocognitive domains, including simple attention, sensory-perceptual and motor functions, language, learning and memory, complex attention, and executive function (Fig. 4.1; Schoenberg & Scott, 2011). Depending on the specific referral question and context, the areas or domains of social cognition, intellectual functioning, symptom validity/effort, and mood and behavior may also be evaluated (Scott, 2011). While some qualitative interpretation of performance on formal neurocognitive tests can inform the overall impression and findings, unlike the bedside neurobehavioral examination, results of formal neuropsychological tests and test batteries are based on population normative information and species-wide behaviors depending on the function or task being examined (Scott, 2011; Strauss et al., 2006). An individual's performance is described in a norm-referenced profile of neurocognitive functions in each domain (i.e., the individual's performance relative to the most appropriate demographic reference group) and is also interpreted within an individual comparison standard, such as estimates of premorbid functioning and personal history (Scott, 2011; Strauss et al., 2006).

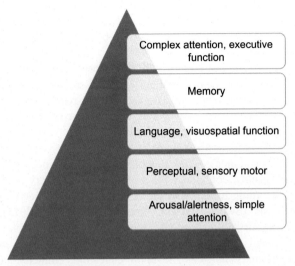

FIGURE 4.1 Hierarchical model of the fundamental structure of neurocognitive evaluation. *Adapted and used with permission from Scott (2011).*

Interpretation of neuropsychological test performance patterns has to consider individual test psychometrics, neuropathological and anatomical correlates, as well as normal or "expected" performance at an individual and group/population level (Parsons & Hammeke, 2014; Schoenberg & Scott, 2011). Normal individuals demonstrate various patterns of cognitive strengths and weaknesses on tests of psychometric intellectual functioning, language ability, and memory. In this context, an important distinction has to be made between what (score/performance) constitutes a clinically significant finding or change versus a statistically significant one (Schretlen, Munro, Anthony, & Pearlson, 2003). To this end, base rates or frequencies in the normative population of a particular score, or difference between scores within a set of tests are used to make this very important distinction. For example, differences of nine points or greater between the Verbal Index and the Performance Index on the Wechsler Adult Intelligence Scale—Third Edition (WAIS-III) are statistically significant at the 0.05 level, yet are found in approximately 43% of the standardization sample for this measure (Smith et al., 2008) and differences that are this common in normal individuals are not very likely to be of clinical significance.

Another challenge in the attempt to define normality is the consideration of the source from which normative data are derived. Demographic variables that are known to be related to cognitive function and performance are age, education, sex, neurodevelopmental milieu, socioeconomic status, and culture (Heaton, Ryan, & Grant, 2009; Smith et al., 2008). Thus, normative data derived from a well-educated, urban population may not generalize to rural patients with lower educational attainment and the normative sample may also influence the distribution (i.e., certain populations may produce truncated distributions or distributions that deviate in other ways from normal distributions and may not generalize to other populations). While the size of the normative group is not the be all and end all of norm selection, it is pertinent. For example, one of the early normative samples provided for a well-known neuropsychological battery (Halstead-Reitan Battery) was derived from a small sample of 26 neurologically normal patients (Russell, Neuringer, & Goldstein, 1970). These early normative samples have improved with time, with more contemporary measures, for example, Wechsler Adult Intelligence Scale 4th Ed. (Wechsler, 2008), including thousands of participants across a wide range of age (i.e., mid-adolescent through late geriatric), education (≤ 8 years through ≥ 16 years), sex, and geographic region. It should be noted, however, that the relationship between various demographic factors and neuropsychological test performance is not static across all measures. For example, verbal reasoning abilities reflecting the acquisition of knowledge over one's life experience tends to be relatively stable despite advancing age, but instead tends to be highly correlated with educational attainment (Heaton et al., 2009). Conversely, measures of fluid reasoning and cognitive proficiency (i.e., attention, working memory, and information processing speed) tend to be more sensitive to the effects of aging as opposed to educational background (Heaton et al., 2009). Ideally, normative data samples are not only described in terms of age, education, and sex, but also the inclusion/exclusion and other sampling methods for which the data were collected. Indeed, if the goal is to assess mild cognitive impairment or prodromal dementia, it would be inappropriate to use datasets where only moderate to severe dementia patients were excluded from the normative sample, for example.

A true normal distribution (Fig. 4.2) is perfectly symmetrical about the mean and has skewness of zero (Urbina, 2004). Positive skew indicates a

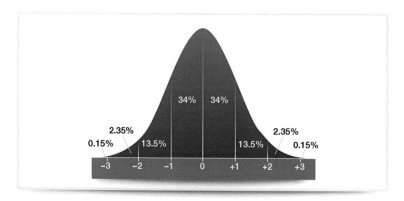

FIGURE 4.2 The normal distribution.

frequency distribution where more scores fall below the mean compared to above the mean. Negative skew refers to distributions where more scores fall above the mean compared to below the mean. Perfect symmetry and zero skew might be more theoretical than practical (Strauss et al., 2006). Many actual distributions of test scores deviate somewhat from the theoretical distribution that we have come to expect (Strauss et al., 2006; Weintraub et al., 2009).

It is not uncommon for a patient to ask—"how did I do?" immediately upon completion of a given neuropsychological task. Unfortunately, a ready answer is typically unavailable as raw scores generally do not carry any interpretive value and require transformation to standard scores based on the demographic factors noted above. Quantitative data may thus be expressed in a variety of ways, including z-scores ($M = 0.0$, $SD = 1.0$), scaled scores ($M = 10.0$, $SD = 3.0$), standard scores ($M = 100.0$, $SD = 15.0$), T-scores ($M = 50.0$, $SD = 10.0$), or percentile ranks.

By examining these transformed scores, a neuropsychologist can make inferences as to how far away from "normal" a given score falls, providing a guide for defining an impairment. It is not uncommon for neuropsychologists to use general rules of thumb with regard to when a score is considered "impaired." For example, scores that are approximately 1.5 standard deviations below the mean are consistent with the 7th percentile rank. In other words, a score at that level indicates that the test taker is performing worse than 93% of an appropriate comparison group (the normative group). Of course, a rigid cutoff is not always appropriate. Indeed, there is a precedent to relax impairment cutoffs for cases in which an individual's baseline abilities are above average to superior. Thus, a score that falls in the low normal range (i.e., one standard deviation below the normative mean) may be considered as a genuine impairment in a high functioning individual (Binder, Iverson, & Brooks, 2009; Jak et al., 2009; Sumowski et al., 2018). In other words, patients may experience decline from previous function without crossing the threshold into "psychometric" impairment, although such decline likely affects real-world functioning (Sumowski et al., 2018). For example, in a test that conforms to a normal distribution, an individual with above-average cognition prior to disease onset (84th percentile) with a decline of 1.5 SD in performance would be categorized within the average or intact range (31st percentile), which conflicts with the individual's experience of cognitive decline (Sumowski et al., 2018).

Complicating the absolute reliance on the normal bell curve, there are certain neurocognitive abilities that are not normally distributed (Weintraub et al., 2009). For example, confrontational visual naming tends to have a negatively skewed distribution, such that one would expect a majority of patients, regardless of demographic background, to perform at a near perfect level. Thus, on skewed tasks such as most available confrontation naming tests, an individual may be considered impaired even when missing only a few items on the test.

It is also of utmost importance to consider normal psychometric variability that is commonplace within neuropsychological batteries. Previous studies (Schretlen et al., 2003) have revealed how wide discrepancies are found in a typical neuropsychological battery, even in healthy individuals. Increasing the number of measures within a given neuropsychological battery increases the probability that abnormal scores on individual measures will emerge (Binder, Iverson, & Brooks, 2009). Thus, the adage of "the more data the better" is not always accurate, with test selection needing to balance sensitivity and specificity specific to the referral question.

CONCLUSION

Neuropsychological testing provides an objective assessment of complex neurocognitive and psychological functions. Cognitive function, mood, and behavior are affected in various neurologic and neuropsychiatric disorders. Comprehensive, objective, and data driven evaluation of neuropsychological function and dysfunction provides a powerful tool that can guide and inform treatment planning and impact health-related outcomes in many neuromedical conditions. Clinical and experimental neuropsychology are continuing to evolve and develop sensitive measures and methods that in combination with modern neuroimaging techniques broaden our understanding of brain–behavior relationships and of the patient experience and related outcomes. Clinical neuropsychologists are uniquely equipped to interpret neurocognitive test results and data and integrate them with other available clinical information to contribute to more comprehensive care and a more nuanced understanding of the individual patient's status and functioning.

REFERENCES

Binder, L. M., Iverson, G. L., & Brooks, B. L. (2009). To err is human: "Abnormal" neuropsychological scores and variability are common in healthy adults. *Archives of Clinical Neuropsychology, 24*(1), 31–46.

Fields, J. A., Ferman, T. J., Boeve, B. F., & Smith, G. E. (2011). Neuropsychological assessment of patients with dementing illness. *Nature Reviews Neurology, 7*, 677–687.

Food and Drug Administration (FDA). (2018). Early Alzheimer's disease: Developing drugs for treatment, guidance for industry *(Revision 1)*. Silver Spring, MD: Office of Communications, Division of Drug Information, Center for Drug Evaluation and Research.

Gläscher, J., Tranel, D., Paul, L. K., Rudrauf, D., Rorden, C., Hornaday, A., et al. (2009). Lesion mapping of cognitive abilities linked to intelligence. *Neuron, 61*(5), 681–691.

Goh, S., Bansal, R., Xu, D., Hao, X., Liu, J., & Peterson, B. S. (2011). Neuroanatomical correlates of intellectual ability across the life span. *Developmental Cognitive Neuroscience, 1*(3), 305–312.

Grant, I., & Adams, K. W. (2009). *Neuropsychological assessment of neuropsychiatric and neuromedical disorders* (3rd ed). New York: Oxford University Press.

Heaton, R. K., Ryan, L., & Grant, I. (2009). Demographic influences and use of demographically corrected norms in neuropsychological assessment. In I. Grant, & K. W. Adams (Eds.), *Neuropsychological assessment of neuropsychiatric and neuromedical disorders* (3rd ed, pp. 127–158). New York: Oxford University Press.

Iverson, G. L., Lange, R. T., Brooks, B. L., & Rennison, V. L. (2010). "Good old days" bias following mild traumatic brain injury. *The Clinical Neuropsychologist, 24*(1), 17–37.

Jak, A. J., Bondi, M. W., Delano-Wood, L., Wierenga, C., Corey-Bloom, J., Salmon, D. P., & Delis, D. C. (2009). Quantification of five neuropsychological approaches to defining mild cognitive impairment. *American Journal of Geriatric Psychiatry, 17*, 368–375.

Keifer, E., & Haut, M. W. (2014). Neurobehavioral examination. In M. W. Parsons, & T. A. Hammeke (Eds.), *Clinical neuropsychology: A pocket handbook for assessment* (3rd ed, pp. 31–52). Washington, DC: American Psychological Association.

Lezak, M. D., Howieson, D. B., Bigler, E. D., & Tranel, D. (2012). *Neuropsychological assessment* (5th ed). New York: Oxford University Press.

Ljunggren, B., Sonesson, B., Saveland, H., & Brandt, L. (1985). Cognitive impairment and adjustment in patients without neurological deficits after aneurysmal SAH and early operation. *Journal of Neurosurgery, 62*(5), 673–679.

Marcotte, T. D., & Grant, I. (2009). *Neuropsychology of everyday functioning*. New York: The Guilford Press.

Martin, R. C., Gerstenecker, A., Nabors, L. B., Marson, D. C., & Triebel, K. L. (2015). Impairment of medical decisional capacity in relation to Karnofsky Performance Status in adults with malignant brain tumor. *Neuro-oncology Practice, 2*(1), 13–19.

Millan, M. J., Agid, Y., Brüne, M., Bullmore, E. T., Carter, C. S., Clayton, N. S., et al. (2012). Cognitive dysfunction in psychiatric disorders: Characteristics, causes, and the quest for therapy. *Nature Reviews Drug Discovery, 11*, 141–168.

Morgan, J. E., & Ricker, J. H. (2017). *Textbook of clinical neuropsychology* (2nd ed). New York: Taylor & Francis.

Nasreddine, Z. S., Phillips, N. A., Bedirian, V., Charbonneau, S., Whitehead, V., Collin, I., ... Chertkow, H. (2005). The Montreal Cognitive Assessment, MoCA: A brief screening tool for mild cognitive impairment. *Journal of the American Geriatrics Society, 53*(4), 695–699.

Nuechterlein, K. H., Green, M. F., Kern, R. S., Baade, L. E., Barch, D. M., Cohen, J. D., et al. (2008). The MATRICS consensus cognitive battery, part 1: Test selection, reliability, and validity. *American Journal of Psychiatry, 165*, 203–213.

Schretlen, D. J., Munro, C. A., Anthony, J. C., & Pearlson, G. D. (2003). Examining the range of normal intraindividual variability in neuropsychological test performance. *Journal of the International Neuropsychological Society, 9*(6), 864–870.

Scott, J. G. (2011). Components of the neuropsychological evaluation. In M. R. Schoenberg, & J. G. Scott (Eds.), *The little black book of neuropsychology: A syndrome based approach* (pp. 127–138). New York: Springer.

Schoenberg, M. R., & Scott, J. G. (2011). *The little black book of neuropsychology: A syndrome based approach*. New York: Springer.

Parsons, M. W., & Hammeke, T. A. (2014). *Clinical neuropsychology: A pocket handbook for assessment* (3rd ed). Washington, DC: American Psychological Association.

Russell, E. W., Neuringer, C., & Goldstein, G. (1970). *Assessment of brain damage: A neuropsychological key approach*. New York: Wiley-Interscience.

Sager, M. A., Dunham, N. C., Schwantes, A., Mecum, L., Halverson, K., & Harlowe, D. (1992). Measurement of activities of daily living in hospitalized elderly: A comparison of self-report and performance-based methods. *Journal of the American Geriatrics Society, 40*(5), 457–462.

Scott, R. B., Eccles, F., Molyneux, A. J., Kerr, R. S., Rothwell, P. M., & Carpenter, K. (2010). Improved cognitive outcomes with endovascular coiling of ruptured intracranial aneurysms: Neuropsychological outcomes from the International Subarachnoid Aneurysm Trial (ISAT). *Stroke*, *41*(8), 1743–1747.

Smith, G. E., Ivnik, R. J., & Lucas, J. (2008). Assessment techniques: Tests, test batteries, norms, and methodological approaches. In J. E. Morgan, & J. H. Ricker (Eds.), *Textbook of clinical neuropsychology* (1st ed, pp. 38–58). New York: Taylor & Francis.

Strauss, E., Sherman, E. M. S., & Spreen, O. (2006). *A compendium of neuropsychological tests: Administration, norms, and commentary* (3rd ed). New York: Oxford University Press.

Strub, R. L., & Black, F. W. (2000). *The mental status examination in neurology* (4th ed). F.A. Davis Company.

Sumowski, J. F., Benedict, R., Enzinger, C., Fillippi, M., Geurts, J. J., & Hamalainen, P. (2018). Cognition in multiple sclerosis: State of the field and priorities for the future. *Neurology*, *90* (6), 1–11.

Sweet, J. J., Benson, L. M., Nelson, N. W., & Moberg, P. J. (2015). The American Academy of Clinical Neuropsychology, National Academy of Neuropsychology, and Society for Clinical Neuropsychology (APA Division 40) 2015 *TCN* Professional Practice and 'Salary Survey': Professional practices, beliefs, and incomes of U.S. neuropsychologists. *The Clinical Neuropsychologist*, *29*(8), 1069–1162.

Urbina, S. (2004). *Essentials of psychological testing* (1st ed.). Hoboken, NJ: Wiley & Sons.

Wechsler, D. (2008). *Wechsler Adult Intelligence Scale* (4th ed.). San Antonio, TX: Pearson.

Weintraub, S., Salomon, D., Mercaldo, N., Ferris, S., Graff-Redford, N. R., Chui, H., et al. (2009). The Alzheimer's Disease Centers' Uniform Data Set (UDS): The neuropsychological tests battery. *Alzheimer's Disease and Associated Disorders*, *23*(2), 91–101.

Section II

Methods

Chapter 5

Functional Neuroimaging in the Presurgical Workup

Leslie C. Baxter

Department of Neurosurgery, Barrow Neurological Institute, St. Joseph's Hospital and Medical Center, Phoenix, AZ, United States

THE SCOPE OF THIS CHAPTER

The intent of this chapter is to provide practical information regarding common, widely available imaging techniques that are used for functional neuroimaging of the presurgical patient. These techniques provide "roadmaps" of the brain regions that subserve behaviors that may be critically affected when resected. Some of the main principles discussed here can apply to many of the neuroimaging techniques, including positron emission tomography (PET), magnetoencephalography (MEG), magnetic resonance spectrometry imaging (MRSI), functional magnetic resonance imaging (fMRI), and diffusion tensor imaging (DTI). However, each of these methods has enough complexity that it is beyond the scope of this chapter to address all the methods in depth. Therefore, this discussion will focus on the most commonly used methods of fMRI and DTI. Further, fMRI will be discussed in more detail than DTI because it focuses on the brain—behavior relationship and is thus well suited to be performed by a neuropsychologist with expertise in understanding potential behavioral confounds that can strongly influence fMRI interpretation (Bobholz, Rao, Saykin, & Pliskin, 2007). With a focus on practical aspects of mapping, this chapter will provide a context explaining how a neuropsychologist can be a key contributor in presurgical neuroimaging mapping, especially fMRI. Neuroimaging mapping benefits from a team approach, given that there are technical, medical, and statistical aspects to the assessments and few professionals have extensive training in every area. Although some institutions choose to use a technician-based model of mapping, this chapter will illustrate why a neuropsychology—neuroradiology team can provide the expertise needed to produce accurate and meaningful results.

Neurosurgical Neuropsychology. DOI: https://doi.org/10.1016/B978-0-12-809961-2.00006-0

This chapter aims to address some of the practical matters that should be considered, illustrating commonly used approaches, to allow each clinical site to tailor and optimize their functional neuroimaging to best meet the needs of their unique neurosurgical practice. There is still debate about how well all the functional neuroimaging techniques can help minimize neurological decline after neurosurgery; however, the techniques discussed in this chapter have merit in presurgical mapping. Numerous excellent reviews of fMRI and DTI have detailed the research supporting their use and clinical validity (Table 5.1) (Bookheimer, 2007; Catani & Mesulam, 2008; Descoteaux, Deriche, Knosche, & Anwander, 2009; James, Rajesh, Chandran, & Kesavadas, 2014; Kekhia, Rigolo, Norton, & Golby, 2011; Official position of the division of clinical neuropsychology (APA division 40) on the role of neuropsychologists in clinical use of fMRI: approved by the Division 40 Executive Committee July 28, 2004, 2004; Stippich et al., 2007). Nonetheless, the reliability and usefulness of neuroimaging techniques may be improved by paying closer attention to some variables that influence the mapping results. Therefore, this chapter focuses on the requirements for implementation, the necessary resources, and the strengths and weaknesses of the mapping techniques that should be considered when using these techniques.

MAPPING ELOQUENT CORTEX: DEFINITIONS AND GENERAL CONSIDERATIONS

Neurosurgeons often use the phrase "eloquent" cortex to describe brain regions that are involved in functions considered to be important for cognition or behavior. This phrase represents more of a general principle than a well-defined set of behaviors or brain regions. Eloquent cortex generally refers to areas of the brain that, if removed, would result in a significant decline in the patient's ability to provide self-care or to be assisted by a caregiver. For most neurosurgeons, motor functioning and language functioning meet these criteria. Vision can also be included if the neurosurgeon or patient believes that a visual deficit will significantly impact current functioning, such as one that would impair driving or playing a sport. With the introduction of tractography, mapping can include both fMRI (cortical activity) and DTI (white matter integrity) mapping of brain regions involved in a given function to ensure best outcomes. Severe language deficits can be devastating. Many epilepsy assessments focus on language lateralization because switched dominance is common (Dym, Burns, Freeman, & Lipton, 2011). Anatomical variability in the location of language within the dominant (usually left) hemisphere occurs in virtually all persons and thus intrahemispheric location can be more important than lateralization of language (Pouratian & Bookheimer, 2010). Motor mapping is also commonly requested, although the motor system is predictably located along the precentral sulcus and the

TABLE 5.1 Published Reviews Supporting the Use and Clinical Validity of fMRI and DTI

Author (Year)	Title	Description
Jones et al. (2013)	White matter integrity, fiber count, and other fallacies: the do's and don'ts of diffusion MRI	Thorough discussion of DTI
Catani and Thiebaut de Schotten (2012)	Atlas of Human Brain Connections	Excellent atlas that includes helpful tract visualization for clinicians
Kekhia et al. (2011)	Special surgical considerations for functional brain mapping	Includes fMRI, DTI, and MEG
Descoteaux et al. (2009)	Deterministic and probabilistic tractography based on complex fiber orientation distributions	Treatise on tractography
Bookheimer (2007)	Presurgical language mapping with functional magnetic resonance imaging	Reviews the literature supporting the use of fMRI for epilepsy, vascular lesions, and brain tumors
Stippich et al. (2007)	Localizing and lateralizing language in patients with brain tumors: feasibility of routine preoperative functional MR imaging in 81 consecutive patients	Large series of patients with good results attributed to good MR methodology
APA Division 40, 2004 ("Official position of the division of clinical neuropsychology (APA division 40) on the role of neuropsychologists in clinical use of fMRI: Approved by the Division 40 Executive Committee July 28, 2004," 2004)	Official position of the Division of Clinical Neuropsychology (APA Division 40) on the role of neuropsychologists in clinical use of fMRI: approved by the Division 40 Executive Committee July 28, 2004	Summary statement regarding the role of neuropsychology in presurgical fMRI

APA, American Psychological Association; *DTI*, diffusion tension imaging; *fMRI*, functional magnetic resonance imaging; *MEG*, magnetoencephalography; *MR*, magnetic resonance; *MRI*, magnetic resonance imaging.

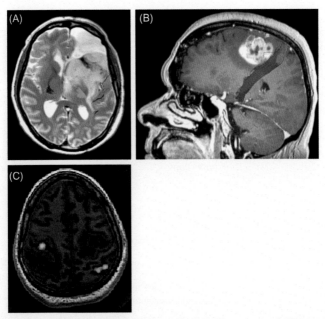

FIGURE 5.1 Examples of (A and C) presurgical mapping of the corticospinal tracts and (C) fMRI of the hand knob area. (A) The fMRI shows a patient with a large left hemisphere tumor causing midline shift. The blue and green corticospinal tracts map through the posterior limb of the internal capsule. The purple tract lateral to the tumor is part of the arcuate fasciculus. (B) The fMRI shows anterior and posterior shift in the corticospinal tract due to tumors. Blue rectangle indicates seeding regions for tractography. (C) The fMRI shows edema near two tumors that have shifted the motor cortex posteriorly in the left hemisphere and anteriorly in the right hemisphere. *Used with permission from Barrow Neurological Institute, Phoenix, Arizona.*

corticospinal tract courses through the internal capsule. However, edema and tumor can cause mass effect or cortical effacement, which distorts the normal gyral and sulcal appearance. For example, in tumor cases (Fig. 5.1), visualizing the motor cortex and tracts as occurring anterior or posterior to the tumor can help determine the risk of resecting a margin of normal-appearing tissue around the tumor. Several studies examining positive outcomes after resection have found that resection margins approximately 1 or 2 cm from the fMRI activation decrease the risk of postoperative deficits (Haglund, Berger, Shamseldin, Lettich, & Ojemann, 1994; Mueller et al., 1996; Wood et al., 2011). Mapping motor function to visualize the location of the supplementary motor area (SMA) is often helpful because resection of the SMA can result in profound motor hemiparesis that usually remits in the months after surgery (Hanakawa et al., 2001). Resection of the pre-SMA (i.e., supplementary speech area), which is anterior to the SMA and is involved in speech-related motor function (Benjamin et al., 2017; Krainik et al., 2003), can result in a temporary speech production deficit; the SMA can be mapped using expressive language tasks. The primary visual cortex shows little or no variability in its location along the calcarine fissure, but fMRI is often helpful in visualizing secondary processing of color, shapes, and motion extending anteriorly and superiorly (the "ventral and dorsal streams"). The use of fMRI to map memory functioning in the hippocampus is extremely desirable for presurgical mapping, especially for patients with mesial temporal lobe epilepsy who may be rendered amnestic if the nonaffected hippocampus is not capable of supporting memory functioning (Milner, Corkin, & Teuber, 1968). Some clinician-researchers have in-house memory tasks to assess memory performance in their patients. Although memory mapping is done by some institutions as an exploratory measure, most do not predict memory outcome using fMRI as an alternative to intracarotid amobarbital testing (Wada, 1949). Similarly, commercial products include vision, language, and motor mapping but none appear yet to have adopted a memory paradigm. A recent position paper reviewing studies of verbal memory outcomes in epilepsy patients concluded that encoding of verbal material, regardless of whether it occurs within the mesial temporal lobe or in language region, declines after left-sided surgery (Szaflarski et al., 2017). Most studies examining fMRI and outcomes in patients with epilepsy use the metric of a lateralizing index, which allows fMRI to be quantified so that correlations with outcome can be done, but calculating lateralizing indices has not been standardized (Seghier, 2008).

The particular expertise of the neuropsychologist is the ability to determine how neurological injury will impact the brain–behavior relationship. Neuropsychologists, therefore, are capable not only of assessing the cognitive function of the preoperative patient but also of providing the treatment team with a perspective of the patient's relative cognitive strengths and weaknesses to help predict the patient's ability to compensate and cope with neurological or cognitive deficits after surgery. This perspective is especially important because many of the conditions requiring neurosurgical resection cannot be treated without some compromise in function, so the relative risks and benefits must be well understood. Patients with brain tumors have a more favorable outcome when resection results in clear margins surrounding the tumor (Brown et al., 2016; Kekhia et al., 2011; Petrella et al., 2006). Vascular lesions such as cavernous malformations have an increased risk of bleeding that can affect more tissue than the resection itself, and epilepsy has quality-of-life trade-offs such that some patients feel that being unable to drive because of seizures is worth risking a potential cognitive deficit. The risk–benefit trade-off philosophy of some neurosurgeons has expanded to other cognitive domains, such as "executive functioning." Their goal is to maximize return to work for patients with more benign lesions or to enable patients to retain musical ability, although mapping these other functions is mostly being done with intraoperative electrical stimulation rather than with neuroimaging techniques. Presurgical brain mapping using fMRI and DTI-derived white matter tractography can complement other mapping procedures, such as intraoperative direct cortical electrical stimulation, or it can be a substitute for patients who have contraindications for intraoperative procedures. Presurgical brain mapping can determine if alternate therapies, such as radiosurgery or laser neurosurgery, are indicated (Fernandez Coello et al., 2013).

WHY ARE SOME FUNCTIONAL IMAGING TECHNIQUES MORE COMMONLY USED THAN OTHERS?

Currently, MRI procedures of fMRI and DTI are the most commonly used neuroimaging methods for presurgical mapping. PET imaging has the benefit of requiring less cooperation from the patient, but the drawbacks include lower spatial and temporal resolution, exposure to radiation, and higher cost. MEG imaging has better temporal resolution than fMRI and its spatial resolution can be superior to that of fMRI because MEG relies on neuron-related electrical discharge instead of blood flow. MEG is often used in epilepsy mapping because it can precisely map cortical focal seizure sites as well as provide information about functional mapping (mostly motor mapping and language lateralization) (Anderson, Carlson, Li, & Raghavan, 2014). Although MRIs are routinely performed at virtually all neurosurgical hospitals, only about 20 clinical sites were using MEG in 2017. MEG imaging has great promise and it is possible that as more centers routinely use MEG and expand language mapping for localization in addition to lateralization, it will become a more widely used neuroimaging mapping procedure. Since obtaining fMRI, DTI, and high-resolution three-dimensional (3D) images compatible with the neuronavigation system can be done in the same imaging session (within days or even hours of a surgery), this combination is currently the most efficient and cost-effective mapping method (Fig. 5.2).

DESCRIPTION OF THE FUNCTIONAL MAGNETIC RESONANCE IMAGING PROCEDURE

In fMRI, an intrinsic contrast signal obtained from a T2-weighted ("T2*") image is used to determine brain regions that are uniquely active during a given behavior. The intrinsic contrast signal is referred to as blood oxygen level—dependent, or BOLD contrast, and it capitalizes on the linked increase in both metabolic activity and degree of blood oxygenation in specific brain regions when the patient performs a cognitive or behavioral function (Huettel, Song, & McCarthy, 2014). Specifically, the T2* signal is sensitive to inhomogeneities in the magnetic field. Unlike oxygenated blood, deoxygenated blood is paramagnetic, thus the T2* signal is suppressed by the tiny local magnetic fields. A fast infusion of oxygenated blood occurs as a response to increased neural activity, which causes an increase in T2* signal.

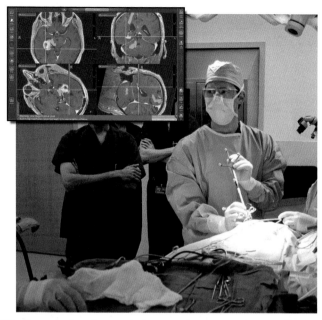

FIGURE 5.2 Intraoperative photograph demonstrating a neurosurgeon utilizing the "wand" portion of the StealthStation (Medtronic, plc) neuronavigation system in a tumor patient. The inset shows the corticospinal tract in green. This system allows the neurosurgeon to obtain presurgical mapping for real-time guidance during surgery. *Used with permission from Barrow Neurological Institute, Phoenix, Arizona.*

For example, the BOLD signal will increase in motor-related brain regions when the patient moves one hand and then it will return to a baseline signal level when that hand is at rest. An important aspect of BOLD-related signal change is that there is no inherent baseline T2* signal, so an fMRI task requires a statistical comparison of the change of signal from a baseline to an activated condition.

The fMRI tasks can be designed in several different ways, but the most common design is a block design task in which a given activity (e.g., hand movement) is alternated with a baseline condition (e.g., rest) to isolate regions that show increased BOLD signal during hand movement compared to the resting condition. For example, with a block design, an fMRI scan acquires a rapid, image of the whole brain in 2 or 3 seconds using echo planar imaging. Each block lasts an average of about 12−15 seconds to acquire a steady-state signal change for that block. Tasks require some repetition of these blocked conditions so that a statistical approach can produce a best fit using a general linear model analysis of the raw data. The data from the scanner are analyzed as a time series of these alternating blocks, and the data analysis compares the signal change among the different conditions (Poldrack, Mumford, & Nichols, 2011). The statistical question answered is: Where in the brain is more signal observed during the active condition compared to the baseline condition, after accounting for factors such as movement and drift in scanner frequency (Fig. 5.3). Each 3D pixel (voxel) of image space is thresholded to reach a user-determined statistical *P*-value and then further constrained by requiring that this *P*-value be reached by a user-determined number of neighboring voxels. The statistical map can be saved as "blobs" that represent these regions of activation associated with a task. This fMRI map must be coregistered to a high-resolution scan so that the functional activity can be related back to the lesion, other brain anatomy, and even the location of vessels that may be impacted by the surgical approach. Smoothing clinical fMRI data is an important factor that will be discussed in more detail later. At the current time, clinical MRIs use single-shot gradient-echo planar images to generate T2* BOLD signal because of the high contrast-to-noise ratio; however, the heightened susceptibility of T2* scans to inhomogeneities in the magnetic field leads to signal loss in brain regions close to air or bone. Thus, signal loss is common in the inferior

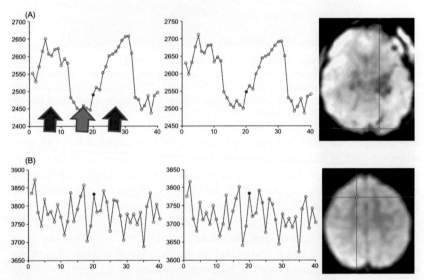

FIGURE 5.3 Example of the BOLD signal change during a block design task. (A) An axial slice of a T2* image at the level of the visual cortex, with a blue cross hair in the occipital cortex. The cross hair includes the two voxels to the left of the image, showing BOLD signal across the 40 volumes collected in the time series. The *x*-axis of the graph represents the time series of brain volumes (collected every 3 s in this case). The *y*-axis is BOLD signal intensity. The task starts with blocks of complex visual color scenes alternating with a cross hair. The first two of the 40 time points showed weak signal, which corresponds to the hemodynamic response function (hrf), or the 6 s it takes for blood to reach the active brain regions. The steady-state signal continues during the picture presentation and drops when the cross hair is presented. (B) The BOLD time series in a brain region not involved in the task, with a random BOLD signal across the brain. *Used with permission from Barrow Neurological Institute, Phoenix, Arizona.*

temporal lobe, which limits interpretation of the contribution of this area to language tasks. As a result, the clinician should inspect the data and the lesion location on the high-resolution structural scans to ensure that they accurately relay any limitations in data collection that affect the interpretation of the results. This signal loss is not visible on high-resolution scans onto which the fMRI "blobs" are projected, so it is important for the fMRI clinician to alert the neurosurgeon to the possible false-negative finding in the inferior temporal lobe (Bookheimer, 2007). Blood products and abnormal blood flow also cause signal loss that can affect the accurate mapping of regions near vascular lesions such as cavernous malformations. Vascular uncoupling describes a change in the typical BOLD response to neuronal activity due to alterations in vasculature, so special attention must be paid to determine whether the underlying BOLD signal near vascular lesions represents a typical hemodynamic response (Kekhia et al., 2011; Ulmer et al., 2003).

The inability of T2* imaging to capture data from some brain regions highlights only one of several reasons why both fMRI and DTI require skilled clinicians involved in data analysis. Because these techniques produce maps from processed data, the clinician must ensure that the underlying caveats for data interpretation can be relayed properly to the treatment team (Bookheimer, 2007; Kekhia et al., 2011). Other factors that also require this kind of oversight include patient cooperation during fMRI and data quality (signal loss as well as excessive motion) signal dropout due to edema in DTI data. Because neuropsychologists have been trained to understand and account for issues such as measurement error and the impact of patient-related factors in data interpretation, they are ideally suited to relay pertinent information to the referral source.

HOW DOES CLINICAL FUNCTIONAL MAGNETIC RESONANCE IMAGING DEVIATE FROM RESEARCH FUNCTIONAL MAGNETIC RESONANCE IMAGING? DATA THRESHOLDING, SMOOTHING, AND TASK DESIGN

Cognitive neuroscientists and neuropsychologists have long studied the brain−behavior relationship using fMRI in cognitive normal populations. Clinical fMRI paradigms are borrowed from research paradigms; however, understanding why research procedures need to be altered for clinical cases (e.g., each patient is essentially a case study) can help to illustrate what to anticipate when mapping patients. For example, these alterations include adjusting the data for different significance levels. Unlike group analyses of data from research participants who are healthy and usually medication free, patients are often ill, fatigued, and taking medications that can affect the robustness of the BOLD reactivity. To be able to choose an appropriate threshold that produces the most informative brain activation pattern, the clinician should be familiar with the typical map expected for a given task. A recent study showed a high degree of consistency for blinded assessment of lateralization and location of six language-related regions when experienced neuropsychologists thresholded maps according to individual judgments of data quality (e.g., strength of response, motion) rather than when they used maps of fixed thresholds. In 2017, Benjamin et al. (Benjamin et al., 2017) found that experienced clinicians varied the maps due to the clinical question, thereby generating maps that were considered to be of higher quality by an independent group of experienced clinicians. This finding emphasizes how knowledge about the clinical question is important for generating presurgical maps that are most helpful to the treatment team.

Applying the proper degree of data smoothing is another fMRI analysis step that requires an experienced clinician for best results. Smoothing refers to the application of an averaging algorithm (Gaussian smoothing kernel) to each voxel to improve the spatial correlation between neighboring voxels. Given the size of a voxel ($2-3$ mm^3), neighboring voxels generally share similar BOLD data. Smoothing helps to address false-positive errors in the data, since rogue voxels that have high signal relative to their neighbors will have their signal averaged with that of their neighbors, which will be less likely to produce false regions of activation. The downside to smoothing is that the data are intentionally "blurred," which reduces the spatial resolution of the fMRI blobs in relation to the high-resolution MRI image. Research studies often use spatial smoothing kernels of 8 mm^3 (about $\geq 2 \times$ the size of a voxel) to protect against false-positive errors, but for clinical neuroimaging mapping of all kinds, false-negative errors pose a much larger problem, since a map that does not include regions that may be subserving a function would not be missing in the presurgical map and may lead to resecting eloquent cortex. One option is to address the uncertainty regarding whether an area is contributing to a given function is to use complementary tasks to find areas of commonality (Bookheimer, 2007; Newman & Twieg, 2001). For example, auditory and visual language comprehension tasks can be used to pinpoint the area that overlaps from both tasks and represents common comprehension areas (Benjamin et al., 2017; Bookheimer, 2007). Motor mapping can be done with little smoothing by mapping several motor regions, and visual inspection shows that the activation maps to the same gyrus. This redundant mapping can allow for minimal (2 mm^3) or no smoothing, resulting in better spatial resolution. Different smoothing kernels can also be used as based on the inherent uncertainty in the function being mapped. Specifically, since there is much less individual variability in motor and vision functioning, smaller smoothing kernels might be used. In contrast, larger smoothing kernels may be needed for language tasks due to the anatomical variability in language location, which increases the degree of spatial uncertainty.

Understanding how to effectively choose and present the fMRI tasks and determine what activation patterns are expected from a given task are key components for producing meaningful results. Although it is possible for a well-trained MRI technologist to administer simple tasks such as motor functioning, an experienced clinician who understands subtle cognitive deficits that can affect data quality is needed to evaluate complex cognitive systems such as language. A typical fMRI evaluation should start with a clinical interview and a task instruction period so that the clinician can determine whether the patient has deficits in attention/concentration, working memory, abstract thinking, or comprehension that require alterations of the fMRI procedure. Patients are often good at "filling in the gaps" when they have comprehension difficulties, which makes them often seem to be more neurologically intact than they actually are. A clinical interview by a neuropsychologist (as well as corroborating information from a neuropsychological evaluation) can help to establish the level of task difficulty that best suits the individual patient. Most established treatment centers have several versions of language tasks that include "step-down" versions that can used for patients who are too impaired for standard administration. In patients with brain tumors or vascular lesions, the patients often are newly diagnosed and thus are very anxious and concerned. They will often have difficulties understanding task instructions because of their anxiety and distraction. It is not uncommon for these patients to be claustrophobic and worried because they have to be in the scanner without sedation.

As a practical guide for optimizing fMRI cooperation, the following points may help the practitioner. Patients often have difficulties with attention or concentration due to fatigue, medications, and other factors. Therefore, it is helpful to begin the tasks with the target condition rather than the control condition. For example, when mapping motor functioning, it is better to have the active movement condition followed by rest, rather than the opposite. This approach is especially appropriate when more than one movement task will be given. Patients with frontal lobe deficits often have difficulty with decision-making. Since they may have difficulty performing language tasks that require a response (e.g., making a decision as to whether a word represents a living or nonliving object, or whether two words are semantically related), alternate language comprehension to semantic processing tasks such as passive reading and auditory comprehension tasks may map better in these patients. Fluency tasks that require verb or letter generation can be too abstract or effortful for some patients, so an easier alternative is category fluency tasks (e.g., silently generating words according to categories such as fruits or birds) (Lezak, Howieson, Bigler, & Tranel, 2012).

Similarly, asking patients to perform bilateral motor movements to map the primary motor cortex can be problematic for several reasons. First, movement from one hemisphere is thought to cause inhibition of movement from the opposite side, so if the target motor area is impaired, bilateral tasks may diminish movement on the affected side. Beyond the practical reason that less movement inside the scanner is better, another practical reason for mapping one side of the brain is a rule of thumb for neuronavigation fMRI maps: The less extraneous activation blobs, the better. Therefore, mapping only the hemisphere of interest makes for less distracting maps in the operating room. It is also helpful to map hand movement as simply as possible; many patients have difficulty making finger-thumb opposition movements, so a simple hand opening and closing movement maps the "hand knob" region accurately and robustly, and even impaired patients usually able to perform this task. Patients are often quite worried about having to remain still for an extended period of time, so it may be helpful to offer anxious or uncomfortable patients the opportunity to move in between scans to keep them from moving excessively during a scan. MRI technologists are trained to ask patients to remain still so that the neuroradiologists will be able to see the same anatomy in images that have gaps in coverage, but since fMRI time series are contiguous and are coregistered to a high-resolution scan that is also contiguous, slight movement in between the scans is not a critical factor. Taking the time to ensure that the patient is comfortable can help avoid patients moving excessively during scans or having patients who are distracted because they feel uncomfortable. Our site starts every scan session with a 2-minute visual task consisting of outdoor scenes (instead of a cross hair baseline), which allows the patient to acclimate to the scanner and often calms claustrophobic patients. The signal in the occipital lobe is robust and allows the clinician to ensure that the patient is showing adequate BOLD responsivity, both at the time of the scan (using real-time fMRI) and during postprocessing.

Task design is always a critical aspect of fMRI studies because the statistical analysis compares two conditions. Thus, it is imperative for the user to have a clear understanding of the expected difference between the two conditions. The goal for clinical studies is to use a task that minimizes irrelevant activation while still producing a robust signal for the intended function. Understanding how the baseline condition affects the statistical design and activation pattern is crucial. For example, using a cross hair or a similarly simple baseline for reading comprehension is likely to show stronger activity in the visual cortex than in language areas. Therefore, most reading comprehension and semantic decision tasks have scrambled words, letters, or other visual stimuli as a baseline (Ashtari et al., 2005). Scans must be relatively short not only to reduce unwanted scanner drift but also to overcome common patient problems of anxiety, fatigue, inattention, and difficulty holding still. Commercial tasks tend to be about 5 minutes long, but good results can be produced from tasks lasting only 3 or 4 minutes, such as the examples provided here. In my experience, block designs are more robust than event-related designs. Block designs can retain a level of robustness due to the sustained BOLD response over several seconds, which is useful when scanning impaired patients. The greatest source of task variability usually comes from the patient; so if a patient has minor difficulty initiating a motor movement or reading a sentence, a block design can accommodate this variability better than an event-related design. For some alternate arguments for event-related designs with longer tasks (Szaflarski et al., 2017).

WHITE MATTER TRACTOGRAPHY FROM DIFFUSION TENSOR IMAGING

Visualizing white matter tracts that are associated with brain functioning is now possible using DTI. Typical tracts include the corticospinal tract (motor), the visual tracts, the classical language tract of the arcuate fasciculus, which is the main connection for frontal and temporoparietal language regions, and a host of additional language tracts that contribute to naming and semantic language processing (e.g., superior longitudinal fasciculus, inferior longitudinal fasciculus, uncinate fasciculus, and inferior fronto-occipital fasciculus) (Catani & Mesulam, 2008). Visualization of the fornix, which is critical for memory, can be helpful for lesions near the pituitary and surrounding structures. White matter mapping is complementary to the cortical information generated by fMRI, and the combination can often provide the best information for determining the surgical path and the extent of resection that is necessary. For example, sparing motor cortex but damaging the corticospinal tract can result in a neurological deficit just as devastating as resection of the primary motor cortex. Tractography relies on DTI, a T2-weighted scan that is based on standard diffusion-weighted imaging (DWI), which is commonly used to determine regions of abnormal restriction of water diffusion due to edema or stroke. Briefly, DTI increases the number of diffusion directions any water molecule may be traveling through any voxel from the typical three directions of a DWI scan to 20 or more directions. This added information allows for computation of an ellipsoid in each voxel that follows the tendency of water to follow a direction within that voxel. The myelin sheath surrounding axons within larger tracts have water molecules that are trapped by these sheaths, and this water is essentially forced to diffuse along the axis of the tract due to restricted diffusion across the myelin barrier. The resulting ellipsoid within each voxel is a measure of both direction and degree of anisotropy (Chung, Chou, & Chen, 2011; Jones & Leemans, 2011; Mori & van Zijl, 2002). DTI maps that contain both degree of anisotropy and directionality are often colored to represent three main directions (red for left−right, blue for inferior−superior, and green for anterior−posterior). The resolution of clinical DTI scans (2 or 3 mm) does not allow for data to be obtained for axons or relatively small tracts but can accurately track larger fiber bundles. Tractography refers to the postprocessing of DTI data using both fractional anisotropy (FA) and direction of the ellipsoid on a voxel-by-voxel basis to isolate white matter that is likely to belong to a specific fiber tract. Postprocessing algorithms are becoming more and more refined, but most of the current methods are based on deterministic algorithms that evaluate whether neighboring voxels share similar directional and FA information, which suggests that they represent the same tract. Most tractography programs currently use a semiautomated technique requiring the user to choose a "seed" region from which a tract is then propagated along some standard FA and angle parameters. Most commercial programs allow users to alter their strategies to test out different strengths of FA and the angular degree allowed for the neighboring ellipsoid to be included in the tract (e.g., most tracts do not make 90-degree directional changes). Figs. 5.1 and 5.5 show tracts generated from a clinical case generated from a commercially available neuronavigational system, with results that are similar to those of tractography generated by other commercial systems. As with fMRI, there are no specific guidelines available for generating tracts, but it is quite likely that the nascent field of clinical tractography will rapidly evolve. Probabilistic methods can evaluate each voxel for factors that contribute to ambiguity, such as low FA values caused by crossing fibers (Mandelli et al., 2014). In the future, it is likely that more commercial tractography programs will provide deterministic data as model-based methods are developed that do not require as excessively long DTI acquisition times (Descoteaux et al., 2009; Mandelli et al., 2014). However, to better understand the limitations and considerations of interpreting tractography and why processing requires an experienced clinician, consider the example of a current commercially available system that shows mapping of the corticospinal tract (Fig. 5.4). In this scenario, the raw DTI data are coregistered to a high-resolution image and the DTIs FA and tensor map is generated that is colorized to highlight

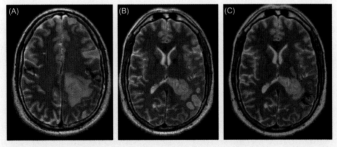

FIGURE 5.4   fMRI and DTI tractography mapping in a tumor case. (A and B) Images show language (blue) and lip motor (red) and (C) corresponding tractography results from the corticospinal tract (green) and the arcuate fasciculus (red). The fMRI and DTI are overlayed on the same T2 interleaved scan that can be used in the neuronavigation system. The lip area maps bilaterally across the central sulcus. Language regions include the left posterior temporal (Wernicke's area) and the left inferior frontal (Broca's area). The fMRI in (A) is superior to those in (B) and (C). Together, these data can assist the neurosurgical team in developing a trajectory to this deep lesion that avoids these eloquent regions. *Used with permission from Barrow Neurological Institute, Phoenix, Arizona.*

the directionality of the tracts in the brain. A "seed" is placed in an area known to include the tract (e.g., above the decussation in the cerebral peduncle), and tracts are automatically generated. An end region can be defined (e.g., the motor cortex) to restrict the computational set and thereby isolate a specific tract (e.g., the corticospinal tract). As with fMRI, several aspects of this procedure require close attention to quality control by the user. First, the colorized directional DTI map should be evaluated to determine whether areas are missing diffusion data because of the presence of a lesion or edema. This step is critical because, as with fMRI data, DTI visualizes the final tracts without reference to the quality of the data underlying the analysis and thus any signal loss that affected the final volume of the tract will be unknown to those who did not create the object. For example, if a corticospinal tract is located near edema, it is difficult to know whether the region of edema invades part of this tract, thereby truncating the boundaries of the tract. Whether the region of edema also includes the tract is impossible to determine with current technology, which must be communicated to the neurosurgeon (Fig. 5.5). Generating a tract requires knowledge of the typical trajectory of the tract as well as a strategy for addressing mass effect and other effects of lesions. As with clinical fMRI, when resection is being guided by the functional map, DTI false-negatives (beta errors) can be much more devastating to a patient neurologically than false-positives (alpha errors). For example, a conservative strategy to isolate the corticospinal tract could be to use a seed region such as the cerebral peduncle, which generates broad tracts across the hemisphere, then to restrict the tracking to include only the white matter that intersects with a region encompassing the motor sensory cortex. In some instances, it may be helpful to delineate a series of seed regions near the area to be resected to determine the various trajectories of white matter around the tumor, thereby determining where resection may impact tracts that may cause a behavioral deficit (Fig. 5.5). This technique can be particularly helpful in patients whose normal anatomy is distorted or when mapping language tracts to account for individual variability in the cortical location of language.

HARDWARE, SOFTWARE, AND CLINICAL FLOW FOR OBTAINING FUNCTIONAL MAGNETIC RESONANCE IMAGING AND DIFFUSION TENSOR IMAGING MAPPING

Establishing a protocol for acquiring data, transferring data from the scanner to postprocessing computers, and transferring postprocessed brain maps to the neuronavigation system will require input from both the information technology and radiology teams but then becomes relatively routine after the initial setup. Scanning DTI and fMRI requires special "keys" that are available for both 1.5 and 3.0 T scanners. Most patients are now cleared to be safely scanned on the 3.0 T scanner, including those with vagal nerve

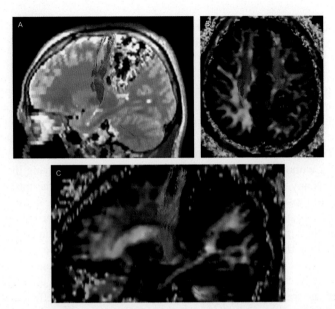

FIGURE 5.5 Example of tractography results from a DTI. Data were collected from 20 directions with 3-mm-thick slices. The number of directions and slice thicknesses acquired in clinically reasonable times (4−7 min) varies from 20 to 32 directions and 2- to 3-mm-thick slices, depending on the scanner. (A) DTI overlays the corticospinal tract (CST) on a T2-weighted scan to show the anterior shift of the CST caused by a large anteriovenous malformation (AVM). (B and C) DTIs show the colorized directional anisotropic map from which the CST tracts are generated (*colorization shows direction as left−right (red), superior−inferior (blue), and anterior−posterior (green)*). Loss of diffusion signal is evident in the region of the AVM. The bright red tract of the left CST shows how it is shifted anteriorly compared to the bright red tract on the right side. However, the possibility that some signal loss is possible along the margin of the lesion is evident when the tracts are displayed on the colorized map, but signal loss is not evident in the final data in (A). Feedback to the neurosurgeon regarding the possible loss of tract along the margin of the lesion is an important part of the workup. *Used with permission from Barrow Neurological Institute, Phoenix, Arizona.*

stimulators or other implants such as deep brain stimulators (see manufacturer details). Some older stimulators require the use of receive-only coils rather than the preferred multichannel coils that should be used for most patients for increased signal-to-noise ratio. For some patients with implants, acquiring some scans may not be possible because the specific absorption rate, which is the degree to which tissue is heated during an MRI sequence, can be too high near the implant and cause tissue burning. In general, higher-strength scanners (3.0 vs 1.5 T), multichannel coils (e.g., 8, 20, 32), and parallel or multiband imaging will improve signal for fMRI and DTI scans. Most scanners also provide real-time fMRI, which allows for a quick view of the BOLD response as the scan is being collected. Real-time tracking of the patient's responsivity allows the clinician to be able to repeat scans or to change the protocol if the expected activity is not present. Most real-time fMRI systems require a block design analysis, which may be another factor for using block design tasks on clinical patients. For the fMRI stimulus to be presented to a patient, the setup must include a computer outside the scanner to deliver the task stimuli and MRI-compatible hardware to transmit the information to the patient during the scan. Several commercial packages provide audio and visual MRI-compatible hardware and response devices. Whether back-projection screens or goggles attached to the head coil are better for presenting visual stimuli is debatable. Rear projectors are cheaper but goggles may help patients overcome claustrophobia better because they help patients to focus more on the computer screen and to see less of the MRI coil. Since the patient has to keep the head still to keep the image in focus in the goggles, the goggles will also help minimize excessive motion. Our site uses goggles and the patients routinely move very little during scans. Many neuropsychologists who also do research use presentation programs and produce their own task designs using stimulus delivery software (e.g., E-Prime, Cogent, Presentation) (James et al., 2014). However, commercial systems have turnkey approaches that include the software, task designs, and hardware needed to perform fMRI, and these are designed to allow centers that do not have dedicated fMRI clinicians to obtain an fMRI using an MRI technician who then relays the data to a radiologist for interpretation. These systems are easy to use but their data analysis may be limited (e.g., limited task choice, number of choices for selecting thresholds or smoothing, or overlaying data on different types of structural scans). In contrast, stimulation presentation hardware (e.g., goggles with a back-projection screen, headphones, response devices) can be used with most stimulation presentation software on a personal computer.

Brain mapping data are most helpful to the treatment team if they can be viewed on the picture archiving and communication system (PACS) in addition to being integrated with the neuronavigation system. Ideally, these data will look identical in both systems, but not all the postprocessing programs that generate fMRI or DTI data show the data identically. For example, some DTI systems show tracts in color in the neuronavigation system, but the images available for the PACS are limited to white tracts that have been "burned in" on the structural image. Data work flow includes transferring the raw Digital Imaging and Communications in Medicine images to a postprocessing computer, then transferring the brain maps to the PACS and the neuronavigation system. Since fMRI or DTI are often completed along with other presurgical scans, the fMRI can be dictated separately from the structural scans if a separate report is generated for the diagnostic examination. Thus, fMRI interpretation is a billable service. The Current Procedural Terminology (CPT; https://www.ama-assn.org/practice-management/cpt-current-procedural-terminology) codes for fMRI are: 96020, 70555, and, 70553 or 70551. The CPT code 96020 is the professional code for the functional

MRI, reflecting the neurofunctional testing selection and administration during noninvasive imaging functional brain mapping, with the test administered entirely by a physician or psychologist who reviews the test results and report. The CPT code 70555 is for MRI of the brain, when the physician or psychologist is administering the neurofunctional testing. These can be billed at the same time as CPT code 70553 for the structural MRI, of the brain with and without contrast material or as CPT code 70551 for imaging without contrast.

Ideally, both fMRI and DTI data should be viewable as an overlay on the structural scan that best images the lesion, usually a T1-weighted, 3D scan with contrast to show high-grade tumors or T2-weighted images that best show low-grade lesions, and cavernous malformations. Some commercial products do not allow the clinician to select the type of structural scan.

SUMMARY

This chapter has presented some of the factors that support the contention that both fMRI and DTI tractography require an experienced clinician to maximize results. This chapter has also outlined the role of the neuropsychologist in carrying out these analyses and contributing to the neurosurgical team. However, many treatment centers do not follow this model, partly for the practical reason that not all neurosurgical patients require presurgical mapping and so this service does not constitute a full-time position. Two factors that influence whether consistent, high-quality data are obtained are (1) the ability to effectively choose and administer the tasks to patients who are often impaired and (2) effective oversight of data acquisition and interpretation for those patients whose brain conditions affect imaging quality. Clinical neuropsychologists are uniquely trained to use assessment tools to understand the anatomical basis of behavior, making them ideally suited for providing high-quality fMRI and tractography maps to assist with neurosurgical interventions.

ACKNOWLEDGMENTS

The authors thank the staff of Neuroscience Publications at Barrow Neurological Institute for assistance with manuscript preparation.

ABBREVIATIONS

BOLD blood oxygen level—dependent
CPT   Current Procedural Terminology
DTI   diffusion tensor imaging
DWI   diffusion-weighted imaging
FA   fractional anisotropy
fMRI functional magnetic resonance imaging

MEG	magnetoencephalography
MRI	magnetic resonance imaging
MRSI	magnetic resonance spectrometry imaging
PACS	picture archiving and communication system
PET	positron emission tomography
SMA	supplementary motor area

REFERENCES

Anderson, C. T., Carlson, C. E., Li, Z., & Raghavan, M. (2014). Magnetoencephalography in the preoperative evaluation for epilepsy surgery. *Current Neurology and Neuroscience Reports*, *14*(5), 446. Available from https://doi.org/10.1007/s11910-014-0446-8.

Ashtari, M., Perrine, K., Elbaz, R., Syed, U., Thaden, E., McIlree, C., ... Ettinger, A. (2005). Mapping the functional anatomy of sentence comprehension and application to presurgical evaluation of patients with brain tumor. *American Journal of Neuroradiology*, *26*(6), 1461−1468.

Benjamin, C. F., Walshaw, P. D., Hale, K., Gaillard, W. D., Baxter, L. C., Berl, M. M., ... Bookheimer, S. Y. (2017). Presurgical language fMRI: Mapping of six critical regions. *Human Brain Mapping*, *38*(8), 4239−4255. Available from https://doi.org/10.1002/hbm.23661.

Bobholz, J. A., Rao, S. M., Saykin, A. J., & Pliskin, N. (2007). Clinical use of functional magnetic resonance imaging: Reflections on the new CPT codes. *Neuropsychology Review*, *17*(2), 189−191. Available from https://doi.org/10.1007/s11065-007-9022-1.

Bookheimer, S. (2007). Pre-surgical language mapping with functional magnetic resonance imaging. *Neuropsychology Review*, *17*(2), 145−155. Available from https://doi.org/10.1007/s11065-007-9026-x.

Brown, T. J., Brennan, M. C., Li, M., Church, E. W., Brandmeir, N. J., Rakszawski, K. L., ... Glantz, M. (2016). Association of the extent of resection with survival in glioblastoma: A systematic review and meta-analysis. *JAMA Oncology*, *2*(11), 1460−1469. Available from https://doi.org/10.1001/jamaoncol.2016.1373.

Catani, M., & Mesulam, M. (2008). The arcuate fasciculus and the disconnection theme in language and aphasia: History and current state. *Cortex*, *44*(8), 953−961. Available from https://doi.org/10.1016/j.cortex.2008.04.002.

Catani, M., & Thiebaut de Schotten, M. (2012). *Atlas of Human Brain Connections. Atlas of Human Brain Connections* (p. 544) New York: Oxford University Press.

Chung, H. W., Chou, M. C., & Chen, C. Y. (2011). Principles and limitations of computational algorithms in clinical diffusion tensor MR tractography. *American Journal of Neuroradiology*, *32*(1), 3−13. Available from https://doi.org/10.3174/ajnr.A2041.

Descoteaux, M., Deriche, R., Knosche, T. R., & Anwander, A. (2009). Deterministic and probabilistic tractography based on complex fibre orientation distributions. *IEEE Transactions on Medical Imaging*, *28*(2), 269−286. Available from https://doi.org/10.1109/TMI.2008.2004424.

Dym, R. J., Burns, J., Freeman, K., & Lipton, M. L. (2011). Is functional MR imaging assessment of hemispheric language dominance as good as the Wada test? A meta-analysis. *Radiology*, *261*(2), 446−455. Available from https://doi.org/10.1148/radiol.11101344.

Fernandez Coello, A., Moritz-Gasser, S., Martino, J., Martinoni, M., Matsuda, R., & Duffau, H. (2013). Selection of intraoperative tasks for awake mapping based on relationships between tumor location and functional networks. *Journal of Neurosurgery, 119*(6), 1380−1394. Available from https://doi.org/10.3171/2013.6.JNS122470.

Haglund, M. M., Berger, M. S., Shamseldin, M., Lettich, E., & Ojemann, G. A. (1994). Cortical localization of temporal lobe language sites in patients with gliomas. *Neurosurgery, 34*(4), 567−576. (discussion 576).

Hanakawa, T., Ikeda, A., Sadato, N., Okada, T., Fukuyama, H., Nagamine, T., ... Shibasaki, H. (2001). Functional mapping of human medial frontal motor areas. The combined use of functional magnetic resonance imaging and cortical stimulation. *Experimental Brain Research, 138*(4), 403−409.

Huettel, S. A., Song, A. W., & McCarthy. (2014). *Functional magnetic resonance imaging* ((3rd ed.)). Sunderland, MA: Sinauer Associates.

James, J. S., Rajesh, P., Chandran, A. V., & Kesavadas, C. (2014). fMRI paradigm designing and post-processing tools. *Indian Journal of Radiology and Imaging, 24*(1), 13−21. Available from https://doi.org/10.4103/0971-3026.130686.

Jones, D. K., Knosche, T. R., & Turner, R. (2013). White matter integrity, fiber count, and other fallacies: The do's and don'ts of diffusion MRI. *Neuroimage, 73*, 239−254. Available from https://doi.org/10.1016/j.neuroimage.2012.06.081.

Jones, D. K., & Leemans, A. (2011). Diffusion tensor imaging. *Methods in Molecular Biology, 711*, 127−144. Available from https://doi.org/10.1007/978-1-61737-992-5_6.

Kekhia, H., Rigolo, L., Norton, I., & Golby, A. J. (2011). Special surgical considerations for functional brain mapping. *Neurosurgery Clinics of North America, 22*(2), 111−132. Available from https://doi.org/10.1016/j.nec.2011.01.004, vii.

Krainik, A., Lehericy, S., Duffau, H., Capelle, L., Chainay, H., Cornu, P., ... Marsault, C. (2003). Postoperative speech disorder after medial frontal surgery: Role of the supplementary motor area. *Neurology, 60*(4), 587−594.

Lezak, M. D., Howieson, D. B., Bigler, E. D., & Tranel, D. (2012). *Neuropsychological assessment* (5th ed.). New York: Oxford University Press.

Mandelli, M. L., Berger, M. S., Bucci, M., Berman, J. I., Amirbekian, B., & Henry, R. G. (2014). Quantifying accuracy and precision of diffusion MR tractography of the corticospinal tract in brain tumors. *Journal of Neurosurgery, 121*(2), 349−358. Available from https://doi.org/10.3171/2014.4.JNS131160.

Milner, B., Corkin, S., & Teuber, H. L. (1968). Further analysis of the hippocampal amnesic syndrome: 14-year follow-up study of H.M. *Neuropsychologia, 6*(3), 215−234. Available from https://doi.org/10.1016/0028-3932(68)90021-3.

Mori, S., & van Zijl, P. C. (2002). Fiber tracking: principles and strategies—A technical review. *NMR in Biomedicine, 15*(7−8), 468−480. Available from https://doi.org/10.1002/nbm.781.

Mueller, W. M., Yetkin, F. Z., Hammeke, T. A., Morris, G. L., 3rd, Swanson, S. J., Reichert, K., ... Haughton, V. M. (1996). Functional magnetic resonance imaging mapping of the motor cortex in patients with cerebral tumors. *Neurosurgery, 39*(3), 515−520. (discussion520-511).

Newman, S. D., & Twieg, D. (2001). Differences in auditory processing of words and pseudowords: An fMRI study. *Human Brain Mapping, 14*(1), 39−47.

Official position of the division of clinical neuropsychology (APA division 40) on the role of neuropsychologists in clinical use of fMRI: Approved by the Division 40 Executive Committee July 28, 2004. (2004). *Clinical Neuropsychology, 18*(3), 349−351. https://doi.org/10.1080/1385404049088718.

Petrella, J. R., Shah, L. M., Harris, K. M., Friedman, A. H., George, T. M., Sampson, J. H., ... Voyvodic, J. T. (2006). Preoperative functional MR imaging localization of language and motor areas: Effect on therapeutic decision making in patients with potentially resectable brain tumors. *Radiology*, *240*(3), 793−802. Available from https://doi.org/10.1148/radiol.2403051153.

Poldrack, R. A., Mumford, J. A., & Nichols, T. E. (2011). *Handbook of functional MRI data analysis*. New York: Cambridge University Press.

Pouratian, N., & Bookheimer, S. Y. (2010). The reliability of neuroanatomy as a predictor of eloquence: A review. *Neurosurgical Focus*, *28*(2), E3. Available from https://doi.org/10.3171/2009.11.FOCUS09239.

Seghier, M. L. (2008). Laterality index in functional MRI: Methodological issues. *Magnetic Resonance Imaging*, *26*(5), 594−601. Available from https://doi.org/10.1016/j.mri.2007.10.010.

Stippich, C., Rapps, N., Dreyhaupt, J., Durst, A., Kress, B., Nennig, E., ... Sartor, K. (2007). Localizing and lateralizing language in patients with brain tumors: Feasibility of routine preoperative functional MR imaging in 81 consecutive patients. *Radiology*, *243*(3), 828−836. Available from https://doi.org/10.1148/radiol.2433060068.

Szaflarski, J. P., Gloss, D., Binder, J. R., Gaillard, W. D., Golby, A. J., Holland, S. K., ... Theodore, W. H. (2017). Practice guideline summary: Use of fMRI in the presurgical evaluation of patients with epilepsy: Report of the Guideline Development, Dissemination, and Implementation Subcommittee of the American Academy of Neurology. *Neurology*, *88*(4), 395−402. Available from https://doi.org/10.1212/WNL.0000000000003532.

Ulmer, J. L., Krouwer, H. G., Mueller, W. M., Ugurel, M. S., Kocak, M., & Mark, L. P. (2003). Pseudo-reorganization of language cortical function at fMR imaging: A consequence of tumor-induced neurovascular uncoupling. *American Journal of Neuroradiology*, *24*(2), 213−217.

Wada, J. (1949). A new method for determination of the side of cerebral speech dominance: A preliminary report on the intracarotid injection of sodium amytal in man. *Igaku to Seibutsugaki*, *14*, 221−222.

Wood, J. M., Kundu, B., Utter, A., Gallagher, T. A., Voss, J., Nair, V. A., ... Prabhakaran, V. (2011). Impact of brain tumor location on morbidity and mortality: A retrospective functional MR imaging study. *American Journal of Neuroradiology*, *32*(8), 1420−1425. Available from https://doi.org/10.3174/ajnr.A2679.

Chapter 6

Wada Testing and Neurosurgical Patients

Caleb M. Pearson[1], Adam Parks[2] and Patrick Landazuri[3]

[1]Departments of Neurology, Neurosurgery, and Psychiatry, University of Kansas Medical Center, Kansas City, KS, United States, [2]Departments of Neurology and Psychiatry, University of Kansas Health System, Kansas City, KS, United States, [3]Department of Neurology, University of Kansas Health System, Kansas City, KS, United States

WADA TESTING AND NEUROSURGICAL PATIENTS

Depending on the study, $\sim 20\% - 40\%$ of individuals with epilepsy do not adequately respond to treatment with anticonvulsant medication (Aicardi & Shorvon, 1997). Uncontrolled seizures not only pose a health risk to patients but are also psychologically and socially burdensome. Further, individuals with uncontrolled seizures are much more likely to be unemployed or underemployed (Hauser & Hesdorffer, 1990). Neurosurgical interventions provide patients with intractable epilepsy the opportunity to drastically reduce, and in some cases eliminate altogether their seizures. However, epilepsy surgery can, in some instances, come at a cost to language, memory, and other cognitive functions. This is especially true in seizure disorders that arise from the temporal lobe, or other areas known to be involved in language or memory functioning. Fortunately, a variety of factors that place patients at risk for postoperative memory or language decline can often be identified prior to surgical intervention. In addition to examining side of surgery (dominant vs nondominant hemisphere resection), the presence of a clear lesion on structural and/or functional imaging studies (mesial temporal sclerosis and/or presence of glucose hypometabolism), and other factors such as age and disease burden, it is essential that the treatment team take into account the functional status of the tissue to be resected. Preoperative neuropsychological assessment is the most common presurgical study that can offer information on the health of a patient's memory and language systems; however, at this time traditional neuropsychological measures are unable to reliably determine language lateralization or assess individual memory circuits. For this, clinicians have often turned to the intracarotid amobarbital test (IAT), which

Neurosurgical Neuropsychology. DOI: https://doi.org/10.1016/B978-0-12-809961-2.00007-2

is otherwise known as the Wada test. The Wada test allows neuropsychologists and other clinicians to selectively anesthetize brain tissue for the purpose of interrogating the functional ability of various brain systems. Not only has this procedure been useful from a clinical standpoint, it has greatly advanced our understanding of the organization of the human language and memory systems. This chapter will examine the Wada test in depth and provide the reader with an understanding of the strengths and weakness of this procedure, as well as practical guidelines for administration and interpretation of Wada test results.

HISTORICAL OVERVIEW

The first known person to apply an anesthetic agent directly to the brain in an attempt to identify essential language cortex was Gardner (1941). At that time, it was generally believed that all right-handed individuals had left hemisphere language dominance and that eloquent language cortex was typically located in the "posterior part of the third left frontal convolution, just in front of the motor center for the face" (Gardner, 1941, p. 1035). It was also thought that most congenitally left-handed individuals possessed right hemisphere dominance for speech functions. Gardner noted that in some cases in which a left-handed individual needed surgery within the right hemisphere, it was difficult to predict for certain where language dominance was localized. He noted that this was especially the case in individuals with intact language functions. To assess this Gardner came up with the bold idea of injecting procaine hydrochloride directly into the brain in an effort to determine the location of language centers.

The first patient to receive such an injection was a 35-year-old man undergoing neurosurgery for the resection of a large right frontal tumor abutting the right motor strip. He presented with a left hemiparesis and simple partial seizures with a classic Jacksonian march beginning in the large toe. A small trephine opening was used to insert a needle just anterior of what was thought to be the facial motor area. Multiple injections of 27 cm^3 of 0.75% Novocain were applied, which resulted in a mild facial droop, but no appreciable aphasia. Given this, Gardner later performed a large resection of the right frontal lobe, which resulted in a left-hand plegia without any noticeable language deficits.

Gardner's second case involved a 27-year-old woman who at the age of 5 years underwent a left frontal craniotomy for a removal of a cyst in an attempt to cure her epilepsy. The seizures persisted, however, and as an adult she elected to undergo a second surgery. In September of 1940 she underwent a second left craniotomy with subsequent injection of 10 cm^3 of 1% Novocain just anterior to the motor face area, which had been identified using a faradic stimulator. This injection again produced no appreciable aphasia, which led to the decision to continue with the cortical excision and ultimately the control of her seizures.

Juhn Wada was the next to advance the localization of eloquent language cortex after he published his paper outlining the IAT (Wada, 1949). His contribution to the development of the IAT test led to a huge advancement in our knowledge of language localization in the brain. Indeed, his name is now synonymous with the procedure. Wada originally developed the IAT to study the evolution of epileptiform discharges between the two hemispheres of the brain in patients receiving electroconvulsive therapy. During the procedure, Wada noted that patient's language abilities would be disrupted following an injection of amobarbital into the patient's language dominant hemisphere. Based on this information, Wada teamed up with Theodore Rasmussen, whom Wada was working with as a fellow at the time, at the Montreal Neurological Institute and began to use the IAT as part of the preoperative evaluation of patients who were candidates for epilepsy surgery. In 1960 the first English language publication describing what is now known as the Wada procedure appeared (Wada & Rasmussen, 1960).

CURRENT USE OF THE WADA TEST

There are generally three purposes for which the Wada test is used in major surgical centers today. First, it allows for the identification of the language dominant hemisphere in patients being considered for neurosurgical intervention. This not only allows for planning the extent to which cerebral tissue can be resected, but also permits the preoperative prediction of postoperative memory impairment (Chelune, Naugle, Lüders, Sedlak, & Awad, 1993) in patients undergoing temporal lobe resections. Second, the Wada test allows the clinician to directly assess memory functioning by simulating the functional disruption of critical memory circuits within the brain. As a result, one can evaluate the memory capacity of the cerebral hemisphere contralateral to the side of injection. The third purpose for the Wada test, specifically for individuals planning to undergo epilepsy surgery, is to help substantiate the presumed hemispheric seizure onset as determined by other invasive and noninvasive procedures designed to lateralize and localize epileptogenic tissue (scalp or invasive electroencephalogram (EEG), structural and functional radiography, video EEG telemetry, magnetoencephalogram (MEG), etc.).

Although the Wada test is considered to be an invasive procedure, which carries a small degree of risk (e.g., stroke, cerebral embolism, infection, etc.), it remains an important component of the workup for many comprehensive surgical epilepsy centers. Despite this, there has been some evidence that suggests less invasive procedures, such as functional magnetic resonance imaging (fMRI) and functional magnetoencephalography for language localization can give similar information and should replace the use of Wada testing in some instances (Binder, 2011). Others have suggested that preoperative neuropsychological testing is sufficient to predict postoperative memory decline when combined with the results of other studies (EEG, structural MRI findings) and demographic and other information (age of seizure onset, chronological age, etc.) (Baxendale, Thompson, Harkness, & Duncan, 2006; Stroup et al., 2003). Likely as a result of these less invasive studies, the use of the Wada test has declined in epilepsy centers across Europe and North America, despite an increase in the number of surgeries performed. In one survey, only 12% of epilepsy centers reported the use of the Wada test on all epilepsy surgical candidates (Baxendale, Thompson, & Duncan, 2008). In contrast to current usage, 85% of centers performed the Wada test on all surgery candidates in 1993 (Rausch et al., 1993). Despite this, the Wada test is still considered the gold standard for lateralizing language function, and in a worldwide survey of 92 epilepsy surgery centers nearly half of all centers reported routinely utilizing the Wada test during presurgical workup of epilepsy surgery candidates (Baxendale et al., 2008). In the United States, the number of centers routinely performing the Wada test is even higher as ~50% of epilepsy surgery centers there report routine use of the Wada test on 75% or more of their patients (Baxendale et al., 2008). Currently, at our own center, we use a combination of neuropsychological testing, fMRI, and Wada studies to predict the risk of temporal lobe epilepsy patients. Specifically, if the patient is right-handed and is to undergo a right temporal lobectomy they will receive presurgical neuropsychological testing and fMRI. If they are found to have right hemisphere language dominance, they are then referred for a Wada procedure. Left temporal lobe epilepsy patients nearly always undergo a Wada procedure in lieu of fMRI given their presumed left hemisphere language dominance.

RATIONALE FOR THE IAT PROCEDURE

The fundamental concept underlying the use of the IAT is to determine the extent to which one cerebral hemisphere can support certain cognitive abilities, such as language and memory, independent of the opposite hemisphere. Disruptions in language and memory during the IAT are assumed to be due to the dispersed anesthetic, which has temporarily "lesioned" an area of one hemisphere, thus the IAT can mimic a hypothetical resection of that neuroanatomical region. An important consideration for the evaluation of language lateralization in epilepsy patients is that this population has a greater incidence of atypical language development, such as right hemispheric or bilateral language representation.

The IAT is unique in its ability to assist with determining the lateralization of presumed cognitive functions due to its deactivation, as opposed to activation, of functional brain tissue (e.g., fMRI, MEG). During the IAT, the anesthetizing agent is typically flushed into the neuroanatomical regions receiving blood supply from the internal carotid artery (ICA; also referred to as the anterior circulation, includes the anterior cerebral artery, middle cerebral artery, and the anterior choroidal artery). The lateral frontal and temporal lobes, and the medial temporal structures including the anterior third of the hippocampus, are all nourished by the vessels of the anterior circulation (Loring et al., 1990).

Sodium amobarbital, a barbiturate, acts as a selective $GABA_A$ agonist and suppresses neuronal excitability in the central nervous system (CNS). Amobarbital is the most widely used anesthetic agent for the IAT as it easily crosses the blood−brain barrier (the selective membrane that separates the blood supply from the brain and extracellular CNS fluid) which allows for a rapid anesthetic effect. This rapid onset as well as the relatively brief wearing off period, with almost complete return of functioning, allows for the assessment of both hemispheres in a short period of time. Other anesthetizing agents, including methohexital, pentobarbital, etomidate, and propofol, are also utilized at some centers. Although methodologies vary across centers, the initial step of the IAT is to perform a cerebral angiogram while the patient is supine. A catheter is inserted into the femoral artery through an incision near the patient's groin. A neuroradiologist, neurosurgeon, or other endovascular specialist then directs the catheter until it reaches the ICA just distal to the common carotid bifurcation contralateral to the hemisphere targeted for study. The amount of anastomotic flow (cross-flow from one hemisphere to the other) is determined using cerebral angiography while also identifying possible atypical vascular formations that may produce unusual results or complications during anesthesia. An example of such atypical formation would be a fetal posterior communicating artery leading to excess anesthetic agent being distributed into the posterior cerebral artery, while a fetal persistent trigeminal artery may lead to agent being dispersed to critical respiratory and cardiac centers in the brain stem, an event which could prove fatal (Loring et al., 1990) (Fig. 6.1).

To provide increased selectivity when "lesioning" portions of the brain, alternative Wada tests utilize microcatheters to reach more circumscribed regions, such as the area supplied by the posterior cerebral artery. One advantage of these alternative Wada tests is that these procedures allow for the assessment of specific cognitive functions, such as memory, without the disruption of other abilities, such as language. The selectivity of this procedure provides the potential for fewer false positive memory test results, that is, fewer memory test failures due to disruption of other cognitive functions (attention, concentration, etc.) rather than a true amnesia, compared to the traditional IAT (Der-Jen et al., 2006). Despite this, many are reluctant to routinely perform such alternative studies due to concerns for possibly increased complication rates or the increased demand for technical expertise. At our center, the selective Wada test is only performed after the validity of the initial IAT results is questioned (e.g., oversedation of the patient during the procedure) or when the results are in contention with other diagnostic testing.

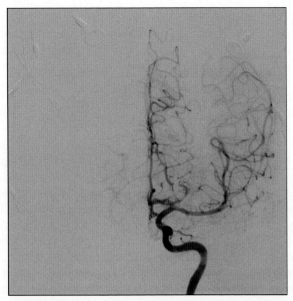

FIGURE 6.1 Cerebral angiogram via the left ICA. Notice the lack of cross-flow of contrast to the contralateral hemisphere. A mild degree of cross-flow is typically not concerning as the injection of amobarbital should be much slower than that of the contrast used during an angiogram. This leads to a lesser degree of cross-flow during the actual Wada procedure itself.

DESCRIPTION OF THE INTERNAL CAROTID ARTERY WADA PROCEDURE AT THE UNIVERSITY OF KANSAS HOSPITAL

At our center, the typical ICA Wada procedure begins with an explanation of the procedure and the presentation of two example items, which the patient is asked to remember. The patient is then told to hold their arms into the air with their palms upward. They are asked to count out loud at which time an injection of between 50 and 150 mg of amobarbital solution is administered over the course of 4−6 seconds by the interventional radiologist or neurosurgeon into the ICA contralateral to the presumed epileptogenic lesion. The dosage recommendations between centers can vary significantly. Some centers use a fixed quantity that does not vary from patient to patient, while others will dose depending on the patient's response during the assessment. Our team typically will inject ∼75 mg, which is then increased by 12.5 mg intervals until the following conditions are met: (1) slowing of the contralateral hemisphere is noted on scalp EEG and (2) contralateral hemiparesis is noted. We find that a typical dose is anywhere between 100 and 125 mg per person. Other symptoms often seen after injection include dysarthria, contralateral hemianopsia, and in some patients, ipsilateral eye version. Speech arrest is often seen following dominant hemisphere injection, but can also be seen in nondominant hemisphere injections as well, presumably due to the anterior cerebral artery's proximity to the presupplemental motor area, which can cause an akinetic mutism-like syndrome.

Immediately following the injection the patient is asked to stick out their tongue or perform some other simple tasks. The purpose of this is to assess comprehension. It is not unusual for the patient to be unable to follow commands following a dominant hemisphere injection, and an inability to do so is often one's first clue as to language dominance. Next, the language and memory portion of the Wada procedure is performed. Our center uses a combination of spoken words and line drawings presented to the patient on a tablet computer, which is held by the neuropsychologist and presented to the patient's ipsilateral visual field so as to avoid problems with any potential contralateral hemianopsia (see Fig. 6.2 for example of Wada content). The patient is asked to either repeat a word, or name a picture depending upon what stimuli is being presented. This allows for the assessment of language and also helps, we believe, to encode the item into memory. Regardless of whether or not the patient can successfully name or repeat the stimuli, the neuropsychologist will name or read the item three times, which again allows for better encoding of the stimuli. Approximately halfway through the presentation of the 16 items in our typical Wada protocol the patient's strength is assessed. This provides the neuropsychologist with information regarding the patient's clinical presentation and helps with interpretation of results. If the patient remains weak in the upper limb contralateral to the side of injection, it can be assumed that they are still under the effects of amobarbital. At this point the epileptologist, or EEG technologist will also communicate the amount of slowing present within the contralateral hemisphere on scalp EEG. Again, this information is crucial to understanding whether the patient is effectively dosed and that the clinical data is valid. After this quick assessment of functioning, and upon completion of the presentation of stimuli, the patient's strength and EEG findings are again examined to ensure proper dosing of amobarbital throughout the examination.

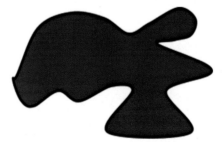

FIGURE 6.2 One of the visual stimuli used during the Wada procedure at the University of Kansas Hospital. Our particular protocol makes use of a tablet computer to display a series of pictures, abstract designs, and words as part of the memory test portion of the study.

Following the presentation of stimuli, which generally takes between 2 and 3 minutes, the patient's speech is assessed. It is expected that the patient's language abilities, if disturbed following injection, will return within ~6−10 minutes following injection. During this time the patient is asked to do a number of tasks, which range from assessing automatic speech (counting to 10, reciting the ABCs, etc.) to repetition and both visual and auditory response naming (e.g., what is an object that tells time, what do I tie to keep my shoes on, etc.). As previously noted, speech arrest can be seen following injection of either hemisphere regardless of language lateralization. For this reason, speech arrest alone should not be considered evidence of language involvement. Instead, abnormalities in speech production such paraphasic errors, which can be seen immediately following the resolution of speech arrest should be used in making one's interpretation about language involvement.

Upon resolution of any speech difficulties, the patient is asked to recall the two example stimuli items, which were presented to them prior to injection of amobarbital. Afterward, they are asked to recall any of the previously presented stimuli in order to assess spontaneous recollection. Oftentimes the patient is able to recall few if any of the stimuli. For this reason, the patient is then presented with a multiple choice recognition format and asked to select the object or word they had previously seen. At our institute we consider anything below the cut score of 9 to represent impaired functioning. Scores between 9 and 10 are considered to be borderline passing scores, and scores of 11 of 16 and greater are considered normal.

At many centers, including our own, a subsequent injection and assessment using similar, but alternative stimuli is performed to assess the opposite hemisphere. This is done for two purposes. First, it allows for the interrogation of functional capacity in the presumed epileptogenic portion of the brain, which can lead to a better understanding of potential deficits following surgery. Second, it allows the neuropsychologist and their team the opportunity to more accurately assess language. This is especially true in the case of bilateral language representation, which can present as either impaired language functioning following each injection, or sometimes even normal language functioning following either injection. Although may centers wait anywhere between 30 minutes and several hours between injections, it has been our experience that starting the second assessment as little as 15 minutes following the original injection can be done without any apparent side effects or validity issues.

POSTERIOR CEREBRAL ARTERY WADA

As previously mentioned, there are some instances in which a posterior cerebral artery Wada is elected to be performed instead of the more traditional internal cerebral artery Wada procedure. In this case, it is important to note some modifications to the Wada protocol. The first consideration is the amount of amobarbital injected into the posterior cerebral artery is much less than the 75−150 mg dosage. At our facility we inject 50 mg of amobarbital solution into the posterior cerebral artery, which is nearly always sufficient. Second, it is important to note that after injection of amobarbital into the posterior cerebral artery the neuropsychologist will not see the traditional contralateral limb weakness. Instead, the most common clinical sign of correct dosage is a contralateral homonymous hemianopsia. Thus, upon injection of amobarbital the patient's visual fields are always tested until a dense visual field cut is seen. Afterward, the stimuli are presented as normal in the unaffected visual field. The memory recognition portion of the study is not completed until after full resolution of any visual field deficit.

INTERPRETATION OF WADA RESULTS

As previously mentioned the goals of the Wada test are threefold. First, the traditional ICA Wada procedure allows for the assessment of language dominance. The second and third goals involve the assessment of memory functioning contralateral to the side of injection. With respect to the first goal, it is important to note that individuals with epilepsy are more likely than normal controls to have atypical language lateralization (Helmstaedter, 2003). This is especially true in individuals who have experienced early brain injuries, have abnormal development of brain tissue in close proximity to eloquent language centers of the brain, and those who develop seizures at an early age. For this reason, it is important to examine language functioning carefully during and after the administration of amobarbital as language dominance is a significant predictor of postoperative memory decline following a traditional temporal lobectomy (Chelune & Najm, 2000). The definition of language dominance varies somewhat between major epilepsy centers, with most major epilepsy centers having their own standardized way of assessment (Benbadis et al., 1998). One method is to compare the length of speech arrest following injection of amobarbital with the side of greater arrest being considered language dominant. As previously mentioned, there are some theoretical problems with this method, as the anterior cingulate is often perfused with amobarbital immediately following injection, which can lead to behavioral arrest that can mimic global aphasia. A second method is to compare scores on a specific battery of language tests given following each hemispheric injection. Despite the multitude of differences in assessing language lateralization between centers, a multicenter study of seven epilepsy centers found a high degree of interrater reliability when assessing language lateralization using the Wada test (Haut et al., 2002).

The rationale behind the memory assessment portion of the Wada test is the anesthetization of the mesial temporal structures, which temporarily simulates the resection of tissue. Although a relatively simple concept, it is important to keep in mind that a neuropsychologist is always assessing the functional ability of the brain tissue contralateral to the site of injection. As previously mentioned, most centers begin with an injection that is ipsilateral to the assumed epileptogenic tissue. This allows for the assessment of what is known as the functional reserve (Chelune, 1995). That is, the memory capacity of the hemisphere proposed to be spared following a traditional temporal lobectomy. If the patient performs poorly following the ipsilateral injection of amobarbital, they are considered to have a low functional reserve and thus may be at higher risk of postoperative memory decline following surgery.

Memory performance of the presumed epileptogenic hemisphere is often referred to as the assessment of functional adequacy (Chelune, 1995). It is not uncommon for the assessment of memory capacity of the epileptogenic temporal lobe to show poor memory, which is a reflection of the epilepsy disease process and its effect on memory structures within the temporal lobe. If the patient's functional adequacy is high (the patient performs well on memory testing following an injection contralateral to the proposed side of surgery) than it can be assumed that surgery would be removing functional tissue and that the patient is at an elevated risk of memory impairment following neurosurgical intervention.

PROBLEMS AND CONSIDERATIONS

The goal of any epilepsy treatment, medical or surgical, is the elimination or improvement of the patient's epilepsy with little or no side effect. As such, the Wada test aims to provide information appraising the risk of memory loss from temporal lobe surgery, whether catastrophic or "merely" clinically significant. However, there remains real concern whether Wada test results can reliably deliver on that promise. Part of this is due to the variability in the way the test is done. As noted above there are multiple methods of performing the Wada test without any clear evidence that one method is superior to the other or that they are even measuring the same specific information. One study examined 21 patients with two different strategies for determining language laterality in comparison to an fMRI paradigm. The study found no correlation between any of the different paradigms in the results (Benbadis et al., 1998). Further muddying the waters for Wada testing is that, as noted previously, there are epilepsy centers that do not routinely perform Wada tests due to the small risks of procedural complication and the belief that the Wada's contribution to preoperative prediction of memory changes do not warrant the risk. Instead, these centers rely on preoperative neuropsychology testing for objective measures of memory function and fMRI for language lateralization. Furthermore, preliminary noninvasive fMRI paradigms testing memory have been published and, if validated and generalizable, may obviate the need for the minimally invasive Wada test (Sidhu et al., 2015).

There is a body of literature that has led these centers to conclude the Wada test is redundant for estimating memory risk and language laterality. In terms of language, fMRI is well regarded in its ability to accurately localize language function (Benjamin et al., 2017; Black et al., 2015). For the purposes of pure language localization, it would appear fMRI offers a suitable noninvasive alternative to the Wada test. However, when done, Wada tests are still often performed on a bilateral basis when localizing language. One paper evaluated its center's Wada tests and found that unilateral Wada procedures accurately predicted language laterality in >80% of its tests (Wellmer et al., 2005). The authors of that paper suggest that bilateral Wada studies could be reserved for surgeries such as corpus callosotomies and could thus decrease the catheter time and the associated small risk of stroke. On a more concerning note, there is a report of a patient with continued language function during left injection who developed expressive aphasia after a functional left hemispherectomy (Loddenkemper et al., 2004). Thus, it would appear that the Wada test is not as reliable as one might hope with regards to language lateralization.

Given the noninvasive method of language lateralization that fMRI offers, currently the Wada test is typically used to explore memory function. However, some studies have demonstrated the results of the Wada test may not always lead to an accurate prediction of postoperative outcomes. Loring et al. (1990) studied ten patients who failed the Wada test in terms of memory scores suggesting higher risk for postoperative anterograde amnesia. None of these ten patients suffered anterograde amnesia. Furthermore, no patient complained of significant subjective decline in memory function. The authors concluded that failing the Wada test need not be an absolute contraindication to surgical consideration. Similarly, Kubu et al. (2000) reviewed ten patients from their center in Ontario, Canada. Like observations by Loring et al., none of the ten patients suffered anterograde amnesia after their temporal lobe resection. Notably, the Ontario cohort assessed patients who had failed memory testing bilaterally. Patients who had some decline were material specific to the resected temporal lobe, i.e., verbal memory with dominant and visual memory with nondominant.

Although there is a great deal that continues to be unknown with respect to the Wada procedure and memory outcomes, there is even greater uncertainty when it comes to functional neuroimaging. For instance, it has been suggested that certain medications commonly used for seizure management may cause changes in how language networks are represented. One such medication, topiramate, has been shown to have a particularly negative impact on cognition and language (Meador et al., 2005). One study (Yasuda, Centeno, & Vollmar, 2013) found that topiramate use was related to a reduction in task-related deactivation on language tasks, specifically verbal fluency, in fMRI. Finally, a recent paper by Wandschneider et al. (2017) compared fMRI activations on a language task between patients taking either topiramate or zonisamide. They found that, while both medications showed similar effects on frontal and parietal cognitive networks, topiramate was also associated with deactivation in language-based cognitive networks. To conclude, certain considerations should be made regarding medication history, particularly topiramate, before an accurate interpretation of language network analysis can be made with functional neuroimaging techniques. In addition, further research needs to be done to better understand how medications may affect the lateralization capability of noninvasive functional neuroimaging studies.

Hormone fluctuations have also been identified as modifying factors impacting the interpretation of hemispheric language dominance analyses. For instance, when comparing language dominance between men and women using functional transcranial Doppler sonography, one group (Helmstaedter, Jockwitz, & Witt, 2015) found that the women showed high intraindividual variability compared to the men, and that this inconsistency was shown to be related to the menstrual cycle. Given this variability, the authors warned against utilizing singular measures of language dominance comparing brain activation patterns between hemispheres in women. Furthermore, this study has not yet been replicated using fMRI and certainly is a direction of future research. Thus, while the Wada test has certainly been a historically useful tool and gold standard for language lateralization and memory testing, there does appear to be a well-founded possibility that noninvasive modalities may provide the future for presurgical memory risk stratification and language lateralization. However, future research should assess the effects of hormonal changes and medication regimens on their accuracy.

CONCLUSION

Refractory epilepsy greatly impacts quality of life, and in many cases surgery must be considered as a treatment option. Unfortunately, postoperative memory deficits are not uncommon following traditional temporal lobe epilepsy surgeries. For that reason, epilepsy surgery centers must be careful to understand the factors that could increase the risk of detrimental cognitive changes. The development of the Wada procedure has been an important contribution to the field in this regard as it allows neurosurgeons, epileptologists, and neuropsychologists the ability to predict the potential impact of neurosurgical interventions on a patient's memory functioning. Although other techniques, such as fMRI have replaced the Wada test in some situations, it remains a valuable tool in the neuropsychologist's arsenal.

REFERENCES

Aicardi, J., & Shorvo, S. D. (1997). *Epilepsy: A comprehensive textbook* (pp. 1325–1331). Philadelphia, PA: Lippincott-Raven.

Baxendale, S., Thompson, P., & Duncan, J. S. (2008). The role of the Wada test in the surgical treatment of temporal lobe epilepsy: An international survey. *Epilepsia, 49*, 715–719.

Baxendale, S., Thompson, P., Harkness, W., & Duncan, J. S. (2006). Predicting memory decline following epilepsy surgery: A multivariate approach. *Epilepsia, 47,* 1887−1894.

Black, D. F., et al. (2015). Retrospective analysis of interobserver spatial variability in the localization of Broca's and Wernicke's areas using three different fMRI language paradigms. *Journal of Neuroimaging, 25*(4), 626−633.

Benbadis, S. R., et al. (1998). Is speech arrest during Wada testing a valid method for determining hemispheric representation of language? *Brain and Language, 65*(3), 441−446.

Benjamin, C. F., et al. (2017). Presurgical language fMRI: Mapping of six critical regions. *Human Brain Mapping, 38*(8), 4239−4255.

Binder, J. R. (2011). fMRI is a valid noninvasive alternative to Wada testing. *Epilepsy & Behavior, 20*(2), 214−222.

Chelune, G. J. (1995). Hippocampal adequacy versus functional reserve: Predicting memory functions following temporal lobectomy. *Archives of Clinical Neurospychology, 10,* 413−432.

Chelune, G. J., & Najm, I. M. (2000). Risk factors associated with postsurgical decrements in memory. In H. O. Luders, & Y. Comair (Eds.), *Epilepsy Surgery* (2nd ed, pp. 497−504). Philadelphia, PA: Lippincott-Raven.

Chelune, G. J., Naugle, R. L., Lüders, H., Sedlak, J., & Awad, I. A. (1993). Individual change after epilepsy surgery: practice effects and base rate information. *Neuropsychology, 1,* 41−52.

Der-Jen, Y., Jiing-Feng, L., Yang-Hsin, S., Ian-Kai, S., Tung-Ping, S., Chien, C., . . . Chun-Hing, Y. (2006). Selective posterior cerebral artery amobarbital test in patients with temporal lobe epilepsy for surgical treatment. *Seizure, 15,* 117−124.

Gardner, J. (1941). Injection of procaine into the brain to locate speech area in left-handed persons. *Archives of Neurology and Psychiatry,* 1035−1038.

Hauser, W. A., & Hesdorffer, D. C. (1990). *Epilepsy: Frequency, causes and consequences.* New York: Epilepsy Foundation of America.

Haut, S. R., Berg, A. T., Shinnar, S., Cohen, H. W., Bazil, C. W., Sperling, M. R., et al. (2002). Interrater reliability among epilepsy centers: Multicenter study of epilepsy surgery. *Epilepsia, 43,* 1396−1401.

Helmstaedter, C. (2003). Neuropsychological aspects of epilepsy surgery. *Epilepsy and Behavior, 5,* S45−S55.

Helmstaedter, C., Jockwitz, C., & Witt, J. (2015). Menstrual cycle corrupts reliable and valid assessment of language dominance: Consequences for presurgical evaluation of patients with epilepsy. *Seizure, 28,* 26−31.

Kubu, C. S., et al. (2000). Does the intracarotid amobarbital procedure predict global amnesia after temporal lobectomy? *Epilepsia, 41*(10), 1321−1329.

Loddenkemper, T., et al. (2004). Aphasia after hemispherectomy in an adult with early onset epilepsy and hemiplegia. *Journal of Neurology, Neurosurgery, and Psychiatry, 75*(1), 149−151.

Loring, D. W., et al. (1990). The intracarotid amobarbital procedure as a predictor of memory failure following unilateral temporal lobectomy. *Neurology, 40*(4), 605−610.

Meador, K. J., Loring, D. W., Vahle, V. J., Ray, P. G., Werz, M. A., Fessler, A. J., . . . Kustra, R. P. (2005). *Neurology, 64*(12), 2108−2114.

Rausch, R., Silfevenius, H., Wieser, H., Dodrill, C., Meador, K., & Jones-Gotman, M. (1993). Intraarterial amobarbital procedures. In J. Engel (Ed.), *Surgical Treatment of the Epilepsies* (2nd ed, pp. 341−357). New York: Raven Press.

Sidhu, M. K., et al. (2015). Memory fMRI predicts verbal memory decline after anterior temporal lobe resection. *Neurology, 84*(15), 1512−1519.

Stroup, E., Langfitt, J., Berg, M., McDermott, M., Pilcher, W., & Como, P. (2003). Predicting verbal memory decline following anterior temporal lobectomy (ATL). *Neurology, 60,* 1266−1273.

Wada, J. (1949). A new method for the determination of the side of the cerebral speech dominance: A preliminary report on the intracarotid injection of sodium amytal in man. *Igaku Seibutsugaku, 14,* 221−222.

Wada, J., & Rasmussen, T. (1960). Intracarotid injection of sodium amytal for the lateralization of cerebral speech dominance: Experimental and clinical observations. *Journal of Neurosurgery, 17,* 266−282.

Wandschneider, B., Burdett, J., Townsend, L., Hill, A., Thompson, P. J., Duncan, J. S., & Matthias, J. (2017). *Koepp Neurology, 88*(12), 1165−1171.

Wellmer, J., et al. (2005). Unilateral intracarotid amobarbital procedure for language lateralization. *Epilepsia, 46*(11), 1764−1772.

Yasuda, C. L., Centeno, M., Vollmar, C., et al. (2013). The effect of topiramate on cognitive fMRI. *Epilepsy Research, 105,* 250−255.

Chapter 7

Awake Craniotomy and Bedside Cognitive Mapping in Neurosurgery

Guillaume Herbet[1,2,3] and Hugues Duffau[1,2,3]

[1]Department of Neurosurgery, Hôpital Gui de Chauliac, Montpellier University Medical Center, Montpellier, France, [2]National Institute for Health and Medical Research (INSERM), U1051, Team "Plasticity of the central nervous system, human stem cells and glial tumors," Institute for Neurosciences of Montpellier, Montpellier University Medical Center, Montpellier, France, [3]University of Montpellier, Montpellier, France

INTRODUCTORY REMARKS AND BRIEF HISTORY

Since the seminal works of eminent neurologists of the late of the 19th century—particularly Jean-Baptiste Bouillaud (1825), Paul Broca (1861), and Paul Wernicke (1874)—cerebral processing was mainly conceived in a rigid, localizationist manner, following the principles previously dictated by the phrenologist school. According to the localizationist principle, the brain is formed by a mosaic of cerebral structures, each being specialized in a given function. As a result, a lesion in one of these structures is presumed to result in a massive and long-lasting deficit. Although the heuristic value of this reductionist approach was decried quickly by authors like Pierre Marie (1906), and more generally by the funders of the Associationist School, the localizationism paradigm continued and remains firmly rooted in the way of thinking, despite regular proposals of more sophisticated neuropsychological and computational models that attempt to include connectivity information and postlesional plasticity (Geschwind, 1965, 1974; Mesulam, 1990, 1998; McClelland & Rogers, 2003). Moreover, it has had important consequences, firstly on the conception of investigation tools in neuropsychology and neuroscience and, consequently, on how the data were interpreted, but more dramatically on the way to take care of patients with cerebral insult—especially in the context of neurosurgery (Vilasboas et al., 2017). Indeed, for many years—and to a lesser extent currently—the decision whether or not to operate on a patient for a brain neoplasm was guided by incorrect anatomical assumptions (e.g., the glioma of a given patient is located in Broca's area;

Neurosurgical Neuropsychology. DOI: https://doi.org/10.1016/B978-0-12-809961-2.00008-4

consequently, this patient will develop severe long-lasting postoperative aphasia-like disorders if he or she is operated on). As a result, the localizationist view has naturally led to a "wait-and-see" attitude, especially in the context of patients harboring a slow-growing tumor where the functional consequences of the tumor remain weak before any oncological treatment (Cochereau, Herbet, & Duffau, 2016; Taphoorn & Klein, 2004). However, the progressive development of awake craniotomy with intraoperative monitoring of cerebral functions by means of direct electrostimulation (DES), first initiated by Penfield (Penfield & Boldrey, 1937) and Ojemann (Ojemann & Whitaker, 1978; Ojemann, 1979) in the context of epilepsy surgery, and then by Berger (Berger & Ojemann, 1992; Berger, Ojemann, & Lettich, 1990) and Duffau (Duffau, Sichez, & Lehéricy, 2000, 2002) in the context of tumor surgery, has radically changed the field. Indeed, thanks to this surgical approach, it was regularly possible to remove impressive neoplastic tumors in the so-called eloquent areas (e.g., Broca's area or Wernicke's area) without causing any severe or long-lasting neurological deficits (Plaza, Gatignol, Leroy, & Duffau, 2009; Sarubbo et al., 2012), demonstrating that cerebral functions are subserved by complex, distributed, and resilient neural networks, and not by discrete and isolated areas.

In parallel, in the past decade, important advances have been accomplished in neuroscience allowing significant progress in understanding the anatomic and functional architecture of the brain. Specifically, the concept of connectome has emerged, the goal of which is to capture the characteristics of spatially distributed, dynamic, neural processes at multiple spatial and temporal scales (Sporns et al., 2005; Sporns, 2001). The new science of connectomics, that aims to map the neural connections of human brain, contributes both to theoretical and computational models of the central nervous system (CNS) as a complex system and, experimentally, to new indices and metrics (e.g., nodes, hubs, efficiency, and modularity), with the goal to characterize and scale the functional organization of the healthy and diseased nervous system (Bressler & Menon, 2010; Bullmore & Bassett, 2011). According to this new theory, the CNS is an ensemble of complex networks which are continually formed and reshaped across the lifespan, and processes information dynamically, opening the window to a huge potential for neuroplasticity, even in adults.

These important advances have found their counterparts in the realm of surgical neuro-oncology. The paradigmatic shift from localizationism to connectionism provides not only a better understanding of the efficient neuroplasticity generally observed in glioma patients (Bonnetblanc, Desmurget, & Duffau, 2006; Desmurget, Bonnetblanc, & Duffau, 2007; Duffau, 2005), but also of the pathophysiological mechanisms that equally constrain this plasticity (Herbet, Lafargue, & Duffau, 2017; Herbet, Moritz-Gasser, & Duffau, 2017; Herbet, Yordanova, & Duffau, 2017; Herbet, Maheu et al., 2016; Herbet, Moritz-Gasser et al., 2016; Ius, Angelini, de Schotten, Mandonnet, & Duffau, 2011). On the other hand, as cortico—subcortical electrical mapping was the only technique allowing to generate real-time anatomo-functional correlations and to interrogate directly the white matter tracts from a functional standpoint (Duffau, 2015a, 2015b), the data gained from "awake surgery" have greatly enhanced our current knowledge about the structural and functional connectivity underpinning a range of brain functions, such as spoken language (Duffau, Herbet, & Moritz-Gasser, 2013; Duffau, Moritz-Gasser, & Mandonnet, 2014) or spatial cognition (Thiebaut de Schotten et al., 2015), taking into consideration the inter-individual variability in the functional topological organization (Duffau, 2017; Tate, Herbet, Moritz-Gasser, Tate, & Duffau, 2014). As a result, such knowledge has enabled the construction of viable, neuropsychological-based, and anatomically-constrained models of human functions.

Although "awake" mapping has historically concerned "overt" functions, and has predominantly been applied at the cortical level, this surgical technique has undergone far-reaching changes in recent years, especially in the context of glioma surgery. In fact, this rapid development is consubstantial with the greater awareness from neurosurgical physicians that preserving the quality of life for patients with a long-survival expectancy (as in low-grade gliomas) cannot not only be reduced to the assessment of functions whose damage leads to overt consequences, such as aphasia or motor defects. As a result, the intraoperative mapping of other important cognitive functions (such as, e.g., sociocognitive functions, spatial cognition, and semantic cognition) in both hemispheres has been advocated in order to give patients the best opportunities of resuming a normal socioprofessional life quickly after surgery (Duffau, 2010; Vilasboas et al., 2017). Moreover, the improvement of our knowledge of subcortical anatomy thanks to both the development of tractography imaging and the revival of anatomical dissection (Basser, Pajevic, Pierpaoli, Duda, & Aldroubi, 2000; Catani, Howard, Pajevic, & Jones, 2002; Martino, Brogna, Robles, Vergani, & Duffau, 2010; Maldonado et al., 2013; Sarubbo et al., 2015; Wang et al., 2016), have enabled a better understanding of the role of anatomical connections in cognition and behavior. This dynamic interaction between awake neurosurgery, neuropsychology, and neuroscience has resulted in the move toward the era of connectomal surgery (Duffau, 2014, 2015a, 2015b).

In this chapter, after defining the main principles of DES mapping in the context of brain tumor, the authors describe the different behavioral paradigms typically used in awake surgery, and how they are selected and modulated to identify and spare critical brain-wide cognitive systems of the CNS. For each described function, a tentative anatomo-functional model based on electrostimulation mapping results is provided. Knowledge gained in this chapter may aid neuropsychologists and neurosurgeons to better cope with both the functional and anatomical connections of the brain, and their reorganizational potential in a network perspective.

PRINCIPLES OF DIRECT ELECTROSTIMULATION MAPPING IN AWAKE SURGERY

During surgery for a tumor generally involving both gray and white matter, it has become common practice to awaken patients with the goal of investigating the functional role of restricted cerebral sites. The surgeon can optimize the extent of resection and, thereby, improve overall survival without causing permanent deficits, owing to individual mapping and preservation of eloquent structures (Duffau, 2014). Consequently, resection is achieved according to functional, as opposed to structural, boundaries. This well-tolerated surgical approach has been proven to drastically reduce the risk of permanent postoperative neurological and cognitive deficits, to optimize tumor removal, even when located within so-called critical areas, and to increase overall survival (De Witt Hamer, Robles, Zwinderman, Duffau, & Berger, 2012). In this way, quality of live is well-preserved, and patients are able to resume a normal social and professional life quickly after surgery.

In practice, patients perform a range of sensorimotor, visuospatial, language and/or social cognition tasks (see below) during surgery, while the surgeon uses DES to temporarily disrupt discrete regions in the vicinity of the tumor. In this approach, a biphasic electrical current creates a "virtual transient lesion," initially at the level of the cortex. At the axonal level, DES is again used to generate a virtual lesion when the electrodes are applied directly in contact with the white matter tracts, leading to a transient disconnection of the brain areas interconnected by the tract. If the patient stops moving or speaking, or produces an inappropriate response (i.e., the patient is unable to make a semantic association), the surgeon can avoid removing the stimulated site at the level of the subcortical connectivity (Duffau, Gatignol, Mandonnet, Capelle, & Taillandier, 2008).

The exact mechanisms of DES remains controversial, but the predominant view is that the stimulation transitorily interferes locally with a small cortical and/or axonal site. Interestingly, Logothetis et al. (2010), demonstrated that as long as the repetitive stimulation is $< 200\,Hz$, y-aminobutyric acid-related inhibition prevents the stimulation from propagating beyond the first synapse. Assuming that there is no propagation of the stimulation to the entire connected network, DES informs us about its impact on a network's function when just a part of that network is stimulated. Thus, DES is able to detect the structures that are essential for brain function by inhibiting a sub-circuit for a few seconds, with the possibility of checking whether the same effects are reproduced when repeated stimulations are applied over the same structure. In surgery, reproducibility is essential to avoid a false positive outcome. It is generally admitted that obtaining similar functional responses during three stimulations over the same site in a nonsequential manner is sufficient to validate that this site is critical.

By gathering all cortical and axonal structures in which DES evokes the same types of error, one can build up a picture of the subnetwork mediating the disrupted subfunction. DES, the validity of which has been recently reestablished (Desmurget, Song, Mottolese, & Sirigu, 2013), has extensively been demonstrated to detect, with great accuracy and reproducibility, the structures that are crucial for cognition functioning in humans, and is currently the only way to directly interrogate the functional role of the white matter fibers. By combining neurological disturbances elicited by DES with anatomical data provided by postoperative magnetic resonance imaging (MRI), it becomes possible to achieve reliable anatomo-functional correlations and to, secondly, model the neural implementation of a given function. Such in vivo correlations have permitted analysis of the anatomical location of the eloquent areas detected by DES—in essence, at the periphery of the surgical cavity, where the resection was stopped according to functional boundaries. This methodology has extensively been validated in various studies (Rech, Herbet, Moritz-Gasser, & Duffau, 2014; Tate et al., 2014; Sarubbo et al., 2015; Kinoshita et al., 2015).

In the next section, we describe the different behavior paradigms typically used in our institution to intraoperatively monitor cognitive functions. When appropriate, we describe the connectionist models that have been built thanks to the data issued from intraoperative brain mapping.

INTRAOPERATIVE MONITORING AND MAPPING OF COGNITIVE PROCESSES IN AWAKE SURGERY

Motor Cognition

Voluntary movement (i.e., the consequence of internal/endogenous activity) engages a set of highly sophisticated, neurocognitive processes grouped under the term of motor cognition (i.e., intention to act, motor planning, motor initiation, and action control, etc.). Impairment of motor cognition can lead to a variety of disabling disorders, such as, for example, disturbance of bimanual coordination or ideomotor motor apraxia. A basic way to intraoperatively monitor all aspects of voluntary movement is to ask the patient to perform a simple, double-motor task engaging the upper limb, such as to lower the arm and open the hand, then to raise the arm and close the hand (the lower limb may be also concerned or both upper and lower limbs at the same time). In addition, in order to control the velocity and the accuracy of the movement during surgery, this task enables mapping crucial areas for motor cognition under stimulation (Schucht et al., 2013). Depending on the structures being stimulated, several manifestations can be observed, for example, stimulation of the supplementary motor area can lead to motor initiation disturbances. Other motor tasks can be performed to map more specific motor abilities. Regularly, we ask patients to make coordinated movements with both hands (Rech et al., 2014, 2016), an ability which is especially crucial for certain professions (e.g., manual work, making music). It is also important to ask the patient to perform more complex movements to assess fine-grained motor abilities, such as drumming or to perform reflexive praxis (e.g., imitation of meaningless movements) to evaluate movement planning.

The use of these motor tasks during awake surgery has provided considerable knowledge about the network subserving motor control—the so-called negative motor network. More specifically, further studies have compiled the stimulation mapping data from patients harboring a tumor around the central region, and have revealed a large neural network originating from the supplementary area, the lateral premotor cortex, and the depth of the precentral gyrus, and passing to anterior to the corticospinal tract. The projections fibers run to the head of the caudate nucleus, corresponding to the frontostriatal tract (Kinoshita et al., 2015). The caudate is one of the inputs of the basal ganglia, which have a crucial role in initiation and execution of voluntary movements, as well as in the inhibition of competing movements (Alexander & Crutcher, 2000). Additional fibers project to the anterior arm of the internal capsule, indicating a likely course toward the spinal cord. Another part of the circuits runs posterior to the primary somatosensory fibers (Almairac, Herbet, Moritz-Gasser, & Duffau, 2014). Precentral and retrocentral pathways are interconnected by U fibers passing beneath the central region. The parietal lobe is known to have strong connections to the precentral areas of the frontal lobe (Filimon, 2010) and to be involved in motor functions, as demonstrated through cortical stimulations in macaques and lesion studies in humans (Fogassi & Lupino, 2005; Jackson et al., 2009).

These stimulation-based findings provide evidence that the network subserving motor control is not centered on the frontal lobe, but extends to other regions, since it includes frontal and parietal white matter fibers, as well as projection pathway with inhibitory and excitatory features. Therefore, the dichotomy between precentral motor and retrocentral sensory structures needs to be revisited with a new connectionist model, which can explain the effects observed when the white matter tracts are stimulated. A revised classification seems necessary to distinguish the motor control fibers with the modulatory influence on movement from the classic cortico-spinal tract originating from the precentral gyrus (Kinoshita et al., 2015).

Basic Visual Processes

Patients with a large visual field defect, such as lateral homonymous hemianopia, generally have a poor functional outcome. In many countries, driving is formally prohibited and a lot of activities, such as reading, become arduous. To map visual connectivity and avoid the occurrence of long-term postoperative visual field defects, we use a simple protocol allowing the assessment of visual fields during surgery (Gras-Combe, Moritz-Gasser, Herbet, & Duffau, 2012). Specifically, with their vision fixed at the center of a screen, patients are asked to name successively two pictures disposed in the two opposite quadrants, knowing that it is absolutely crucial to preserve the inferior quadrant (the superior being compensable in daily life). The position of the pictures is determined by the laterality of the lesion. Although DES of visual pathways, especially the optic radiations, generally evokes a range a phenomena subjectively described by the patients themselves (either "inhibitory phenomena" such as blurred vision or impression of shadow, or "excitatory phenomena," such as phosphenes), the described task allows the benefit from a more objective confirmation of the transitory visual disturbance induced by electrostimulation (i.e., the patient cannot name the picture presented in the inferior quadrant contralateral to the lesion). Some indicators must also be taken into consideration during the assessment of visual fields, most notably the amplitude of visual saccades or the possible increase of naming response time in the visual field under scrutiny. It is also very important to regularly manually check the extent of the visual field of the patient.

Visual Cognition

The inferolateral occipito-temporal cortex is reputed to broadcast critical information in the service of object recognition (Bar et al., 2001). Damage to this neural system may lead to visual agnosia—the disabling inability to recognize objects with visual modality—as repeatedly observed in the neurological population, especially when the lesion is bilateral or involves corpus callosal (Karnath, Rüter, Mandler, & Himmelbach, 2009). A simple way to map these high-order visual processes during awake surgery is to use a picture naming task. If a disturbance of object recognition is induced during electrostimulation, the patient generally commits a nonsemantically related "visual" paraphasia.

Our group has previously shown that axonal stimulation of the inferior longitudinal fasciculus (ILF) leads to contralateral visual hemiagnosia (Mandonnet, Gatignol, & Duffau, 2009; Coello, Duvaux, De Benedictis, Matsuda, & Duffau, 2013), demonstrating the vital role of this white matter connectivity in conveying high-level visual information from primary occipital structures to more posterior structures located in the temporal lobe, in particular the fusiform gyrus and the anterior temporal lobe. It is worth mentioning that visual agnosia is generally generated in the right hemisphere and not in the left hemisphere where stimulation or lesion-related disconnection of the ILF leads to alexia (Zemmoura et al., 2015) or lexical retrieval difficulties (Herbet, Maheu et al., 2016; Herbet, Moritz-Gasser et al., 2016). Note that the bilateral disconnection of ILF can lead to other forms of visual agnosia, in particular prosopagnosia (Corrivetti, Herbet, Moritz-Gasser, & Duffau, 2017)—an inability to recognize familiar faces. This is consistent with evidence that this contingent of white matter fibers are specifically atrophied in patients with congenital prosopagnosia (Thomas et al., 2009; Grossi et al., 2014).

Visuospatial Cognition

Unilateral spatial neglect is a debilitating neurological condition characterized by a failure to explore and allocate attention in the space contralateral to the damaged hemisphere. It may occur after lesion of many cortical territories of the brain, especially in the right hemisphere (Molenberghs, Cunnington, & Mattingley, 2012; Molenberghs, Sale, & Mattingley, 2012). This cognitive impairment has a major impact on quality of life by depriving the patient to resume a normal social and professional life (e.g., neglect patients cannot drive a car).

Decades of neuropsychological studies have pinpointed a wide range of cortical areas associated with spatial neglect, including mainly parietal (angular gyrus, supramarginal gyrus, superior parietal lobule and temporo-parietal junction (TPJ)) (Heilman, Bowers, & Watson, 1983; Vallar & Perani, 1986; Mort et al., 2003; Chechlacz et al., 2010; Karnath, Rennig, Johannsen, & Rorden, 2010; Thiebaut de Schotten et al., 2012), temporal (superior and middle temporal gyri) (Samuelsson et al., 1997; Karnath, Fruhmann Berger, Küker, & Rorden, 2004), and posterior frontal areas (inferior and middle frontal gyri, premotor cortex) (Binder, Marshall, Lazar, Benjamin, & Mohr, 1992; Husain & Kennard, 1996; Committeri et al., 2007; Verdon et al., 2009; Rengachary et al., 2011). As spatial neglect is widely viewed as a complex polymorph syndrome, the involved cortical regions (and associated subnetworks) are likely to vary according to the type of neglects observed. For example, certain studies have evidenced some kinds of anatomo-functional dissociations between neglect due to disturbance of visuo-exploratory processes (posterior frontal lesions) and that due to disturbance of more perceptual processes (posterior parietal lesions) (Binder et al., 1992).

A classical test to evaluate spatial neglect is the bisection line task. For surgery, we have adapted this task in a touch-screen environment. The patient is asked to separate a line into two identical segments (i.e., find the true center of the line). The length of the line is 18 centimeters. If, during the time of stimulation, a significant rightward deviation is induced (typically 7 mm or slightly more if the patient presents with a behavioral variability), the brain areas under scrutiny are considered as eloquent for visuospatial cognition (Bartolomeo et al., 2017).

The line bisection task is useful to map the posterior parietal cortex, including the inferior and superior parietal lobule, but also to a lesser extent the posterior temporal cortex (Thiebaut de Schotten et al., 2005; Roux et al., 2011; Vallar et al., 2014). Most importantly, it also enables the identification of the dorsal white matter connectivity, especially layer II of the superior longitudinal fasciculus (SLF). This white matter tract, by interconnecting the superior and inferior parietal lobules (and the cortex within the intraparietal sulcus) with the posterior middle frontal gyrus (Makris et al., 2004; Schmahmann et al., 2007; Thiebaut de Schotten et al., 2011) is thought to maintain information exchange between the ventral and the dorsal attention systems (Bartolomeo, De Schotten, & Chica, 2012), and is strongly associated to spatial neglect in the case of neurological injury (Thiebaut de Schotten et al., 2012).

Using the same behavioral paradigm, further DES studies have shown that the right inferior fronto-occipital fasciculus (IFOF), a brain-wide white matter tract providing connections between the occipital, parietal, and temporal lobes and the prefrontal cortex (Catani et al., 2002; Sarubbo et al., 2013; Caverzasi, Papinutto, Amirbekian, Berger, & Henry, 2014) may also play a role in spatial cognition (Herbet, Lafargue et al., 2017; Herbet, Moritz-Gasser et al., 2017; Herbet, Yordanova et al., 2017), as previously hypothesized, but not clearly behaviorally evidenced (Sarubbo et al., 2013). This is consistent with the emerging view that the IFOF may support a multilayer organization. More specifically, the IFOF may be organized in two (Sarubbo et al., 2013; Hau et al., 2016) or even more streams (i.e., five) (Wu et al., 2016) having projections to cortical structures known to be involved in spatial cognition, such as the posterior dorsolateral prefrontal cortex, the superior parietal lobule, the angular gyrus, and the posterior part of the temporal gyrus. This multilayer structural organization is in agreement with current experimental evidence suggesting an involvement of the right IFOF in at least three cerebral functions—nonverbal semantic cognition (Herbet, Lafargue et al., 2017; Herbet, Moritz-Gasser et al., 2017; Herbet, Yordanova et al., 2017) mentalizing (Yordanova et al., 2017), and basic emotion recognition (Philippi, Mehta, Grabowski, Adolphs, & Rudrauf, 2009). Thus, it is possible that one, or maybe more, of the IFOF layers are specifically devoted to some aspects of spatial cognition/attention, especially layers connecting the ventrolateral prefrontal (i.e., IFOF-III from Wu et al., 2016) or the posterior dorsolateral prefrontal cortex (i.e., deep layer from Sarubbo et al., 2013, or IFOF-IV from Wu et al., 2016) to the posterior parietal cortex (i.e., superior parietal lobule and angular gyrus). In view of the large panel of connections of the IFOF in brain regions belonging to both the dorsal (i.e., posterior dorsolateral prefrontal cortex, superior parietal lobule) and the ventral (i.e., posterior inferior frontal gyrus, angular gyrus) attention network, this associative fasciculus might, in principle, have the same physiological capacities as the SLFII in governing between-system integration in the attention network.

In brief, the use of a simple line, bisection task enables to map and preserve critical structures for spatial attention. A further study performed by our group has shown that, although approximately half of the patients with a tumor experienced a transitory neglect in the immediate postoperative phase, none of them presented persistent spatial neglect—demonstrating the effectiveness of the intraoperative mapping (Charras et al., 2015).

Dual-Tasking and Multitasking

Numerous activities in daily life necessitate processing different matters at the same time. This crucial multitasking ability requires maintaining in working memory several tasks to be performed and concurrently allocating attention among them. During surgery, this higher capacity can be assessed by asking the patient to simultaneously perform a regular movement of the upper limb and a naming task or a semantic association task. A multitasking disturbance is observed when the patient is no longer able to perform both tasks at the same time while the realization of each task separately remains possible. This impairment, to a lesser extent, may be manifested in a temporary desynchronization/lack of coordination between the two tasks.

Social Cognition

Human beings are especially talented in understanding and predicting others' behavior. This complex ability, commonly referred to as "mentalizing" or "Theory of Mind" (Premack & Woodruff, 1978), is one of the main foundations on which social cognition is built (Frith & Frith, 2012; Gallagher & Frith, 2003; Lieberman, 2007). Unsurprisingly, it is disrupted in a wide range of neuropsychiatric conditions, especially those in which social communication is highly problematic, such as in autism spectrum disorders and schizophrenia. Mentalizing is not a unitary process, but rather encompasses a variety of more basic as well as some nonspecific subprocesses, such as emotion processing, inferential reasoning, understanding of causality, and self/other distinction (Schaafsma et al., 2015). Patients with social cognition disorders have difficulties in understanding their social environment leading to abnormal behaviors and lack of social flexibility. Current literature is subserved by an extensive network of spatially distributed cortical areas, mainly including the medial prefrontal cortex, the IFG, the TPJ along with the pSTG, the posterior medial parietal cortex (i.e., precuneus and posterior cingulate gyrus), and the temporal pole (Frith & Frith, 2003; Amodio & Frith, 2006; Van Overwalle, 2009; Mar, 2011; Schurz et al., 2014; Molenberghs et al., 2016). Some of these areas, namely the medial prefrontal cortex and the bilateral TPJ, are consistently engaged irrespective of the mental state under consideration (e.g., intention, emotion, desire, and belief, etc.) and task modalities (Mitchell, 2009; Schurz et al., 2014), whereas the involvement of other regions seems to be more task-dependent (Carrington & Bailey, 2009; Molenberghs et al., 2016).

To avoid long-term postoperative social cognition impairments, during awake surgery we use an adapted version of a well-used mentalizing task (i.e., The Read the Mind in the Eyes Task) (Baron-Cohen, Wheelwright, Hill, Raste, & Plumb, 2001). The original version of this behavioral task consists of the presentation of 36 photographs depicting the eye region of human faces. For each of them four affective states are suggested and participants are asked to select the one that best describes what the person on the photograph is feeling or thinking. This behavioral task enables us to tap the lower, prereflective, and automatic (as opposed to high-level, reflective, inferential) aspect of mentalizing. We have shown that patients with a resection of the pars opercularis of the right inferior frontal gyrus did not completely recover after surgery (Herbet, Lafargue, Bonnetblanc, Moritz-Gasser, & Duffau, 2013; Herbet, Lafargue, Bonnetblanc et al., 2014; Herbet, Lafargue, De Champfleur et al., 2014), clinically justifying the use of a new intraoperative task. Consistent with previous literature using face-based mentalizing tasks (Schurz et al., 2014), crucial cortical epicenters are regularly identified, however, with some degree of interindividual variability in the posterior inferior frontal gyrus, the dorsolateral prefrontal cortex, and the posterior superior temporal gyrus (Herbet, Lafargue, & Duffau, 2015; Herbet, Lafargue, Moritz-Gasser, & Bonnetblanc et al., 2015; Herbet, Lafargue, Moritz-Gasser, & de Champfleur et al., 2015; Herbet, Latorre, & Duffau, 2015; Yordanova et al., 2017). More importantly, critical sites are also pinpointed along the inferior IFOF and within the white matter fibers supplying the DLPFC (Yordanova et al., 2017), suggesting that functional integrity of both the direct ventral connectivity and the dorsal connectivity via the SLF are required for accurately inferring complex mental states from human faces.

The combination of these intraoperative results with the results from behavioral examinations performed before and after surgery, have allowed the demonstration that mentalizing is made possible by the parallel functioning of two, or even three, neurocognitive subsystems (Fig. 7.1). The first subcircuit, the mirror system, processes low-level perceptual aspects of mentalizing and is subserved by the arcuate fasciculus/SLF complex in the right hemisphere (Herbet, Lafargue, Bonnetblanc et al., 2014; Herbet, Lafargue, De Champfleur et al., 2014; Herbet, Lafargue et al., 2015; Herbet, Lafargue, Moritz-Gasser, & Bonnetblanc et al., 2015; Herbet, Lafargue, Moritz-Gasser, & de Champfleur et al., 2015; Herbet, Latorre et al., 2015). This observation is in agreement with anatomical dissection and DTI-based studies which have shown that the anterior part of the mirror system (including the postero-inferior frontal gyrus and the ventral premotor cortex) is connected to posterior temporoparietal areas by the perisylvian network (Makris et al., 2004; Schmahmann et al., 2007; de Schotten et al., 2011). Moreover, damage to this connectivity has previously been associated with poor social and emotional intelligence and, in healthy subjects, interindividual variability in

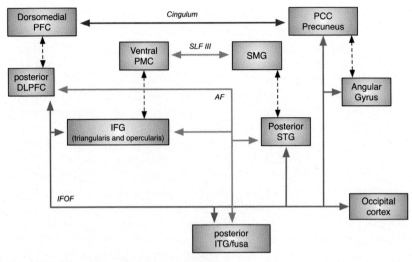

◄─► Medial stream: High-level, inference-based mentalizing

◄─► Dorsal stream: Low-level, perceptive-based mentalizing

◄─► Ventral stream: Facial emotion recognition

FIGURE 7.1 A triple-stream model of mentalizing derived from multimodal (intraoperative and lesion) mapping in diffuse low-grade glioma patients. *PFC*, prefrontal cortex; *PCC*, posterior cingulate cortex; *DLPFC*, dorsolateral prefrontal cortex; *IFG*, inferior frontal gyrus; *ITG*, inferior temporal gyrus; *PMC*, premotor cortex; *SMG*, supramarginal gyrus; *STG*, superior temporal gyrus; *SLF*, superior longitudinal fasciculus; *AF*, arcuate fasciculus; and *IFOF*, inferior fronto-occipital fasciculus.

low-level emotional empathy—a basic socio-affective process known to engage the mirror network—has been correlated with DTI parameters in some parts of the SLF/arcuate fascicle (AF), especially in the right hemisphere (Parkinson & Wheatley, 2014).

The second subcircuit, the mentalizing system per se, is essential for high-level, inference-based, mentalizing processing (i.e., the ability to overtly attribute mental states to others) and is supported by the cingulum which interconnects two crucial cortical structures of the mentalizing network on the medial face of the brain—namely the medial prefrontal cortex and the precuneus. Tumor-related disconnection of the right and the left cingulum have been shown to be associated in glioma patients with a disturbance of intention attribution (Herbet, Lafargue, Bonnetblanc et al., 2014; Herbet, Lafargue, De Champfleur et al., 2014) and cognitive empathy (a synonymous term for mentalizing) (Herbet, Lafargue et al., 2015; Herbet, Lafargue, Moritz-Gasser, & Bonnetblanc et al., 2015; Herbet, Lafargue, Moritz-Gasser, & de Champfleur et al., 2015; Herbet, Latorre et al., 2015), respectively.

Finally, the third subcircuit is supported by the right, inferior, fronto-occipital fasciculus, and may be concerned in processes important for face-based mentalizing, such as facial emotion processing (Yordanova et al., 2017). Damage to this white matter tract in the right hemisphere has been associated with strong emotion recognition deficits (Philippi et al., 2009; Genova et al., 2015) while inter-individual variation in the microstructure of the IFOF has been shown to predict facial emotion recognition and face memory performance (Unger et al., 2016). In view of these findings, and the fact that the IFOF provides direct connections between cortical areas known to be involved in emotion processing, in particular occipito-temporal visual association areas and prefrontal areas (Adolphs, 2002), its disconnection during stimulation may induce transient disturbance of processes involved in decoding facial emotional cues, and are necessary for accurate, affective mentalizing.

Language

To map language processes, the use of a naming task remains the gold standard. This task is easy to implement during surgery and especially adapted to patient positioning constraints, and it is very sensitive to all levels of processing. During electrostimulation, different kinds of impairments may be observed: speech arrest, dysarthria (disturbance of motor programming), anomia (disturbance of lexical retrieval: the patient is unable to name the object), phonological paraphasia (disturbance of phonological encoding: production of a word with phonological deviations; e.g., phelephant for elephant), semantic paraphasia (disturbance of semantic processing, the production of a word semantically related to the target word; e.g., cow for horse) or perseveration (disturbance of inhibitory control mechanisms, repetition of a prior word before a new picture) (Duffau et al., 2002). Beyond classical language-related cortical areas, the naming task enables us to map the main associative connectivities (i.e., the arcuate fasciculus for phonological processes, the lateral SLF for articulatory processes, the IFOF for semantic control processes (see below) and the ILF for lexical retrieval), as well as certain intralobar tracts, such as the frontal aslant tract (speech initiation and control) (Duffau et al., 2013, 2014).

To intraoperatively assess the nonverbal semantic system (the naming tasks enables only an assessment of verbal semantics), we also routinely use a semantic association task (i.e., the Pyramids and Palm Trees Test) (Howard & Patterson, 1992). This task consists of 52 black and white pictures. For each target picture, two new pictures are proposed and the patient is asked to match one with the target one according to the semantic link by pointing it out. We have previously shown that this task is useful to map and preserve the direct ventral connectivity, especially the IFOF in the left hemisphere (Moritz-Gasser, Herbet, & Duffau, 2013), but also in the right hemisphere (Herbet, Maheu et al., 2016; Herbet, Moritz-Gasser et al., 2016). When the tumor concerns the left occipito-temporal cortex, especially the visual word form area and its underlying white matter connectivity, it is necessary to map the different subprocesses involved in reading aloud. To this end, we typically use a reading task in which the patient is asked to read aloud different word categories, including regular and irregular words, and pseudowords (Zemmoura et al., 2015). Finally, to map brain areas involved in the motor implementation of automatic speech production, we routinely use a counting task consisting in counting aloud from 1 to 10 in loop.

On the basis of electrostimulation mapping findings, our group have proposed a dual stream model for visual language processing (Duffau et al., 2013, 2014; Duffau, 2015a, 2015b), with a ventral stream involved in mapping visual information to meaning (semantics) and a dorsal stream dedicated to mapping visual information to articulation through visuo-phonological conversion (Fig. 7.2). Compared to other neurocognitive or computational models (e.g., Hickok & Poeppel, 2004), this model has the great advantage to incorporate anatomical constraints, especially white matter connectivity information.

The first step of this model is object recognition. Electrostimulation of the ILF, especially in the right hemisphere, typically elicits (hemi-)visual

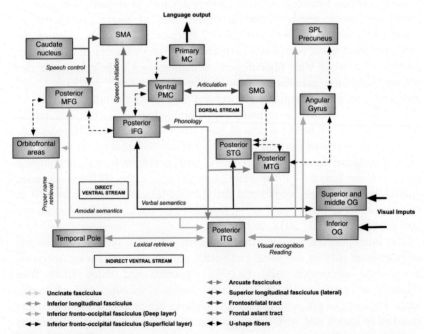

FIGURE 7.2 A dual-stream model of language based on the results from intraoperative stimulation mapping in awake surgery. *MFG*, middle frontal gyrus; *IFG*, Inferior frontal gyrus; *PMC*, premotor cortex; *MC*, motor cortex; *MFG*, middle frontal gyrus; *ITG*, inferior temporal gyrus; *MTG*, middle temporal gyrus; *STG*, superior temporal gyrus; *SMG*, supramarginal gyrus; *AG*, angular gyrus; and *OG*, occipital gyrus.

paraphasia, demonstrating that the patient is no longer able to correctly recognize the observed picture (Mandonnet et al., 2009). These deficits of visual cognition are generally generated by DES of the posterior part of the ILF, which interconnects the primary visual cortex with the visual word form area located (WVFA) in the fusiform gyrus. It is worth mentioning that stimulation (or surgical resection) of the same white matter fibers in the left hemisphere, especially those interconnecting the primary visual cortex and the VWFA, leads to pure alexia (i.e., an incapacity to read letters and words) (Zemmoura et al., 2015). Disruption of the VWFA per se induces different forms of alexia depending the site of stimulation (i.e., posterior, dorsal, or anterordorsal).

The language network is supported by two main pathways that work in parallel and also interact: the phonological dorsal pathway and the semantic ventral pathway (Saur et al., 2008; Friederici, 2009; Kümmerer et al., 2013). A double dissociation between phonemic and semantic errors has been observed during axonal stimulation, demonstrating that the two processes are performed in a synergetic manner (Duffau et al., 2014). The dorsal pathway is subserved by the SLF and consists of two subparts (Catani et al., 2015). The deep part is the classic AF, which connects the posterior temporal structures (mostly the middle and inferior gyri) to the inferior frontal gyrus (essentially the pars opercularis). Stimulation of this subpart results in conduction aphasia, that is, a combination of phonemic paraphasia (supporting a role for the subpart in phonological processing) and repetition disorders, but without semantic paraphasia. Interestingly, the posterior cortical origin of this tract corresponds to the visual object form area (Mandonnet et al., 2009), a functional epicenter involved both in phonological processing dedicated to visual material and in semantics (Vigneau et al., 2006).

The superficial portion of the dorsal stream is subserved by the lateral part of the SLF (also called the SLF III), stimulation of which induces anarthria (i.e., articulatory disorders) (Geemen, Herbet, Moritz-Gasser, & Duffau, 2014). This lateral operculo—opercular component of the SLF is involved in articulation by connecting the junction between the posterior part of the superior temporal gyrus (which receives feedback information from somatosensory and auditory areas) and the supramarginal gyrus with the ventral premotor cortex (which receives afferents bringing phonological—phonetic information to be translated into articulatory motor programs) (Duffau, Gatignol, Denvil, Lopes, & Capelle, 2003). During word repetition, this loop also enables the conversion of auditory input, which is processed in the verbal working memory system, into phonological and articulatory representations within the ventral premotor cortex. This observation is in agreement with data obtained from cortical DES (probabilistic atlas based on 771 stimulation sites), which demonstrated that Broca's area is not the speech output region (Tate et al., 2014), thereby challenging the classic theories on language.

The ventral pathway is divided into a direct bundle (the IFOF) and an indirect pathway formed by the anterior ILF and the uncinate fascicle (UF). These fascicles relay information to one another in the temporal pole (Duffau et al., 2013). The IFOF connects the posterior occipital lobe and the fusiform gyrus to anterior cortical frontal areas, including the inferior frontal gyrus and dorsolateral prefrontal cortex (Martino et al., 2010; Sarubbo et al., 2013). Stimulation of this white matter connectivity demonstrated a major role in verbal semantics, because its stimulation reproducibly elicited semantic paraphasias. However, further studies have shown that this tract is also involved in nonverbal semantic cognition in the left (Moritz-Gasser et al., 2013) *and* in the right (Herbet, Lafargue et al., 2017; Herbet, Moritz-Gasser et al., 2017; Herbet, Yordanova et al., 2017) hemisphere—demonstrating that the nonverbal semantic system is bilaterally distributed.

The indirect ventral semantic pathway has a relay at the level of the temporal pole, which represents a semantic "hub" that allows plurimodal integration of the multiple signals emanating from the unimodal systems (Holland & Lambon Ralph, 2010; Lambon Ralph, 2014). This indirect ventral stream includes the anterior part of the ILF, which connects the fusiform gyrus with the temporal pole, and information is then relayed by the UF, which links the temporal pole with the pars orbitalis of the inferior frontal gyrus. Stimulation of this indirect pathway does not generate semantic paraphasia (Mandonnet et al., 2009), but may nonetheless induce nonverbal semantic disorders (Papagno et al., 2010), in agreement with other studies that have provided evidence in support of a semantic role for the UF (Han et al., 2013).

The ventral semantic pathway seems to contribute to repetition of real words or pseudowords, and may also participate in proper name retrieval (Papagno et al., 2010). The difference between UF and IFOF stimulation on picture naming might reflect their different frontal terminations. Basic picture naming requires minimal levels of semantic control (Jefferies & Lambon Ralph, 2006), which might explain the negligible effects of UF DES on this process. Naming at a more specific level (including naming people) might be somewhat more executively demanding and, thus, UF DES does have an effect.

Of note, the middle longitudinal fascicle, which connects the angular gyrus with the superior temporal gyrus up to the temporal pole, could also be part of the ventral semantic route (Makris et al., 2013). However, subcortical DES of this fascicle failed to induce any naming disorders (De Witt Hamer, Moritz-Gasser, Gatignol, & Duffau, 2011), so its exact functional role in the language network is still unclear.

Beyond picture naming, DES demonstrated that syntactic processing was subserved by delocalized cortical regions (left inferior frontal gyrus and posterior middle temporal gyrus) connected by a subpart of the left SLF. Axonal stimulation of the SLF can elicit specific disorders of grammatical gender, thus disrupting one subfunction without interfering with the others. Interestingly, this subcircuit interacts with, but is independent of, the subnetwork involved in naming, as demonstrated by double dissociation between disruption of syntactic and naming processes during DES. These findings support parallel rather than serial theory, calling into question the concept of the "lemma"—an abstract lexical representation of a word before its phonological properties are assigned (Vidorreta et al., 2011).

Conscious Information Processing

Based on the patients' behavior, electrostimulation can be used to identify critical cerebral structures involved in the maintenance of conscious information processing. For example, our group has shown, in a single patient, that DES of the white matter emanating from the dorsal posterior cingulate cortex induces a breakdown in conscious experience characterized by a transient behavioral unresponsiveness with loss of external connectedness (retrospectively, the patient described himself as a dream, outside the operating room) (Herbet, Lafargue, Bonnetblanc et al., 2014; Herbet, Lafargue, De Champfleur et al., 2014). This finding was replicated in three other patients (Herbet, Maheu et al., 2016; Herbet, Moritz-Gasser et al., 2016). Taken as a whole, these observations suggest that functional integrity of the posterior cingulate connectivity is absolutely necessary to maintain consciousness of the environment.

Other Cognitive Tasks

Patients may have a strong expertise in some cognitive domains due to their occupation or their hobbies (e.g., numerical cognition in a mathematician expert, working memory in a management assistant, and so forth). In such cases, we can implement some specific tests to ensure the patient that he or she will recover their normal professional life after surgery. For example, we regularly use a two-back task to assess working memory. This task consists of naming the picture viewed two trials before. For high-level patients, we can exceptionally increase cognitive load by asking the patient to name the picture viewed three trials before. This kind of behavioral paradigms may be useful to identify and preserve the frontal—parietal network mediated by the SLF. Alternatively, a classic span task can be used (Papagno et al., 2017).

On several occasions, we have also used mathematical cognition tasks to assess basic mathematical operations. This is useful when the lesion is located in the left parietal lobules. For example, we can ask the patient to mentally resolve simple operations such as multiplication or subtraction. These operations can be visually displayed on a computer screen.

THE FUTURE OF AWAKE COGNITIVE MAPPING

An issue frequently raised by actors of awake surgery concerns the use of other cognitive/language tasks for the intraoperative mapping such as more fine-grained linguistic or memory tasks (Rofes & Miceli, 2014), or specific executive function tasks (Wager et al., 2013). However, the choice to implement new tasks in standard practice involves several considerations. Firstly, the cognitive paradigms described in this chapter appear to be sufficient to map both the cortical epicenters and the main white matter connectivity fibers which are reluctant to brain plasticity (Herbet, Maheu et al., 2016; Herbet, Moritz-Gasser et al., 2016). The tasks used must be simple and easily workable given the constraints inherent to the electrostimulation procedure (stimulation duration is four seconds maximum), clinical contexts (intraoperative mapping cannot be too long), and surgery theater constraints (patient position). In connection with this, high-level functions such as, for example, certain executive functions particularly distributed at the anatomical scale, are probably very difficult to map under stimulation. For example, cognitive flexibility required the simultaneous contribution of multiple subnetworks (Cocchi, Zalesky, Fornito, & Mattingley, 2013). Third, the best oncofunctional balance must be found: the first goal towards the surgery is to optimize the extent of resection while preserving quality of life. Adding too many tasks might eventually affect the effectiveness of surgery. Supplying an objective answer to this issue necessitates longitudinally studying (i.e., before and after the surgery) patients' cognitive and language performances on a variety of behavioral paradigms. If patients do not recover sufficiently and are impeded in their daily life functioning, it seems reasonable to think of the implementation of new well-controlled tasks. However, before doing this, we have to comprehend the pathophysiological mechanisms of the lack of recovery. Indeed, a number of clinical or sociodemographic factors can explain a lack of recovery after surgery, such as the degree of infiltration of white matter connectivity (Herbet, Lafargue, Bonnetblanc et al., 2014; Herbet, Lafargue, De Champfleur et al., 2014; Herbet, Lafargue et al., 2015; Herbet, Lafargue, Moritz-Gasser, & Bonnetblanc et al., 2015; Herbet, Lafargue, Moritz-Gasser, & de Champfleur et al., 2015; Herbet, Latorre et al., 2015), the growth rate of the tumor, the preoperative functional status, the inter-individual variability in the neuroplasticity potential (Duffau, 2017), the socio-educational level, and the patients' personality (Minsky, 1933).

CONCLUSION AND FUTURE DIRECTIONS

Although the collaboration between neuropsychology and neurosurgery has already had a long and fruitful history, this interaction has never been as strong as is today. For the first time in the history of cognitive neurosciences and neuropsychology, stimulation mapping in awake surgery offers a unique opportunity to investigate the functional anatomy of the human brain, especially the white matter pathways mediating information exchange within brain-wide neurocognitive networks. This methodology, in which real-time anatomo-functional correlations are performed, is providing critical information on the functional connectivity of a vast array of human brain functions, especially visuospatial and social cognition as well as language—enabling a deep reappraisal of the classic neuropsychological models mainly derived from the behavioral performances of stroke patients.

Beyond this fundamental implication, the results gained from stimulation mapping have led to a better understanding of the anatomic factors governing the potential for neuroplasticity. Specifically, a probabilistic atlas of functional plasticity derived from both anatomic MRI results and intraoperative mapping data on 231 patients having undergone surgery for diffuse low-grade glioma was generated (Herbet, Maheu et al., 2016; Herbet, Moritz-Gasser et al., 2016). This work showed that cortical plasticity is generally high in the cortex (except in primary unimodal areas and in a small set of neural hubs) and rather low in connective tracts. Such results provide critical insights into the topological organization of critical (noncompensable) neural systems and give compelling support for the previously developed idea of "minimal common brain" (Ius et al., 2011), that is, a set of critical structures indispensable for the proper functioning of basic cognitive functions. They may also be useful in predicting the likelihood of recovery (as a function of lesion topology) in glioma patients, but also in various neuropathological conditions in which cerebral disconnection is a key pathophysiological feature such as in strokes (Corbetta et al., 2015). Finally, these kinds of results may be critical for identifying patients who require cognitive rehabilitation and for providing appropriate language, cognitive of behavioral rehabilitation programs.

REFERENCES

Adolphs, R. (2002). Neural systems for recognizing emotion. *Current Opinion in Neurobiology*, *12*(2), 169–177.

Almairac, F., Herbet, G., Moritz-Gasser, S., & Duffau, H. (2014). Parietal network underlying movement control: Disturbances during subcortical electrostimulation. *Neurosurgical Review*, *37*(3), 513–517.

Amodio, D. M., & Frith, C. D. (2006). Meeting of minds: The medial frontal cortex and social cognition. *Nature Reviews. Neuroscience*, *7*(4), 268.

Bar, M., Tootell, R. B., Schacter, D. L., Greve, D. N., Fischl, B., Mendola, J. D., ... Dale, A. M. (2001). Cortical mechanisms specific to explicit visual object recognition. *Neuron*, *29*(2), 529–535.

Baron-Cohen, S., Wheelwright, S., Hill, J., Raste, Y., & Plumb, I. (2001). The "Reading the Mind in the Eyes" Test revised version: A study with normal adults, and adults with Asperger syndrome or high-functioning autism. *The Journal of Child Psychology and Psychiatry and Allied Disciplines*, *42*(2), 241–251.

Bartolomeo, P., De Schotten, M. T., & Chica, A. B. (2012). Brain networks of visuospatial attention and their disruption in visual neglect. *Frontiers in Human Neuroscience*, *6*, 110.

Basser, P. J., Pajevic, S., Pierpaoli, C., Duda, J., & Aldroubi, A. (2000). In vivo fiber tractography using DT-MRI data. *Magnetic Resonance in Medicine*, *44*(4), 625−632.

Berger, M. S., & Ojemann, G. A. (1992). Intraoperative brain mapping techniques in neurooncology. *Stereotactic and Functional Neurosurgery*, *58*(1-4), 153−161.

Berger, M. S., Ojemann, G. A., & Lettich, E. (1990). Neurophysiological monitoring during astrocytoma surgery. *Neurosurgery Clinics of North America*, *1*(1), 65−80.

Binder, J., Marshall, R., Lazar, R., Benjamin, J., & Mohr, J. P. (1992). Distinct syndromes of hemineglect. *Archives of Neurology*, *49*(11), 1187−1194.

Bonnetblanc, F., Desmurget, M., & Duffau, H. (2006). Low grade gliomas and cerebral plasticity: Fundamental and clinical implications. *Medecine Sciences: M/S*, *22*(4), 389−394.

Bouillaud, J. (1825). Recherches cliniques propres à démontrer que la perte de la parole correspond à la lésion des lobules antérieures du cerveau, et à confirmer l'opinion de M. Gall sur le siège de l'organe du language articulé. *Archives Généralede Médecine*, *3*, 25−45.

Bressler, S. L., & Menon, V. (2010). Large-scale brain networks in cognition: Emerging methods and principles. *Trends in Cognitive Sciences*, *14*(6), 277−290.

Broca, P. (1861). Remarques sur le siège de la faculté du langage articulé, suivies d'une observation d'aphémie (perte de la parole). *Bulletin et Memoires de la Societe anatomique de Paris*, *6*, 330−357.

Bullmore, E. T., & Bassett, D. S. (2011). Brain graphs: Graphical models of the human brain connectome. *Annual Review of Clinical Psychology*, *7*, 113−140.

Carrington, S. J., & Bailey, A. J. (2009). Are there theory of mind regions in the brain? A review of the neuroimaging literature. *Human Brain Mapping*, *30*(8), 2313−2335.

Catani, M., Howard, R. J., Pajevic, S., & Jones, D. K. (2002). Virtual in vivo interactive dissection of white matter fasciculi in the human brain. *Neuroimage*, *17*(1), 77−94.

Caverzasi, E., Papinutto, N., Amirbekian, B., Berger, M. S., & Henry, R. G. (2014). Q-ball of inferior fronto-occipital fasciculus and beyond. *PLoS One*, *9*(6), e100274.

Charras, P., Herbet, G., Deverdun, J., De Champfleur, N. M., Duffau, H., Bartolomeo, P., & Bonnetblanc, F. (2015). Functional reorganization of the attentional networks in low-grade glioma patients: A longitudinal study. *Cortex*, *63*, 27−41.

Chechlacz, M., Rotshtein, P., Bickerton, W.-L., Hansen, P. C., Deb, S., & Humphreys, G. W. (2010). Separating neural correlates of allocentric and egocentric neglect: Distinct cortical sites and common white matter disconnections. *Cognitive Neuropsychology*, *27*(3), 277−303.

Cocchi, L., Zalesky, A., Fornito, A., & Mattingley, J. B. (2013). Dynamic cooperation and competition between brain systems during cognitive control. *Trends in Cognitive Sciences*, *17* (10), 493−501.

Cochereau, J., Herbet, G., & Duffau, H. (2016). Patients with incidental WHO grade II glioma frequently suffer from neuropsychological disturbances. *Acta Neurochirurgica*, *158*(2), 305−312.

Coello, A. F., Duvaux, S., De Benedictis, A., Matsuda, R., & Duffau, H. (2013). Involvement of the right inferior longitudinal fascicle in visual hemiagnosia: A brain stimulation mapping study: Case report. *Journal of Neurosurgery*, *118*(1), 202−205.

Corrivetti, F., Herbet, G., Moritz-Gasser, S., & Duffau, H. (2017). Prosopagnosia induced by a left anterior temporal lobectomy following a right temporo-occipital resection in a multicentric diffuse low-grade glioma. *World Neurosurgery*, *97*, 756. e1−756. e5.

de Schotten, M. T., Bizzi, A., Dell'Acqua, F., Allin, M., Walshe, M., Murray, R., . . . Catani, M. (2011). Atlasing location, asymmetry and inter-subject variability of white matter tracts in the human brain with MR diffusion tractography. *Neuroimage*, *54*(1), 49−59.

De Witt Hamer, P. C., Moritz-Gasser, S., Gatignol, P., & Duffau, H. (2011). Is the human left middle longitudinal fascicle essential for language? A brain electrostimulation study. *Human Brain Mapping*, *32*(6), 962−973.

De Witt Hamer, P. C., Robles, S. G., Zwinderman, A. H., Duffau, H., & Berger, M. S. (2012). Impact of intraoperative stimulation brain mapping on glioma surgery outcome: A meta-analysis. *Journal of Clinical Oncology*, *30*(20), 2559−2565.

Desmurget, M., Bonnetblanc, F., & Duffau, H. (2007). Contrasting acute and slow-growing lesions: A new door to brain plasticity. *Brain*, *130*(4), 898−914.

Desmurget, M., Song, Z., Mottolese, C., & Sirigu, A. (2013). Re-establishing the merits of electrical brain stimulation. *Trends in Cognitive Sciences*, *17*(9), 442−449.

Duffau, H. (2005). Lessons from brain mapping in surgery for low-grade glioma: Insights into associations between tumour and brain plasticity. *The Lancet Neurology*, *4*(8), 476−486.

Duffau, H. (2010). Awake surgery for nonlanguage mapping. *Neurosurgery*, *66*(3), 523−529.

Duffau, H. (2014). Surgical neurooncology is a brain networks surgery: A "Connectomic" perspective. *World Neurosurgery*, *82*(3), e405−e407.

Duffau, H. (2015a). Awake mapping of the brain connectome in glioma surgery: Concept is stronger than technology. *European Journal of Surgical Oncology*, *41*(9), 1261−1263.

Duffau, H. (2015b). Stimulation mapping of white matter tracts to study brain functional connectivity. *Nature Reviews Neurology*, *11*(5), 255−265.

Duffau, H. (2017). A two-level model of interindividual anatomo-functional variability of the brain and its implications for neurosurgery. *Cortex*, *86*, 303−313.

Duffau, H., Capelle, L., Sichez, N., Denvil, D., Lopes, M., Sichez, J.-P., . . . Fohanno, D. (2002). Intraoperative mapping of the subcortical language pathways using direct stimulations: An anatomo-functional study. *Brain*, *125*(1), 199−214.

Duffau, H., Gatignol, P., Denvil, D., Lopes, M., & Capelle, L. (2003). The articulatory loop: Study of the subcortical connectivity by electrostimulation. *Neuroreport*, *14*(15), 2005−2008.

Duffau, H., Gatignol, P., Mandonnet, E., Capelle, L., & Taillandier, L. (2008). Intraoperative subcortical stimulation mapping of language pathways in a consecutive series of 115 patients with Grade II glioma in the left dominant hemisphere.

Duffau, H., Herbet, G., & Moritz-Gasser, S. (2013). Toward a pluri-component, multimodal, and dynamic organization of the ventral semantic stream in humans: Lessons from stimulation mapping in awake patients. *Frontiers in Systems Neuroscience*, *7*, 44.

Duffau, H., Moritz-Gasser, S., & Mandonnet, E. (2014). A re-examination of neural basis of language processing: Proposal of a dynamic hodotopical model from data provided by brain stimulation mapping during picture naming. *Brain and Language*, *131*, 1−10.

Duffau, H., Sichez, J.-P., & Lehéricy, S. (2000). Intraoperative unmasking of brain redundant motor sites during resection of a precentral angioma: Evidence using direct cortical stimulation. *Annals of Neurology*, *47*(1), 132−135.

Filimon, F. (2010). Human cortical control of hand movements: Parietofrontal networks for reaching, grasping, and pointing. *The Neuroscientist*, *16*(4), 388−407.

Fogassi, L., & Luppino, G. (2005). Motor functions of the parietal lobe. *Current Opinion in Neurobiology*, *15*(6), 626−631.

Friederici, A. D. (2009). Pathways to language: Fiber tracts in the human brain. *Trends in Cognitive Sciences*, *13*(4), 175−181.

Frith, C. D., & Frith, U. (2012). Mechanisms of social cognition. *Annual Review of Psychology*, *63*, 287−313.

Gallagher, H. L., & Frith, C. D. (2003). Functional imaging of 'theory of mind'. *Trends in Cognitive Sciences*, *7*(2), 77−83.

Geemen, K., Herbet, G., Moritz-Gasser, S., & Duffau, H. (2014). Limited plastic potential of the left ventral premotor cortex in speech articulation: Evidence from intraoperative awake mapping in glioma patients. *Human Brain Mapping*, *35*(4), 1587−1596.

Genova, H. M., Rajagopalan, V., Chiaravalloti, N., Binder, A., Deluca, J., & Lengenfelder, J. (2015). Facial affect recognition linked to damage in specific white matter tracts in traumatic brain injury. *Social Neuroscience*, *10*(1), 27−34.

Geschwind, N. (1965). Disconnexion syndromes in animals and man. *Brain*, *88*(3), 585-585.

Geschwind, N. (1974). *Disconnexion syndromes in animals and man. Selected papers on language and the brain* (pp. 105−236). New York: Springer.

Gras-Combe, G., Moritz-Gasser, S., Herbet, G., & Duffau, H. (2012). Intraoperative subcortical electrical mapping of optic radiations in awake surgery for glioma involving visual pathways. *Journal of Neurosurgery*, *117*(3), 466−473.

Grossi, D., Soricelli, A., Ponari, M., Salvatore, E., Quarantelli, M., Prinster, A., & Trojano, L. (2014). Structural connectivity in a single case of progressive prosopagnosia: The role of the right inferior longitudinal fasciculus. *Cortex*, *56*, 111−120.

Han, Z., Ma, Y., Gong, G., He, Y., Caramazza, A., & Bi, Y. (2013). White matter structural connectivity underlying semantic processing: Evidence from brain damaged patients. *Brain*, *136* (10), 2952−2965.

Hau, J., Sarubbo, S., Perchey, G., Crivello, F., Zago, L., Mellet, E., ... Tzourio-Mazoyer, N. (2016). Cortical terminations of the inferior fronto-occipital and uncinate fasciculi: Anatomical stem-based virtual dissection. *Frontiers in Neuroanatomy*, *10*, 58.

Heilman, K. M., Bowers, D., & Watson, R. T. (1983). Performance on hemispatial pointing task by patients with neglect syndrome. *Neurology*, *33*(5), 661.

Herbet, G., Lafargue, G., Bonnetblanc, F., Moritz-Gasser, S., & Duffau, H. (2013). Is the right frontal cortex really crucial in the mentalizing network? A longitudinal study in patients with a slow-growing lesion. *Cortex*, *49*(10), 2711−2727.

Herbet, G., Lafargue, G., Bonnetblanc, F., Moritz-Gasser, S., Menjot de Champfleur, N., & Duffau, H. (2014). Inferring a dual-stream model of mentalizing from associative white matter fibres disconnection. *Brain*, *137*(3), 944−959.

Herbet, G., Lafargue, G., De Champfleur, N. M., Moritz-Gasser, S., Le Bars, E., Bonnetblanc, F., & Duffau, H. (2014). Disrupting posterior cingulate connectivity disconnects consciousness from the external environment. *Neuropsychologia*, *56*, 239−244.

Herbet, G., Lafargue, G., & Duffau, H. (2015). The dorsal cingulate cortex as a critical gateway in the network supporting conscious awareness. *Brain*, *139*(4), e23.

Herbet, G., Lafargue, G., & Duffau, H. (2017). Un atlas du potentiel neuroplastique chez les patients cérébrolésés. *Médecine/Sciences*, *33*(1), 84−86.

Herbet, G., Lafargue, G., Moritz-Gasser, S., Bonnetblanc, F., & Duffau, H. (2015). Interfering with the neural activity of mirror-related frontal areas impairs mentalistic inferences. *Brain Structure and Function*, *220*(4), 2159−2169.

Herbet, G., Lafargue, G., Moritz-Gasser, S., de Champfleur, N. M., Costi, E., Bonnetblanc, F., & Duffau, H. (2015). A disconnection account of subjective empathy impairments in diffuse low-grade glioma patients. *Neuropsychologia*, *70*, 165−176.

Herbet, G., Latorre, J. G., & Duffau, H. (2015). The role of cerebral disconnection in cognitive recovery after brain damage. *Neurology*, *84*(14), 1390−1391.

Herbet, G., Maheu, M., Costi, E., Lafargue, G., & Duffau, H. (2016). Mapping neuroplastic potential in brain-damaged patients. *Brain*, *139*(3), 829−844.

Herbet, G., Moritz-Gasser, S., Boiseau, M., Duvaux, S., Cochereau, J., & Duffau, H. (2016). Converging evidence for a cortico-subcortical network mediating lexical retrieval. *Brain*, *139*(11), 3007−3021.

Herbet, G., Moritz-Gasser, S., & Duffau, H. (2017). Direct evidence for the contributive role of the right inferior fronto-occipital fasciculus in non-verbal semantic cognition. *Brain Structure and Function*, *222*(4), 1597−1610.

Herbet, G., Yordanova, Y. N., & Duffau, H. (2017). Left spatial neglect evoked by electrostimulation of the right inferior fronto-occipital fasciculus. *Brain Topography*, *30*(6), 747−756.

Hickok, G., & Poeppel, D. (2004). Dorsal and ventral streams: A framework for understanding aspects of the functional anatomy of language. *Cognition*, *92*(1), 67−99.

Holland, R., & Lambon Ralph, M. A. (2010). The anterior temporal lobe semantic hub is a part of the language neural network: Selective disruption of irregular past tense verbs by rTMS. *Cerebral Cortex*, *20*(12), 2771−2775.

Howard, D., & Patterson, K. E. (1992). *The pyramids and palm trees test: A test of semantic access from words and pictures*. UK: Thames Valley Test Company.

Husain, M., & Kennard, C. (1996). Visual neglect associated with frontal lobe infarction. *Journal of Neurology*, *243*(9), 652−657.

Ius, T., Angelini, E., de Schotten, M. T., Mandonnet, E., & Duffau, H. (2011). Evidence for potentials and limitations of brain plasticity using an atlas of functional resectability of WHO grade II gliomas: Towards a "minimal common brain.". *Neuroimage*, *56*(3), 992−1000.

Jackson, S. R., Newport, R., Husain, M., Fowlie, J. E., O'Donoghue, M., & Bajaj, N. (2009). There may be more to reaching than meets the eye: Re-thinking optic ataxia. *Neuropsychologia*, *47*(6), 1397−1408.

Jefferies, E., & Lambon Ralph, M. A. (2006). Semantic impairment in stroke aphasia vs semantic dementia: A case-series comparison. *Brain*, *129*(8), 2132−2147.

Karnath, H.-O., Fruhmann Berger, M., Küker, W., & Rorden, C. (2004). The anatomy of spatial neglect based on voxelwise statistical analysis: A study of 140 patients. *Cerebral Cortex*, *14*(10), 1164−1172.

Karnath, H.-O., Rennig, J., Johannsen, L., & Rorden, C. (2010). The anatomy underlying acute vs chronic spatial neglect: A longitudinal study. *Brain*, *134*(3), 903−912.

Karnath, H.-O., Rüter, J., Mandler, A., & Himmelbach, M. (2009). The anatomy of object recognition—visual form agnosia caused by medial occipitotemporal stroke. *Journal of Neuroscience*, *29*(18), 5854−5862.

Kinoshita, M., de Champfleur, N. M., Deverdun, J., Moritz-Gasser, S., Herbet, G., & Duffau, H. (2015). Role of fronto-striatal tract and frontal aslant tract in movement and speech: An axonal mapping study. *Brain Structure and Function*, *220*(6), 3399−3412.

Kümmerer, D., Hartwigsen, G., Kellmeyer, P., Glauche, V., Mader, I., Klöppel, S., ... Saur, D. (2013). Damage to ventral and dorsal language pathways in acute aphasia. *Brain*, *136*(2), 619−629.

Lambon Ralph, M. A. (2014). Neurocognitive insights on conceptual knowledge and its breakdown. *Philosophical Transactions of the Royal Society of London B: Biological Sciences*, *369*(1634), 20120392.

Lieberman, M. D. (2007). Social cognitive neuroscience: A review of core processes. *Annual Review of Psychology*, *58*, 259−289.

Logothetis, N. K., Augath, M., Murayama, Y., Rauch, A., Sultan, F., Goense, J., ... Merkle, H. (2010). The effects of electrical microstimulation on cortical signal propagation. *Nature Neuroscience*, *13*(10), 1283−1291.

Makris, N., Kennedy, D. N., McInerney, S., Sorensen, A. G., Wang, R., Caviness, V. S., Jr, & Pandya, D. N. (2004). Segmentation of subcomponents within the superior longitudinal fascicle in humans: A quantitative, in vivo, DT-MRI study. *Cerebral Cortex, 15*(6), 854−869.

Makris, N., Preti, M. G., Wassermann, D., Rathi, Y., Papadimitriou, G. M., Yergatian, C., ... Kubicki, M. (2013). Human middle longitudinal fascicle: Segregation and behavioral-clinical implications of two distinct fiber connections linking temporal pole and superior temporal gyrus with the angular gyrus or superior parietal lobule using multi-tensor tractography. *Brain Imaging and Behavior, 7*(3), 335−352.

Maldonado, I. L., Champfleur, N. M., Velut, S., Destrieux, C., Zemmoura, I., & Duffau, H. (2013). Evidence of a middle longitudinal fasciculus in the human brain from fiber dissection. *Journal of Anatomy, 223*(1), 38−45.

Mandonnet, E., Gatignol, P., & Duffau, H. (2009). Evidence for an occipito-temporal tract underlying visual recognition in picture naming. *Clinical Neurology and Neurosurgery, 111*(7), 601−605.

Marie, P. (1906). Revision de la question de l'aphasie: La troisième circonvolution frontale gauche ne joue aucun rôle spécial dans la fonction du langage. *Semaine Méd (Paris), 26*, 241−247.

Martino, J., Brogna, C., Robles, S. G., Vergani, F., & Duffau, H. (2010). Anatomic dissection of the inferior fronto-occipital fasciculus revisited in the lights of brain stimulation data. *Cortex, 46*(5), 691−699.

McClelland, J. L., & Rogers, T. T. (2003). The parallel distributed processing approach to semantic cognition. *Nature Reviews Neuroscience, 4*(4), 310.

Mesulam, M. (1990). Large-scale neurocognitive networks and distributed processing for attention, language, and memory. *Annals of Neurology, 28*(5), 597−613.

Mesulam, M.-M. (1998). From sensation to cognition. *Brain: A Journal of Neurology, 121*(6), 1013−1052.

Molenberghs, P., Cunnington, R., & Mattingley, J. B. (2012). Brain regions with mirror properties: A *meta*-analysis of 125 human fMRI studies. *Neuroscience & Biobehavioral Reviews, 36*(1), 341−349.

Molenberghs, P., Sale, M. V., & Mattingley, J. B. (2012). Is there a critical lesion site for unilateral spatial neglect? A *meta*-analysis using activation likelihood estimation. *Frontiers in Human Neuroscience, 6*, 78.

Moritz-Gasser, S., Herbet, G., & Duffau, H. (2013). Mapping the connectivity underlying multimodal (verbal and non-verbal) semantic processing: A brain electrostimulation study. *Neuropsychologia, 51*(10), 1814−1822.

Mort, D. J., Malhotra, P., Mannan, S. K., Rorden, C., Pambakian, A., Kennard, C., & Husain, M. (2003). The anatomy of visual neglect. *Brain, 126*(9), 1986−1997.

Ojemann, G. A. (1979). Individual variability in cortical localization of language. *Journal of Neurosurgery, 50*(2), 164−169.

Ojemann, G. A., & Whitaker, H. A. (1978). Language localization and variability. *Brain and Language, 6*(2), 239−260.

Papagno, C., Comi, A., Riva, M., Bizzi, A., Vernice, M., Casarotti, A., ... Bello, L. (2017). Mapping the brain network of the phonological loop. *Human Brain Mapping, 38*(6), 3011−3024.

Papagno, C., Miracapillo, C., Casarotti, A., Romero Lauro, L. J., Castellano, A., Falini, A., ... Bello, L. (2010). What is the role of the uncinate fasciculus? Surgical removal and proper name retrieval. *Brain, 134*(2), 405−414.

Penfield, W., & Boldrey, E. (1937). Somatic motor and sensory representation in the cerebral cortex of man as studied by electrical stimulation. *Brain: A Journal of Neurology, 60*(4), 389−443.

Philippi, C. L., Mehta, S., Grabowski, T., Adolphs, R., & Rudrauf, D. (2009). Damage to association fiber tracts impairs recognition of the facial expression of emotion. *Journal of Neuroscience*, *29*(48), 15089−15099.

Plaza, M., Gatignol, P., Leroy, M., & Duffau, H. (2009). Speaking without Broca's area after tumor resection. *Neurocase*, *15*(4), 294−310.

Premack, D., & Woodruff, G. (1978). Does the chimpanzee have a theory of mind? *Behavioral and Brain Sciences*, *1*(4), 515−526.

Rech, F., Herbet, G., Moritz-Gasser, S., & Duffau, H. (2014). Disruption of bimanual movement by unilateral subcortical electrostimulation. *Human Brain Mapping*, *35*(7), 3439−3445.

Rech, F., Herbet, G., Moritz-Gasser, S., & Duffau, H. (2016). Somatotopic organization of the white matter tracts underpinning motor control in humans: An electrical stimulation study. *Brain Structure and Function*, *221*(7), 3743−3753.

Rengachary, J., He, B. J., Shulman, G. L., & Corbetta, M. (2011). A behavioral analysis of spatial neglect and its recovery after stroke. *Frontiers in Human Neuroscience*, *5*, 29.

Rofes, A., & Miceli, G. (2014). Language mapping with verbs and sentences in awake surgery: A review. *Neuropsychology Review*, *24*(2), 185−199.

Roux, F.-E., Dufor, O., Lauwers-Cances, V., Boukhatem, L., Brauge, D., Draper, L., ... Démonet, J.-F. (2011). Electrostimulation mapping of spatial neglect. *Neurosurgery*, *69*(6), 1218−1231.

Samuelsson, H., Jensen, C., Ekholm, S., Naver, H., & Blomstrand, C. (1997). Anatomical and neurological correlates of acute and chronic visuospatial neglect following right hemisphere stroke. *Cortex*, *33*(2), 271−285.

Sarubbo, S., De Benedictis, A., Maldonado, I. L., Basso, G., & Duffau, H. (2013). Frontal terminations for the inferior fronto-occipital fascicle: Anatomical dissection, DTI study and functional considerations on a multi-component bundle. *Brain Structure and Function*, *218*(1), 21−37.

Sarubbo, S., Le Bars, E., Moritz-Gasser, S., & Duffau, H. (2012). Complete recovery after surgical resection of left Wernicke's area in awake patient: A brain stimulation and functional MRI study. *Neurosurgical Review*, *35*(2), 287−292.

Saur, D., Kreher, B. W., Schnell, S., Kümmerer, D., Kellmeyer, P., Vry, M.-S., ... Abel, S. (2008). Ventral and dorsal pathways for language. *Proceedings of the National Academy of Sciences*, *105*(46), 18035−18040.

Schaafsma, S. M., Pfaff, D. W., Spunt, R. P., & Adolphs, R. (2015). Deconstructing and reconstructing theory of mind. *Trends in Cognitive Sciences*, *19*(2), 65−72.

Schmahmann, J. D., Pandya, D. N., Wang, R., Dai, G., D'arceuil, H. E., de Crespigny, A. J., & Wedeen, V. J. (2007). Association fibre pathways of the brain: Parallel observations from diffusion spectrum imaging and autoradiography. *Brain*, *130*(3), 630−653.

Schucht, P., Moritz-Gasser, S., Herbet, G., Raabe, A., & Duffau, H. (2013). Subcortical electrostimulation to identify network subserving motor control. *Human Brain Mapping*, *34*(11), 3023−3030.

Schurz, M., Radua, J., Aichhorn, M., Richlan, F., & Perner, J. (2014). Fractionating theory of mind: A *meta*-analysis of functional brain imaging studies. *Neuroscience & Biobehavioral Reviews*, *42*, 9−34.

Sporns, O., Tononi, G., & Kötter, R. (2005). The human connectome: A structural description of the human brain. *PLoS Computational Biology*, *1*(4), e42.

Taphoorn, M. J., & Klein, M. (2004). Cognitive deficits in adult patients with brain tumours. *The Lancet Neurology*, *3*(3), 159−168.

Tate, M. C., Herbet, G., Moritz-Gasser, S., Tate, J. E., & Duffau, H. (2014). Probabilistic map of critical functional regions of the human cerebral cortex: Broca's area revisited. *Brain, 137*(10), 2773−2782.

Thiebaut de Schotten, M., Tomaiuolo, F., Aiello, M., Merola, S., Silvetti, M., Lecce, F., ... Doricchi, F. (2012). Damage to white matter pathways in subacute and chronic spatial neglect: A group study and 2 single-case studies with complete virtual "in vivo" tractography dissection. *Cerebral Cortex, 24*(3), 691−706.

Thomas, C., Avidan, G., Humphreys, K., Jung, K., Gao, F., & Behrmann, M. (2009). Reduced structural connectivity in ventral visual cortex in congenital prosopagnosia. *Nature Neuroscience, 12*(1), 29−31.

Unger, A., Alm, K. H., Collins, J. A., O'Leary, J. M., & Olson, I. R. (2016). Variation in white matter connectivity predicts the ability to remember faces and discriminate their emotions. *Journal of the International Neuropsychological Society, 22*(2), 180−190.

Vallar, G., & Perani, D. (1986). The anatomy of unilateral neglect after right-hemisphere stroke lesions. A clinical/CT-scan correlation study in man. *Neuropsychologia, 24*(5), 609−622.

Van Overwalle, F. (2009). Social cognition and the brain: A meta-analysis. *Human Brain Mapping, 30*(3), 829−858.

Verdon, V., Schwartz, S., Lovblad, K.-O., Hauert, C.-A., & Vuilleumier, P. (2009). Neuroanatomy of hemispatial neglect and its functional components: A study using voxel-based lesion-symptom mapping. *Brain, 133*(3), 880−894.

Vidorreta, J. G., Garcia, R., Moritz-Gasser, S., & Duffau, H. (2011). Double dissociation between syntactic gender and picture naming processing: A brain stimulation mapping study. *Human Brain Mapping, 32*(3), 331−340.

Vigneau, M., Beaucousin, V., Herve, P.-Y., Duffau, H., Crivello, F., Houde, O., ... Tzourio-Mazoyer, N. (2006). *Meta*-analyzing left hemisphere language areas: Phonology, semantics, and sentence processing. *Neuroimage, 30*(4), 1414−1432.

Vilasboas, T., Herbet, G., & Duffau, H. (2017). Challenging the myth of right « non-dominant » hemisphere: Lessons from cortico-subcortical stimulation mapping in awake surgery and surgical implications. *World Neurosurgery*. Available from https://doi.org/10.1016/j.wneu.2017.04.021.

Wager, M., Du Boisgueheneuc, F., Pluchon, C., Bouyer, C., Stal, V., Bataille, B., ... Gil, R. (2013). Intraoperative monitoring of an aspect of executive functions: Administration of the Stroop test in 9 adult patients during awake surgery for resection of frontal glioma. *Neurosurgery, 72*(2), ons169−180.

Wang, X., Pathak, S., Stefaneanu, L., Yeh, F.-C., Li, S., & Fernandez-Miranda, J. C. (2016). Subcomponents and connectivity of the superior longitudinal fasciculus in the human brain. *Brain Structure and Function, 221*(4), 2075−2092.

Wu, Y., Sun, D., Wang, Y., & Wang, Y. (2016). Subcomponents and connectivity of the inferior fronto-occipital fasciculus revealed by diffusion spectrum imaging fiber tracking. *Frontiers in Neuroanatomy, 10*, 88.

Yordanova, Y. N., Duffau, H., & Herbet, G. (2017). Neural pathways subserving face-based mentalizing. *Brain Structure and Function, 222*(7), 3087−3105.

Zemmoura, I., Herbet, G., Moritz-Gasser, S., & Duffau, H. (2015). New insights into the neural network mediating reading processes provided by cortico-subcortical electrical mapping. *Human Brain Mapping, 36*(6), 2215−2230.

Section III

Applications

Chapter 8

Neuropsychology in Adult Epilepsy Surgery

Jeffrey Cole and Marla J. Hamberger

Department of Neurology, Columbia University Medical Center, New York, NY, United States

Despite the availability of numerous antiseizure medications, seizures remain refractory to pharmacological treatment in approximately 30% of epilepsy patients (Berg, 2009; Kwan & Brodie, 2000). For patients with focal epilepsy, surgical resection or ablation of the region of seizure onset offers the possibility of seizure freedom. The challenge of epilepsy surgery is to remove a sufficient amount of epileptogenic brain tissue to eliminate seizure activity, yet, at the same time, minimize the amount of brain tissue that is removed or disrupted in order to reduce the likelihood of compromise to cognitive function, postoperatively.

HISTORICAL PERSPECTIVE

The imperative for collaboration between neurosurgery and neuropsychology became apparent in the early days of epilepsy surgery, unfortunately, at the expense of several patients who suffered severe postoperative cognitive decline. The now well-established knowledge of temporal lobe mediation of memory came to light when the renowned patient, HM, developed profound anterograde amnesia following resection of both the left and right temporal lobes (Milner, 1958; Scoville & Milner, 1957). Although his seizures were successfully eliminated, HM was no longer able to form new (episodic) memories. Several other patients suffered severe memory declines following unilateral temporal lobe resection due to significant bilateral disease that was not detected in the presurgical evaluation. In these patients, the seemingly healthy temporal lobe contralateral to the resection was unable to support the formation of new memories (Chelune, 1995).

These devastating cases highlighted the need to determine the functionality of the tissue to be resected and perhaps, most importantly, the functional integrity of the remaining tissue that, postoperatively, would carry the full burden of supporting function. Two invasive procedures, Wada testing

Neurosurgical Neuropsychology. DOI: https://doi.org/10.1016/B978-0-12-809961-2.00009-6

141

(Milner, Branch, & Rasmussen, 1962a) and electrocortical stimulation mapping (Penfield, 1959), were developed to, essentially, mimic the effects of the surgery by temporarily inactivating the brain regions to be removed. Both of these procedures exploit the lesion method, enabling brief functional assessment of the nonaffected (i.e., nonanesthetized or nonstimulated) brain areas, thereby simulating the functional outcome if the inactivated region were to be removed.

As temporal lobe epilepsy (TLE) is the most common type of localization-related refractory epilepsy, it is among the most common epilepsy syndromes treated via neurosurgical approach (Engel, 1998). Due largely, to the critical role of the temporal lobe in memory and language, there has been a general movement toward greater precision and reductions in size of the temporal lobe regions removed or disrupted in the service of minimizing damage to eloquent brain areas. The earliest resections tended to remove approximately 8 cm of lateral neocortex from the temporal lobe, in addition to removal of medial structures (Scoville & Milner, 1957). Not surprisingly, results of several reported series showed a relation between extent of temporal lobe removal and extent of cognitive decline, with more marked decline associated with larger resections (Graydon, Nunn, Polkey, & Morris, 2001; Katz, et al., 1989). On the other hand, however, several other studies failed to find a similar relation (McIntosh, Wilson, & Berkovic, 2001; Wolf et al., 1993), possibly reflecting that multiple factors influence the presence and extent of postoperative decline, including but not limited to preoperative status of the medial temporal region (Davies et al., 1998; Hermann, Wyler, Somes, Berry, & Dohan, 1992; Seidenberg et al., 1996, 1997), age at time of surgery (Langfitt & Rausch, 1996), utilization of preresection cortical mapping (Alpherts et al., 2008; Hermann et al., 1999), and baseline level of functioning (Jokeit et al., 1997).

A more consistent and robust finding that emerged from the early decades of temporal lobe surgery, was that postoperative seizure freedom was more likely to be achieved with complete removal of medial temporal structures (Katz et al., 1989). Thus, the next phase of resections, which came to be known as standard anteromesial temporal lobe resection ("SAMTLR" or "standard temporal lobe resection") was generally defined as radical removal of hippocampus and amygdala, with limited cortical resection, defined as 3−4.5 cm of lateral temporal neocortex (Engel, Van Ness, Rasmussen, & Ojemann, 1993). In some surgery programs, resections were marginally smaller on the dominant (~3.5) versus nondominant (~4.5 cm) hemisphere to preserve postoperative speech and language (Spencer, Spencer, Mattson, Williamson, & Novelly, 1984).

SAMTLR is very much in use today, particularly for patients who show evidence of unilateral mesial temporal sclerosis (MTS) on preoperative magnetic resonance imaging (MRI), with electroencephalography (EEG) patterns consistent with ipsilateral medial temporal seizure onset. However, additional, less invasive or potentially less disruptive procedures have been developed as well for patients with seizures arising from the temporal lobe. Selective amygdalohippocampectomy (SAH), as the name implies, involves complete removal of the hippocampus and amygdala, yet, theoretically, spares neocortical tissue (Wieser, 1988). Although SAH was expected to reduce morbidity to language and memory, postoperative results have been inconsistent. Whereas some studies support better cognitive outcome with SAH compared to SAMTLR (Helmstaedter, Elger, & Hufnagel, 1996), others have found no significant difference (Gleissner, Helmstaedter, Schramm, & Elger, 2004; Lutz, Clussman, Elger, Schramm, & Helmstaedter, 2004). It has been speculated that despite limited removal of temporal neo-cortex, insult occurs nevertheless due to "collateral damage" incurred by the surgical approach to mesial structures (Helmstaedter et al., 2004).

More recently, neurosurgery has introduced two less invasive ablative techniques into the arsenal against seizures. Both gamma knife surgery (Regis et al., 2004) and MRI-guided stereotactic laser amygdalohippocam-potomy (SLAH) (Willie et al., 2014) target mesial temporal structures using laser technology. However, within the United States, the use of gamma knife for epilepsy has declined (in favor of SLAH) due to the unwanted radiation exposure, frequent post procedure swelling, and the 1−2 years delay before a therapeutic response is clear (Spencer, 2008). Neuropsychological reports on possible cognitive effects of gamma knife surgery have been inconsistent, with two small studies finding a postoperative deficit in verbal memory (McDonald, Norman, Tecoma, Alksne, & Iragui, 2004; Srikijvilaikul et al., 2004) and one larger study finding memory to be stable or improved (Regis et al., 2004). Although current evidence falls short of supporting gamma knife surgery as an equally efficacious alternative to resection, it remains an option for patients who are disinclined toward or cannot undergo general anesthesia, and has been suggested as a second-line treatment for cases of "failed" surgical treatment of seizures, especially in those averse to a second invasive surgery (Lee et al., 2015). On the other hand, SLAH avoids craniot-omy, yet uses direct thermal ablation instead of radiation, and holds the promise of more immediate results. SLAH is relatively new and not yet con-sistently available outside the United States. Although results are considered preliminary, to date, seizure outcome data are considered close to that obtained with SAMTLR (Curry, Gowda, McNichols, & Wilfong, 2012; Willie et al., 2014). Cognitive outcome studies are underway, with prelimi-nary result suggestive of better language and memory outcome relative to SAMTLR (Drane et al., 2015).

Finally, implantation of stimulation devices is another type of surgical intervention that might be offered to patients who are not candidates for resective or ablative surgery. These neuromodulatory devices, which include responsive neurostimulation, vagus nerve stimulation, and deep brain stimu-lation, are considered palliative, as they are much less likely to result in seizure freedom (Chang, Englot, & Vadera, 2015; Hamberger, Williams, & Schevon, 2011). From a cognitive perspective, however, stimulation devices are not associated with cognitive morbidity; rather, there is some evidence of improvement in cognitive function and quality of life (Loring, Kapur, Meador, & Morrell, 2015; Meador et al., 2015).

ROLE OF THE NEUROPSYCHOLOGIST

While there may be general trends in the field for types of surgical treatment favored at any given time, the surgical technique ultimately selected for an individual patient is based on the clinical presentation and needs of the individual. For some patients, this will be evident in the early stages of the workup. However, for many patients the path to a particular treatment will unfold as relevant clinical data, including neuropsychological results, become available. Moreover, in addition to providing the surgical team with detailed information regarding the patient's cognitive profile, the neuropsychologist is also actively involved in deciding whether and which additional procedures might be necessary to provide critical functional information toward considering a patient's surgical candidacy, the type and potential limits of the surgery, and the likely risks or benefits to cognitive and psychological functioning given a particular patient, a particular set of diagnostic results and the surgical procedure under consideration. Thus, it is necessary for the neuropsychologist to be familiar with (epilepsy related) neuroimaging, EEG, and seizure semiology, and to stay abreast of the growing literature on cognition and epilepsy surgery.

CURRENT PRACTICE

Neuropsychological Evaluation

Clinical neuropsychology in epilepsy is primarily concerned with the effects of seizures and seizure-related abnormalities on cognitive and emotional functioning. In surgical epilepsy, one of the main goals of the neuropsychological evaluation is to identify dysfunction that may correspond with other indicators of brain abnormalities (e.g., EEG, MRI, positron emission tomography (PET), single-photon emission computed tomography (SPECT), and magnetoencephalography (MEG) results) related to the epileptogenic region. In some cases, all or most of these data points implicate the same area, and the neuroanatomic target for surgery is relatively clear-cut. However, many cases are less straightforward, rendering the neuropsychological component of the workup particularly salient. The neuropsychological evaluation typically assesses multiple domains including intellectual capacity, verbal/language functions, nonverbal/visual—spatial functions, attention/working memory, processing speed, executive functions, learning and memory for both verbal and visual information, fine motor skills, mood, personality, and quality of life. A battery of tests may require up to 5 or 6 hours, and can be completed in either an outpatient or an inpatient setting. The neuropsychologist uses this quantifiable information, along with a structured clinical interview and behavioral observations, to derive a detailed profile for the examinee. In a sense, the profile serves as a basic functional-neuroanatomical "map," as a relative deficit implies that there is an underlying abnormality in the brain region associated with a given function.

Evaluations are a critical component of comprehensive presurgical workups, to help establish a patient's baseline level of functioning and identify potential risks and benefits of a proposed surgery. As mentioned earlier, the main goal of epilepsy surgery is to maximize the chance of seizure freedom while simultaneously minimizing the chance of any negative functional outcomes. Generally speaking, good neuropsychological performance in the proposed surgical region is typically associated with higher risk for postoperative cognitive decline because it implies that the to-be-resected or to-be-ablated tissue is functionally intact, even if it is the source of seizure onset. In contrast, poor neuropsychological performance in the proposed surgical region is usually associated with lower risk for postoperative cognitive decline because it implies that the tissue is functionally compromised and, therefore, there is "less to lose."

In nonsurgical cases, evaluations can also be requested if an individual is experiencing cognitive difficulties in their everyday life, which can potentially be related to seizure activity, side effects from antiepileptic medications, psychiatric comorbidities, or other complicating factors.

WADA TESTING

In addition to traditional paper-and-pencil or computer-based testing, neuropsychologists in epilepsy surgery programs often engage in more invasive procedures that further assist with surgical planning. In 1955, Dr. Juhn Wada introduced a technique of selective hemispheric anesthesia at the Montreal Neurological Institute. The procedure was originally developed to determine language representation in patients who were suspected to have atypical cerebral language laterality (Wada & Rasmussen, 1960). Dr. Brenda Milner subsequently included memory testing to identify patients who might be at risk for postoperative amnesia due to good functional adequacy on the side of the proposed surgery and/or poor functional reserve in the opposite hemisphere (Milner, Branch, & Rasmussen, 1962b). As the procedure remains unstandardized across centers, "Wada" testing refers not to a specific paradigm, but rather, to a group of procedures involving the assessment of cognitive function during temporary anesthesia of one cerebral hemisphere. With the injection of an anesthetizing agent into the internal carotid artery, Wada testing enables assessment of language ability in the ipsilateral hemisphere while simultaneously allowing for assessment of memory capacity in the contralateral hemisphere. In this way, it serves as a crude and reversible analog of a proposed resection or ablation.

For many years, the anesthetic agent utilized was sodium amobarbital (it came to be called intracarotid amobarbital testing). However, spurred by amobarbital shortages in the 2000s, other medications such as methohexital and etomidate, which tend to be shorter-acting and less sedating, have come into common use. The specifics of the procedure vary among epilepsy centers. However, hemispheric anesthesia is usually verified by unilateral slowing on scalp EEG ipsilateral to the injection and upper extremity hemiplegia contralateral to the injection. Hemispheric language assessment typically involves automatic speech (e.g., counting or reciting the alphabet), execution of verbal commands, object naming, repetition, and reading. As a "disruption" technique, language dominance is inferred when a patient is unable to perform these tasks during the period of temporary anesthesia. Memory assessment usually includes the spontaneous recall or recognition of previously presented commands, pictures, and objects after the anesthesia has worn off and the patient is back to their baseline level of functioning. Accuracy and subjective confidence can both be measured to determine the strength of a patient's memory.

One advantage of Wada testing is its power in determining language representation in one hemisphere without influence from the other. However, language is rarely a strictly unilateral function (Springer et al., 1999) and, moreover, Wada creates a diffuse unilateral lesion that does not allow for the more precise assessment of intrahemispheric language organization. Recent advances in technology and neuroimaging have yielded other techniques, most notably functional MRI (fMRI), which is less invasive and can more precisely evaluate language functions in both cerebral hemispheres at the same time. However, another advantage of Wada testing is the ability to measure memory capacity in each hemisphere individually, which cannot be accomplished with traditional neuropsychological testing and, although numerous fMRI studies demonstrate strong correlation with Wada memory results (Detre et al., 1998; Janszky et al., 2005; Rabin et al., 2004) none have yet been accepted for clinical use. Moreover, neuroimaging techniques such as fMRI reveal activation, that is, areas that are involved in function; however, activation techniques fail to indicate which brain areas are *necessary* for function, as disruptive methods do. For this reason, Wada remains the gold standard for preoperative language *and* memory assessment in many comprehensive epilepsy surgery centers.

ELECTROCORTICAL STIMULATION (LANGUAGE) MAPPING

When surgery involves the dominant hemisphere, it can be important to identify critical language areas in the region of the proposed resection in order to prevent postoperative language impairments. This is accomplished with extra- or intraoperative language mapping, another invasive practice used by neuropsychologists in comprehensive epilepsy surgery programs. Electrical stimulation-based mapping was largely pioneered by Dr. Wilder Penfield for clinical purposes (Penfield & Roberts, 1959), though the extensive work of Dr. George Ojemann has further advanced our understanding of more precise structure—function relations within the language dominant hemisphere (Ojemann & Mateer, 1979; Ojemann, Ojemann, Lettich, & Berger, 1989; Ojemann & Whitaker, 1978). Although unstandardized across centers, these techniques typically involve the performance of speech and language tasks while electrical stimulation is briefly (2—5 seconds) applied to the brain surface or, to the surrounding brain parenchyma in the context of depth electrodes, effectively creating a reversible functional lesion to a discrete cortical area (Hamberger, 2011).

Across surgery programs, many language tasks have been used; however, the vast majority of programs include object naming, most often via pictures, yet also using auditory descriptions (Hamberger, 2007; Hamberger, Williams, & Schevon, 2014). Electrical stimulation is applied immediately before stimulus presentation, or intermittently during other tasks such as spontaneous speech and sentence reading. Similar to Wada testing, electrical stimulation mapping is a "disruptive" technique. Therefore, unlike the identification of sensory and motor cortex, in which stimulation evokes positive responses such as a subjective sensation (e.g., tingling) or observable movement (e.g., twitching), the stimulation of language cortex results in negative responses (e.g., anomia, aphasia), or other interference with the task at hand. If a particular location is associated with language deficit during the period of stimulation, it is considered a positive site, and it is typically spared from resection or ablation with the goal of preserving function postoperatively. Research in patients with epilepsy has shown that stimulation-identified language sites can extend well beyond traditional Broca's and Wernicke's areas due to intrahemispheric reorganization (Devinsky et al., 2000; Devinsky, Perrine, Llinas, Luciano, & Dogali, 1993; Ojemann, 1979), indicating that traditional landmarks cannot be relied upon to predict the localization of essential language cortex in this population, and underscoring the need for stimulation-based mapping on a case-by-case basis.

The advantage of language mapping is its precision in identifying specific brain areas that are *necessary* for language within a given hemisphere, not merely supplemental areas, which is very important for the planning of accurate surgical margins. However, it is an invasive procedure with several practical constraints. With extraoperative mapping, testing is often spatially limited by the location of preplaced electrodes which may not cover all aspects of eloquent cortex. Also, although intraoperative mapping allows for greater precision and control with regard to location of stimulation, testing can be adversely affected by patient anxiety, fatigue, and discomfort, and time is restricted due to concerns of infection associated with prolonged brain exposure.

CASE CONFERENCE

In a comprehensive epilepsy surgery program, individual patients are typically presented at a case conference, which is an interdisciplinary meeting that helps to determine a patient's candidacy for surgery and expected outcome with regard to seizure relief or freedom. The meeting typically includes epileptologists, radiologists, surgeons, nurse practitioners, and neuropsychologists, and it involves the careful and systematic review of multiple sources of information including medical history, seizure semiology and EEG, MRI, PET, SPECT, MEG, neuropsychological assessment results, and Wada and language mapping findings. The neuropsychologist is a vital participant as he/she reviews the data and weighs the risks and/or benefits associated with a given type of surgery.

From a neuropsychological standpoint, those at high risk for postoperative cognitive decline are typically older at the time of seizure onset and surgery because their brains are already well formed and have lower chance for reorganization. They also have negative pathology on neuroimaging (e.g., no MTS), indicating the absence of a structural lesion that may be the source of seizure onset. In addition, they have high scores on traditional neuropsychological testing, suggesting that their brains are functionally intact. Furthermore, they have good ipsilateral memory adequacy but poor contralateral memory reserve on Wada testing, which means that the side of seizure onset (and proposed surgery) is supporting memory more than the side away from seizure onset. Finally, they have positive language sites in the region of the proposed resection or ablation, implying that they will potentially lose this ability if the region is surgically altered. In contrast, those at low risk for postoperative cognitive decline are typically younger at the time of seizure onset and surgery because their brains have more plasticity and are more capable of reorganizing. They have positive pathology or an identifiable lesion on neuroimaging that is presumed to be involved in seizure onset. They have low scores on traditional neuropsychological testing, suggesting that their brains are already functionally compromised. They have poor memory adequacy but good memory reserve on Wada testing, which means that the side of seizure onset (and proposed surgery) does not support memory as well as the side away from seizure onset. Finally, they do not have positive language sites in the region of the proposed resection or ablation, indicating that they will not lose any ability if the area is surgically modified (Hamberger, Seidel, McKhann, Perrine, & Goodman, 2005; Helmstaedter, Kurthen, Lux, Reuber, & Elger, 2003; Hermann et al., 1992; Loring et al., 1995; et al., 2005).

CASE EXAMPLES

Case #1

S. was a 20-year-old, right-handed, woman with a history of epilepsy since age 7. She was undergoing presurgical evaluation because her seizures were refractory to more traditional pharmacological treatment. Her EEGs were consistently abnormal with interictal abnormalities and seizure onset in the left anterior inferior temporal region. Brain PET showed decreased glucose metabolism in the left anterior and medial temporal lobe. Brain MRI demonstrated clear left MTS but no other structural abnormalities.

As a routine part of her presurgical workup, she underwent comprehensive neuropsychological assessment, which demonstrated average intellectual functioning but weak verbal memory among the findings. Wada testing was recommended to assess for language dominance and left versus right hemisphere memory support. The Wada showed that she had left hemisphere language dominance and poor memory support from the left medial temporal region (ipsilateral to the seizure focus) yet good memory support from the right medial temporal region (contralateral to the seizure focus). The findings were presented in a case conference, and it was determined that she was a good candidate for an SAH on the left, sparing lateral cortex that would otherwise be included in a standard anterior temporal lobe resection. She was advised that she had a high chance for seizure freedom but a low risk for significant cognitive decline from this type of surgery. Her seizures were well controlled after the procedure. Also, as predicted, she did not experience a significant decline in verbal memory postoperatively (delayed recall on an auditory verbal learning test was at the 6th percentile both before and after surgery). She did not notice a significant change in her daily functioning and remained in college as a full-time student.

Case #2

D. was a 52-year-old, right-handed, woman with a history of epilepsy since age 46. She was undergoing presurgical evaluation because her seizures were refractory to pharmacological treatment. Her EEGs were abnormal due to left temporal slowing and epileptiform discharges (time 1), left temporal slowing without epileptiform discharges (time 2), and left > right bitemporal slowing with four recorded seizures of left temporal onset (time 3). Brain PET was grossly normal with no focal areas of decreased glucose metabolism. Brain MRI revealed no evidence of MTS but was notable for status postright pterional craniotomy for clipping of a right middle cerebral artery aneurysm.

As part of her presurgical workup, baseline neuropsychological assessment demonstrated average intellectual functioning with strong verbal memory among the findings. Wada testing was then recommended to assess for language dominance and for more specific assessment of left versus right hemisphere memory support. The Wada showed that she had left hemisphere language dominance and good memory support from the left medial temporal region (ipsilateral to the seizure focus) but poor memory support from the right medial temporal region (contralateral to the seizure focus). Unfortunately, subsequent stereo-EEG with depth electrodes showed that her seizures were starting in the left mesial temporal structures, despite the fact that they were both structurally and functionally intact. The findings were presented in a case conference, and the surgical team agreed that she should be advised that she had a high risk for postoperative verbal memory decline. She was nevertheless determined to proceed because her seizures were quite debilitating, and she opted for surgery involving stereotactic laser ablation of the left amygdala and hippocampus (SLAH). Her seizures were well controlled after the procedure. However, as predicted, she experienced a significant verbal memory decline postoperatively (delayed recall on an auditory verbal learning test went from the 90th percentile to the 6th percentile). This impacted her daily functioning and she had to resign from several educational boards on which she previously served due to her suboptimal memory.

These cases highlight the important role of neuropsychological test data and Wada findings in predicting postoperative cognitive decline, specifically involving memory for TLE patients. For many reasons, patient #1 was a better candidate for surgery than patient #2. First, she was younger at the time of seizure onset and at the time of surgery. She also had a sclerotic left hippocampus on MRI and corresponding hypometabolism on PET imaging. Neuropsychological assessment and Wada testing additionally suggested that her memory structures on the left were functionally compromised. All results consistently pointed to the same region: Left mesial temporal. Additionally, baseline testing suggested she had already "taken the hit" with regard to verbal memory, reflected by preoperative relative impairment. For these reasons, she had a much lower risk for postoperative memory changes, and although they both performed at the same level after surgery, only patient #2 experienced a significant *decline* from her baseline functioning.

Case #3

J. was a 34-year-old, right-handed, man with a history of epilepsy since age 5. He was undergoing presurgical evaluation because his seizures were refractory to pharmacological treatment. His EEGs showed that his seizures were arising from the left frontal—temporal region. Brain MRI revealed probable cortical dysplasia in the left anterior quadrant, corresponding with a focal area of PET hypometabolism.

As part of his presurgical workup, baseline neuropsychological testing showed that he had high average intellectual functioning with good verbal fluency and naming among the findings. This was somewhat unexpected, and Wada testing was recommended to determine if his language functions had reorganized to the right hemisphere. However, despite the possibility of interhemispheric reorganization, the Wada showed that his left hemisphere was indeed language dominant. Because of this, he underwent subdural electrode grid and strip implantation to better delineate the area of seizure onset, and for language mapping purposes. Ictal recordings indicated that his seizures were primarily starting in the left inferior frontal gyrus (i.e., Broca's area) with subsequent spread to the left temporal lobe. However, when electrical stimulation was applied to the same electrodes for testing purposes, it did not interfere with his ability to perform language-based tasks (though positive language sites away from this traditional landmark signified apparent intrahemispheric reorganization). The findings were reviewed in a surgical case conference, and he was advised that he would not experience significant language problems if he had a focal resection of the primary epileptogenic zone. He therefore had surgery to remove the dysfunctional tissue. He no longer had seizures after the procedure and he did not complain of any word-finding difficulty or other language deficits afterwards.

This case highlights the importance of language mapping in helping to identify eloquent cortex and defining the boundaries of a surgical resection. If patient #3 had positive language sites in the area of seizure onset, he would not have been considered a good candidate for surgery, and a more palliative approach would most likely have been recommended. However, language mapping determined that it would be safe to remove an area of his brain that would otherwise be assumed to regulate expressive language.

CLOSING COMMENTS

In this chapter, we reviewed the history and current practice of the neuropsychologist's role in comprehensive epilepsy surgery centers. The goal was to familiarize the reader with different responsibilities and opportunities in these highly specialized settings. Rather than an exhaustive review, we aimed to highlight some of the most important contributions of neuropsychology in the context of surgical treatment for epilepsy.

REFERENCES

Alpherts, W. C., Vermeulen, J., van Rijen, P. C., da Silva, F. H., van Veelen, C. W., & Dutch Collaborative Epilepsy Surgery, Program. (2008). Standard versus tailored left temporal lobe resections: Differences in cognitive outcome? *Neuropsychologia*, *46*(2), 455–460. Available from https://doi.org/10.1016/j.neuropsychologia.2007.08.022.

Berg, A. T. (2009). Identification of pharmacoresistant epilepsy. *Neurologic Clinics*, *27*(4), 1003–1013. Available from https://doi.org/10.1016/j.ncl.2009.06.001.

Chang, E. F., Englot, D. J., & Vadera, S. (2015). Minimally invasive surgical approaches for temporal lobe epilepsy. *Epilepsy & Behavior*, *47*, 24–33. Available from https://doi.org/10.1016/j.yebeh.2015.04.033.

Chelune, G. J. (1995). Hippocampal adequacy versus functional reserve: Predicting memory functions following temporal lobectomy. *Archives of Clinical Neuropsychology, 10*(5), 413−432.

Curry, D. J., Gowda, A., McNichols, R. J., & Wilfong, A. A. (2012). MR-guided stereotactic laser ablation of epileptogenic foci in children. *Epilepsy & Behavior, 24*(4), 408−414.

Davies, K., Bell, B., Bush, A., Hermann, B., Dohan, F. C., & Japp, A. S. (1998). Naming decline after left anterior temporal lobectomy correlates with pathological status of resected hippocampus. *Epilepsia, 39*(4), 407−419.

Detre, J. A., Maccotta, L., King, D., Alsop, D. C., Glosser, G., D'Esposito, M., . . . French, J. A. (1998). Functional MRI lateralization of memory in temporal lobe epilepsy. *Neurology, 50* (4), 926−932.

Devinsky, O., Perrine, K., Hirsch, J., McMullen, W., Pacia, S., & Doyle, W. (2000). Relation of cortical language distribution and cognitive function in surgical epilepsy patients. *Epilepsia, 41*(4), 400−404.

Devinsky, O., Perrine, K., Llinas, R., Luciano, D. J., & Dogali, M. (1993). Anterior temporal language areas in patients with early onset of TLE. *Annals of Neurology, 34*, 727−732.

Drane, D. L., Loring, D. W., Voets, N. L., Price, M., Ojemann, J. G., Willie, J. T., . . . Gross, R. E. (2015). Better object recognition and naming outcome with MRI-guided stereotactic laser amygdalohippocampotomy for temporal lobe epilepsy. *Epilepsia, 56*(1), 101−113. Available from https://doi.org/10.1111/epi.12860.

Engel, J., Jr. (1998). Etiology as a risk factor for medically refractory epilepsy: A case for early surgical intervention. *Neurology, 51*(5), 1243−1244.

Engel, J., Jr., Van Ness, P., Rasmussen, T., & Ojemann, L. (1993). Outcome with respect to epileptic seizures. In J. Engel, Jr. (Ed.), *Surgical treatment of the epilepsies* (2nd ed., pp. 609−621). New York: Raven Press.

Gleissner, U., Helmstaedter, C., Schramm, J., & Elger, C. E. (2004). Memory Outcome after selective amygdalohippocampectomy in patients with temporal lobe epilepsy: One-year follow-up. *Epilepsia, 45*(8), 960−962.

Graydon, F. J., Nunn, J. A., Polkey, C. E., & Morris, R. G. (2001). Neuropsychological outcome and the extent of resection in the unilateral temporal lobectomy. *Epilepsy & Behavior, 2*(2), 140−151. Available from https://doi.org/10.1006/ebeh.2001.0163.

Hamberger, M. J. (2007). Cortical language mapping in epilepsy: A critical review. *Neuropsychology Review, 4*, 477−489.

Hamberger, M. J. (2011). Cortical mapping. In J. K. B. Caplan, & J. DeLuca (Eds.), *Encyclopedia of clinical neuropsychology* (pp. 719−721). New York: Springer Science.

Hamberger, M. J., Seidel, W. T., McKhann, G. M., Perrine, K., & Goodman, R. R. (2005). Brain stimulation reveals critical auditory naming cortex. *Brain, 128*(11), 2742−2749.

Hamberger, M.J., Williams, A.C., & Schevon, C.A. (2011). *Results of an international extraoperative neurostimulation mapping survey*. Paper presented at the American Epilepsy Society Annual Meeting, Baltimore, MD.

Hamberger, M. J., Williams, A. C., & Schevon, C. A. (2014). Extraoperative neurostimulation mapping: Results from an international survey of epilepsy surgery programs. *Epilepsia, 55* (6), 933−939. Available from https://doi.org/10.1111/epi.12644.

Helmstaedter, C., Elger, C. E., & Hufnagel, A. (1996). Differential effects of left anterior temporal lobectomy, selective amygdalohippocampectomy, and temporal cortical lesionectomy on verbal learning, memory and recognition. *Journal of Epilepsy, 9*, 39−45.

Helmstaedter, C., Kurthen, M., Lux, S., Reuber, M., & Elger, C. E. (2003). Chronic epilepsy and cognition: A longitudinal study in temporal lobe epilepsy. *Annals of Neurology, 54*(4), 425−432.

Helmstaedter, C., Van Roost, D., Clussman, H., Urbach, H., Elger, C. E., & Schramm, J. (2004). Collateral brain damage, a potential source of cognitive impairment after selective surgery for control of mesial temporal lobe epilepsy. *Journal of Neurology, Neurosurgery & Psychiatry, 75*(2), 323–326.

Hermann, B. P., Perrine, K., Chelune, G. J., Barr, W., Loring, D. W., Strauss, E., ... Westerveldt, M. (1999). Visual confrontation naming following left ATL: A comparison of surgical approaches. *Neuropsychology, 13*(1), 3–9.

Hermann, B. P., Wyler, A. R., Somes, G., Berry, A., & Dohan, F. C. (1992). Pathological status of the mesial temporal lobe predicts memory outcome after left anterior temporal lobectomy. *Neurosurgery, 31*, 653–657.

Janszky, J., Jokeit, H., Kontopoulou, K., Mertens, M., Ebner, A., Polhlmann-Eden, B., & Woermann, F. G. (2005). Functional MRI predicts memory performance after right mesio-temporal epilepsy surgery. *Epilepsia, 46*(2), 244–250.

Jokeit, H., Ebner, A., Holthausen, H., Markowitsch, H. J., Moch, A., Pannek, H., ... Tuxhorn, I. (1997). Individual prediction of change in delayed recall of prose passages after left-sided anterior temporal lobectomy. *Neurology, 49*(2), 481–487.

Katz, A., Awad, I. A., Kong, A. K., Chelune, G. J., Naugle, R. I., Wyllie, E., ... Luders, H. (1989). Extent of resection in temporal lobectomy for epilepsy. II. Memory changes and neurologic complications. *Epilepsia, 30*(6), 763–771.

Kwan, P., & Brodie, M. J. (2000). Early identification of refractory epilepsy. *New England Journal of Medicine, 342*(5), 314–319. Available from https://doi.org/10.1056/NEJM200002033420503.

Langfitt, J., & Rausch, R. (1996). Word-finding deficits persist after anterotemporal lobectomy. *Archives of Neurology, 53*, 72–76.

Lee, E. M., Kang, J. K., Kim, S. J., Hong, S. H., Ko, T. S., Lee, S. A., ... Lee, J. K. (2015). Gamma knife radiosurgery for recurrent or residual seizures after anterior temporal lobectomy in mesial temporal lobe epilepsy patients with hippocampal sclerosis: Long-term follow-up results of more than 4 years. *Journal of Neurosurgery, 123*(6), 1375–1382. Available from https://doi.org/10.3171/2014.12.JNS141280.

Loring, D. W., Kapur, R., Meador, K. J., & Morrell, M. J. (2015). Differential neuropsychological outcomes following targeted responsive neurostimulation for partial-onset epilepsy. *Epilepsia, 56*(11), 1836–1844. Available from https://doi.org/10.1111/epi.13191.

Loring, D. W., Meader, K. J., Lee, G. P., King, D. W., Nichols, M. E., Park, A. M., ... Smith, J. R. (1995). Wada memory asymmetries predict verbal memory decline after anterior temporal lobectomy. *Neurology, 45*, 1329–1333.

Lutz, M. T., Clussman, H., Elger, C. E., Schramm, J., & Helmstaedter, C. (2004). Neuropsychological outcome after selective amygdalohippocampectomy with transsylvian versus transcortical approach: A randomized prospective clinical trial of surgery for temporal lobe epilepsy. *Epilepsia, 45*(7), 809–816.

McDonald, C. R., Norman, M. A., Tecoma, E., Alksne, J., & Iragui, V. (2004). Neuropsychological change following gamma knife surgery in patients with left temporal lobe epilepsy: A review of three cases. *Epilepsy & Behavior, 5*(6), 949–957. Available from https://doi.org/10.1016/j.yebeh.2004.08.014.

McIntosh, A. M., Wilson, S. J., & Berkovic, S. F. (2001). Seizure outcome after temporal lobectomy: Current research practice and findings. *Epilepsia, 42*(10), 288–307.

Meador, K. J., Kapur, R., Loring, D. W., Kanner, A. M., Morrell, M. J., & Investigators, R. N. S. System Pivotal Trial. (2015). Quality of life and mood in patients with medically intractable epilepsy treated with targeted responsive neurostimulation. *Epilepsy & Behavior,* *45,* 242–247. Available from https://doi.org/10.1016/j.yebeh.2015.01.012.

Milner, B. (1958). Psychological defects produced by temporal lobe excision. *Research* *Publications—Association for Research in Nervous & Mental Disease, 36,* 244–257.

Milner, B., Branch, C., & Rasmussen, T. (1962a). Study of short term memory after intracarotid injection of sodium amobarbital. *Transactions of the American Neurological Association,* *91,* 306–308.

Milner, B., Branch, C., & Rasmussen, T. (1962b). Study of short term memory after intracarotid injection of sodium amytal. *Transactions of the American Neurological Association, 87,* 224–226.

Ojemann, G. A. (1979). Individual variability in the cortical localization of language. *Journal of* *Neurosurgery, 50,* 164–169.

Ojemann, G. A., & Mateer, C. (1979). Human language cortex: Localization of memory, syntax and sequential motor-phoneme sequencing. *Science, 205,* 1401–1403.

Ojemann, G. A., Ojemann, J., Lettich, E., & Berger, M. (1989). Cortical language localization in left-dominant hemisphere: An electrical stimulation mapping investigation in 117 patients. *Journal of Neurosurgery, 71,* 316–326.

Ojemann, G. A., & Whitaker, H. (1978). Language localization and variability. *Brain and* *Language, 6,* 239–260.

Penfield, W. (1959). Mapping the speech area. In L. R. W. Penfield (Ed.), *Speech and brain* *mechanisms* (pp. 103–118). Princeton, NJ: Princeton University Press.

Penfield, W., & Roberts, L. (1959). Evidence from cortical mapping. In W. Penfield (Ed.), *Speech and brain mechanisms* (pp. 119–137). Princeton, NJ: Princeton University Press.

Rabin, M. L., Narayan, V. M., Kimberg, D. Y., Casasanto, D. J., Glosser, G., Tracy, J. I., ... Detre, J. A. (2004). Functional MRI predicts post-surgical memory following temporal lobectomy. *Brain, 127,* 2286–2298.

Regis, J., Rey, M., Bartolomei, F., Vladyka, V., Liscak, R., Schrottner, O., & Pendl, G. (2004). Gamma knife surgery in mesial temporal lobe epilepsy: A prospective multicenter study. *Epilepsia, 45*(5), 504–515. Available from https://doi.org/10.1111/j.0013-9580.2004.07903. x.

Scoville, W. B., & Milner, B. (1957). Loss of recent memory after bilateral hippocampal lesions. *Journal of Neurology, Neurosurgery & Psychiatry, 20*(1), 11–21.

Seidenberg, M., Hermann, B. P., Dohan, F. C., Jr., Wyler, A. R., Perrine, A., & Schoenfeld, J. (1996). Hippocampal sclerosis and verbal encoding ability following anterior temporal lobectomy. *Neuropsychologia, 34*(7), 699–708.

Seidenberg, M., Hermann, B. P., Schoenfeld, J., Davies, K., Wyler, A. R., & Dohan, F. C. (1997). Reorganization of verbal memory function in early onset left temporal lobe epilepsy. *Brain and Cognition, 35,* 132–148.

Spencer, D. D., Spencer, S. S., Mattson, R. H., Williamson, P. D., & Novelly, R. A. (1984). Access to the posterior medial temporal lobe structures in the surgical treatment of temporal lobe epilepsy. *Neurosurgery, 15,* 667–671.

Spencer, S. S. (2008). Gamma knife radiosurgery for refractory medial temporal lobe epilepsy: Too little, too late? *Neurology, 70*(19), 1654–1655. Available from https://doi.org/10.1212/ 01.wnl.0000311272.33720.6b.

Springer, J. A., Binder, J. R., Hammeke, T. A., Swanson, S. J., Frost, J. A., Bellgowan, P. S., . . . Mueller, W. M. (1999). Language dominance in neurologically normal and epilepsy subjects: A functional MRI study. *Brain, 122(Pt, 11,* 2033–2046.

Srikijvilaikul, T., Najm, I., Foldvary-Schaefer, N., Lineweaver, T., Suh, J. H., & Bingaman, W. E. (2004). Failure of gamma knife radiosurgery for mesial temporal lobe epilepsy: Report of five cases. *Neurosurgery*, *54*(6), 1395−1402. (discussion 1402-1394).

Wada, J., & Rasmussen, T. (1960). Intracarotid injection of sodium amytal for the lateralization of cerebral speech dominance: Experimental and clinical observations. *Journal of Neurosurgery*, *17*, 266−282.

Wieser, H. G. (1988). Selective amygdalahippocampectomy for temporal lobe epilepsy. *Epilepsia*, *29*, 100−113.

Willie, J. T., Laxpati, N. G., Drane, D. L., Gowda, A., Appin, C., Hao, C., . . . Gross, R. E. (2014). Real-time magnetic resonance-guided stereotactic laser amygdalohippocampotomy for mesial temporal lobe epilepsy. *Neurosurgery*, *74*(6), 569−585.

Wolf, R. L., Ivnik, R. J., Hirschorn, K. A., Sharbrough, F. W., Cascino, G. D., & Marsh, W. R. (1993). Neurocognitive efficiency following left temporal lobectomy: Standard versus limited resection. *Journal of Neurosurgery*, *79*(1), 76−83. Available from https://doi.org/10.3171/jns.1993.79.1.0076.

Chapter 9

Neuropsychology in the Neurosurgical Management of Primary Brain Tumors

Kyle Noll[1], David Sabsevitz[2], Sujit Prabhu[3] and Jeffrey Wefel[1]

[1]*Department of Neuro-Oncology, The University of Texas M.D. Anderson Cancer Center, Houston, TX, United States,* [2]*Department of Neurology, Medical College of Wisconsin, Milwaukee, WI, United States,* [3]*Department of Neurosurgery, The University of Texas M.D. Anderson Cancer Center, Houston, TX, United States*

BRIEF HISTORY

Primary brain tumors account for 1.4% of all cancers with an incidence of 21.4 cases per 100,000 (Ostrom et al., 2014). These relatively rare tumors may arise from the meninges or brain parenchyma, including neuronal and supportive glial cells. In adults, primary malignant brain tumors are most frequently located in the frontal and temporal lobes, while childhood brain tumors are more often infratentorial. Glial tumors, or glioma, represent nearly 30% of all cerebral tumors and 80% of all malignant (WHO grade II−IV) brain tumors. Of all glioma, WHO grade IV glioblastoma (GBM) represent the most common (54%) and most aggressive, with a median overall survival of 14.6 months despite standard of care treatment (Tran & Rosenthal, 2010). This is in stark contrast to low grade glioma (WHO grade II), which have a median overall survival of 13.8 years with standard care. Of note, survival also differs widely within histopathological tumor type, as various genetic markers (e.g., IDH mutations, 1p/19q deletions) are strong prognostic indicators. In fact, recognizing the prognostic and predictive significance of tumor molecular genetics, the diagnostic criteria for brain tumors was revised in 2016 to include both classical histopathological features and molecular signatures yielding an "integrated" diagnosis (Louis, Ohgaki, Wiestler, & Cavenee, 2016). Despite differences in prognosis across histopathological and molecular subtypes, most malignant glioma require

Neurosurgical Neuropsychology. DOI: https://doi.org/10.1016/B978-0-12-809961-2.00010-2

neurosurgical intervention as a first line treatment, with adjuvant therapies varying according to the specific diagnosis.

Patients with primary brain tumors often initially present with seizures (especially in lower grade tumors), headaches, focal neurological signs, neurocognitive changes, or some combination thereof. Prior to the availability of modern neuroimaging techniques, neuropsychologists were often consulted to help determine the etiology of such signs and symptoms, including localization of suspected intracranial lesions (Anderson & Ryken, 2008). However, current neuroimaging protocols allow for far more precise characterization of lesion size, location, and associated features than can be inferred from neuropsychological testing. Scans utilizing contrast enhancing agents (e.g., gadolinium) and advanced brain tumor imaging (e.g., magnetic resonance spectroscopy) can even suggest preliminary tumor histology in the absence of tissue biopsy (Bulik, Jancalek, Vanicek, Skoch, & Mechl, 2013). Accordingly, neuropsychological assessment is unnecessary for detection and localization of cerebral pathology in patients with primary brain tumors. Rather, the role of the neuropsychologist in the modern era of brain tumor management encompasses the quantification of neurocognitive deficits, determination of the extent of cerebral dysfunction, facilitation of other neuromedical procedures such as brain mapping, and identification of targets for individualized neuropsychological interventions.

The following provides an overview of neuropsychological research and practice related to the neurosurgical management of patients with primary brain tumors. Particular focus is paid to neuropsychological assessment and brain mapping in adult patients with glioma, which represent the most common primary brain malignancies. Discussion also centers on neuropsychology within the context of the typical open neurosurgical procedure with the intent to maximize extent of tumor resection. Less attention is paid to other neurosurgical procedures (e.g., biopsy only, laser interstitial thermal therapy, stereotactic radiosurgery) and surgical adjuvants (e.g., intraoperative imaging, ultrasound, fluorescing dyes).

MODERN FUNCTIONAL NEUROSURGERY

Following identification of the underlying structural brain lesion via neuroimaging, referral to neurosurgery is generally made for consideration of primary management options. In the majority of primary brain tumor cases, especially suspected infiltrating glioma, the neurosurgeon will recommend surgical intervention. The minimal aim of neurosurgery is to obtain adequate tissue to establish a diagnosis, with an optimal goal of maximizing extent of safe resection to increase survival and reduce neurologic symptoms and other problems associated with the tumor (Duffau & Mandonnet, 2013).

Onco-Functional Balance

Successful neurosurgical intervention requires not only maximal extent of resection for tumor control and improved survival, but also minimization of surgically acquired neurologic and neurocognitive deficits that can profoundly impact patients' daily lives. Essential to this "onco-functional balance" is delineation of both tumor and functional tissue boundaries. Tumor boundaries are somewhat vague for infiltrating tumors like glioma, as tumor cells tend to penetrate the parenchyma beyond the bounds of the primary mass. Nonetheless, modern magnetic resonance imaging (MRI) sequences, including gadolinium contrast enhanced scans, can provide detailed high-resolution structural images that assist in identification of the bulk of the tumor (Mabray, Barajas, & Cha, 2015). Intraoperative MRI and microsurgical techniques incorporating fluorescence image guidance can also illuminate structural boundaries to guide resection as surgery proceeds. Other structural imaging techniques also have a role in identifying functional boundaries. For instance, diffusion tensor magnetic resonance imaging (DTI), utilized both extra- and intraoperatively, has proven to be a reliable means of delineating important functional boundaries via anatomical landmarks, such as the visual pathways and long motor tracts (Buchmann et al., 2011). However, given substantial interindividual variability, structural anatomical imaging alone is insufficient for identifying the cortical and subcortical areas supporting higher level cognitive functions like speech and language (Duffau, 2017).

Contributions of Neuropsychology

Various techniques have emerged for individualized mapping of higher neurocognitive functions for neurosurgical planning. Neuropsychologists have been instrumental in the development of many of these methods and remain involved in their implementation in both clinical and research settings. These techniques include noninvasive preoperative tools such as navigated transcranial magnetic stimulation (nTMS) and functional magnetic resonance imaging (fMRI), which are now routinely used in brain mapping for epilepsy and brain tumor surgical planning (Austermuehle et al., 2017; Papanicolaou et al., 2017). However, these methods are not without limitations and delineation of functional boundaries for tumor resection continues to rely heavily upon direct cortical stimulation (DCS), an invasive technique often implemented intraoperatively in close collaboration with neuropsychologists.

While the role of neuropsychology in brain mapping of patients with brain tumors is in many ways similar to that of epilepsy, relevant considerations involved in the surgical decision-making process can differ substantially between the two populations. For example, optimal onco-functional balance in the tumor setting often requires sacrificing eloquent tissue to obtain an adequate resection to provide greatest survival benefit. That is, avoidance of eloquent tissue is not always possible or may be a secondary goal, particularly given data suggesting that gross total resection (of at least 98% of enhancing tumor) for GBM is necessary to maximize survival (Lacroix et al., 2001). Nonetheless, identification of eloquent tissue at the periphery of the tumor boundary can have significant impact on neurosurgical decision-making, such as whether to proceed with supratotal resection (i.e., extension of the surgical margin beyond the contrast enhancing portion of the tumor), which may risk compromising important neurocognitive functions.

In addition to facilitating pre- and intraoperative brain mapping procedures, neuropsychologists have been instrumental in monitoring and managing neurocognitive dysfunction in patients with brain tumors. Primary brain tumors represent a dynamic disease in which neurocognitive functioning may drastically differ across the disease course according to varying tumor, treatment, and individual characteristics. Unfortunately, tumor and treatment effects almost invariably include deficits in neurocognitive functioning, despite utilization of state of the art brain mapping, microsurgical techniques, and well-tolerated adjuvant therapies (Wefel & Schagen, 2012; Wu et al., 2011). Neurocognitive impairment is also associated with reductions in patient health-related quality of life (Noll, Bradshaw, Weinberg, & Wefel, 2017b), functional independence (Noll, Bradshaw, Weinberg, & Wefel, 2017a), and even survival (Johnson, Sawyer, Meyers, O'neill, & Wefel, 2012; Meyers, Hess, Yung, & Levin, 2000). Accordingly, monitoring of neurocognitive functioning is not only essential during the perioperative period, but also important during the adjuvant treatment phase and throughout the entire survivorship period.

CURRENT PRACTICE

Neuropsychological Assessment

Prior to proceeding with neurosurgery, the neurosurgical team is tasked with gathering data necessary to determine surgical risks and help optimize the onco-functional balance. An integral part of this data collection process includes presurgical neuropsychological assessment. Identification of presenting neurocognitive impairment can indicate lesion proximity to functional networks, risk of postoperative deficits, and potential for functional improvement. At time of initial presentation, up to 90% of patients with glioma exhibit neurocognitive impairment (Taphoorn & Klein, 2004; Tucha, Smely, Preier, & Lange, 2000). Impairment may be found within a variety of domains, with memory and executive functioning comprising the most frequently impacted, though considerable interindividual variability exists. Numerous clinical and demographic factors are known to contribute to variability in neurocognitive functioning across patients with glioma. These include lesion size and location, tumor histology, tumor molecular characteristics, germ line genetic factors, and cognitive reserve or level of preillness functioning (Wefel, Noll, & Scheurer, 2016).

While focal patterns on neurocognitive testing may be evident (e.g., aphasic syndromes in cases with left perisylvian tumors), impairment tends to be more nonspecific in patients with primary brain tumors as compared to those with stroke (Anderson, Damasio, & Tranel, 1990). This relates, at least in part, to differences in pathophysiology. Ischemic stroke typically results in rapid and focal destruction of generally discrete neuronal populations. In contrast, primary brain tumors grow more slowly, often initially displacing structures rather than directly and immediately causing tissue damage. This helps explain false lateralizing signs sometimes observed in patients with brain tumors, as mass effect of large tumors may result in cerebral dysfunction in the hemisphere contralateral to tumor location. The slower rate of neuronal injury in patients with brain tumors can also allow for reorganization of function, mitigating the impairment that might be expected based solely upon lesion location and size. Even across glioma histologies, variation in growth kinetics appears to influence neurocognitive outcomes. Specifically, patients with a more rapidly proliferating genetic subtype of malignant glioma (i.e., IDH1 wild type) tend to exhibit broad neurocognitive impairment at a greater severity than their less aggressive counterparts (Wefel, Noll, Rao, & Cahill, 2016). Further, this appears related to differences in the extent of disruption to the structural connectome across disease subtypes, suggesting that the more rapidly growing lesions may allow for less neuroplastic reorganization of function (Kesler, Noll, Cahill, Rao, & Wefel, 2017). Despite these considerations, infiltrating glioma tend to disrupt broad networks given that tumor cells can invade and disrupt circuits both proximal and distal to the tumor mass.

In addition to determining the pattern of neurocognitive impairment and extent of cerebral dysfunction related to the presence of primary brain tumor, preoperative neuropsychological assessment is also useful in determining the feasibility of other preoperative neuromedical procedures involved in surgical planning, as well as in the planning of intraoperative brain mapping itself. Patients presenting with severe neurocognitive deficits (e.g., profound aphasia) or altered level of alertness may be questionable candidates for preoperative and intraoperative mapping, and paradigms utilized in these procedures must be carefully selected and/or modified in such cases. Preoperative testing is also necessary to establish a baseline point of comparison for follow-up examination in the postoperative period, which is particularly important in patients with glioma given that most present with impairment at time of initial presentation.

Neurocognitive decline following neurosurgical resection of brain tumors is common, with indication that over 60% of patients show worsening in the near postoperative period (Noll et al., 2015; Satoer et al., 2012). Surgical resection of brain tumors may damage healthy tissue surrounding the tumor, which can result in more focal dysfunction than may be present at the time of initial preoperative presentation. Indeed, resections near or involving eloquent regions important to speech and memory tend to be associated with declines in these functions. In addition, tumors in the left hemisphere tend to harbor greater risk of postoperative worsening than right hemisphere lesions (Noll et al., 2015). However, postoperative decline is not uncommon following resection of right hemisphere glioma, even in functions more associated with the contralateral hemisphere, such as language and verbal memory. These more diffuse neurocognitive changes likely relate to disruption of distributed networks and potential nonspecific effects of surgery such as postoperative edema.

Neuropsychological testing can also assist neurosurgeons in understanding surgical risks to neurocognitive functioning. Specifically, preoperative neuropsychological testing is a strong predictor of postoperative outcomes in specific domains (Gehring, Sawyer, Etzel, Lang, & Wefel, 2011a; Gehring, Sawyer, Etzel, Lang, & Wefel, 2011b). Patients with greater preoperative levels of memory and language functioning exhibit increased risk of postoperative decline in these domains. Postoperative neurocognitive functioning also harbors significant prognostic value. Overall and progression free survival time is highly associated with degree of neurocognitive impairment (Armstrong et al., 2013; Johnson et al., 2012; Noll, Garbarino, Turner, Verhaak, & Wefel, 2014).

Long-term neuropsychological follow-up and mitigation of neurocognitive impairment is imperative throughout the disease course. While some spontaneous recovery of functioning tends to occur after initial postoperative decline, a sizeable proportion of patients do not return to baseline preoperative levels and most patients with glioma will experience deterioration related to progression of disease and/or adjuvant treatments received. In patients with GBM, around 30% exhibit decline during treatment with temozolomide, despite radiographically and clinically stable disease (Wefel & Schagen, 2012). Radiation therapy for central nervous system neoplasms also comprises a unique set of risks. Some patients may experience acute but transient neurocognitive decline following radiotherapy, while others can develop subacute or even late neurotoxicity in the months to years after completion of radiation (Keime-Guibert, Napolitano, & Delattre, 1998). Interestingly, worsening on formal neuropsychological assessment tends to precede evidence of radiographic progression (Meyers & Hess, 2003). Impairment also frequently entails consequences upon social and occupational functioning, with evidence indicating that neurocognitive impairment impacts health-related quality of life more than physical or other neurological symptoms (Henriksson, Asklund, & Poulsen, 2011).

Emotional functioning represents another important consideration in the management of patients with primary brain tumors. While psychiatric comorbidities are common in various cancer populations, patients with glioma have the greatest risk of affective complications, with depression representing the most prevalent problem (Acquaye, Vera-Bolanos, Armstrong, Gilbert, & Lin, 2013). Rates vary widely across studies, though estimates suggest that over 40% (Arnold et al., 2008), and perhaps as many as 90% of patients with glioma (Pelletier, Verhoef, Khatri, & Hagen, 2002) exhibit significant symptoms of depression during the disease course. Increased incidence of depression is associated with being newly diagnosed and higher tumor grade, with GBM patients showing the most frequent depressive symptoms of all histologies. In addition to comprising the single largest influence upon quality of life in patients with malignant glioma (Noll et al., 2015), evidence indicates that depression also has a detrimental impact upon survival (Gathinji et al., 2009; Litofsky et al., 2004; Noll et al., 2014). Identification of neurocognitive and emotional symptoms via neuropsychological evaluation may facilitate early intervention, which may improve quality of life and potentially even prolong survival, though research regarding efficacious neuropsychological, psychosocial, and rehabilitative interventions in this population is lacking.

Brain Mapping

As noted earlier, the ability to localize eloquent or functional brain regions during presurgical planning and surgical intervention is essential for risk assessment and optimizing outcome. As noted previously, there are a number of localization techniques available, including nTMS, fMRI, and DCS. Effective use of these tools requires a high level of technical expertise, specialized knowledge about the conceptual and functional neuroanatomical underpinnings of the systems being mapped, and understanding of the methodological and interpretative limitations unique to each mapping method. A multidisciplinary approach to presurgical mapping is ideal with neuroradiologists, neurosurgeons, neurologists, and neuropsychologists each contributing specialized knowledge to the process. Neuropsychologists can offer a sophisticated view of the conceptual and anatomical correlates to higher order neurocognitive functions (e.g., language) to allow for more informed task selection and interpretation of the mapping results. Presurgical neuropsychological assessment plays an essential role in assessing the ability level of the patient and determining suitability for participation in different mapping efforts. Neuropsychologists also play an important role in the operating room, with recent evidence indicating higher rates of gross total resection, shorter duration of surgery, and lower rates of unexpected residual tumor when neuropsychologists are used for intraoperative mapping (Kelm et al., 2017).

Motor, sensory, and language functions are the primary focus of brain mapping for surgical purposes in patients with primary brain tumors. Mapping motor and sensory systems are relatively straightforward and predictable with respect to anatomic localization, while language is more complex and variable. An accurate and thorough understanding of the neuro-cognitive processes and underlying neuroanatomy involved in language is necessary to effectively design sound mapping paradigms, make informed decisions on task choice, and accurately interpret mapping results. A brief neurofunctional review of language is provided, followed by a more focused discussion of research regarding specific mapping methods. Deliberate focus is paid to fMRI and DCS, which represent the methods most often involving active participation of neuropsychologists in the development of paradigms, task selection, and clinical implementation. While clinicians are increasingly mapping broader neurocognitive domains (e.g., visuoperception, executive functions), the vast majority of pre- and intraoperative mapping cases involve the localization of language functions given the availability of language mapping paradigms and the importance of language to maintaining independence and functional capacity. Accordingly, the following discussion of brain mapping centers upon language, though the general considerations regarding practical application remain pertinent to other mapping situations.

Functional Neuroanatomy of Language

Understanding of language and its functional neuroanatomy advanced considerably in the time since the first localization model was proposed by Carl Wernicke in the early 19th century. The classical model localizes language to two main areas: (1) an area in the left inferior frontal lobe (now known as Broca's area) where the motor images responsible for speech production are stored and (2) an area in the left posterior temporal lobe (now known as Wernicke's area) where the sound images responsible for speech perception are stored (Tremblay & Dick, 2016). While the basic principles and predictions of this classic model remain true, it is now considered oversimplified and inadequate in explaining how language is represented in and produced by the brain.

Language involves a far more distributed network than just Broca's and Wernicke's areas in the left hemisphere, with distributed areas within a broader language network specializing in the processing of specific aspects of language. Fig. 9.1 illustrates the major left hemisphere regions involved in language and their respective functions. Orthography, or the analysis of written letter combinations, involves an area of the brain in the ventral occipitotemporal fusiform region known as the visual word form area. This region is reliably activated in response to viewing words or word-like non-word stimuli (Cohen et al., 2002; Dehaene et al., 2001, 2004; Polk & Farah, 2002) and shows a graded activation pattern in response to orthographic regularity or letter sequencing probability (Binder, Medler, Westbury, Liebenthal, & Buchanan, 2006). Damage to this region results in a pure alexia whereby patients can identify single letters but are unable to read

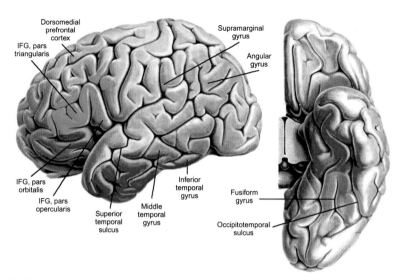

FIGURE 9.1 The major left hemisphere regions involved in language and their respective functions. Yellow, phoneme and auditory word form perception areas; red, semantic storage and retrieval systems; blue, phonological access and phonological output systems; light blue, visual word form perception area; green, general verbal retrieval, selection, and working memory functions. *Reproduced with permission from Binder (2011).*

letter strings. Phonology or the processing of speech sounds localizes along the superior temporal gyrus (STG). In fMRI studies, the more anterior STG activates preferentially to speech sounds compared to nonspeech sounds (De Witt & Rauschecker, 2012). Bilateral damage to this area can result in pure word deafness in which basic hearing is preserved but speech sound recognition is impaired. More posteriorly in the STG are areas involved in the selection of phoneme codes for word assembly. fMRI studies show activation in the posterior STG in response to a wide range of tasks requiring phoneme retrieval (Binder, 2015). Damage to the posterior STG can result in impairments in selecting and ordering phonemes resulting in frequent phonemic paraphasic errors, though basic phoneme perception and auditory comprehension are unaffected. Semantics (i.e., the processing of meaning) involves a distributed dominant hemisphere (i.e., typically left lateralized) network including the posterior inferior parietal lobe (angular gyrus), lateral and ventral temporal lobe, and areas in the dorsolateral and ventromedial prefrontal cortex (Binder, Desai, Graves, & Conant, 2009). Lesions to these areas can cause a wide variety of semantic impairments, though in some cases damage to aspects of the temporal lobe can cause highly circumscribed impairments (e.g., affecting specific object categories such as fruits and vegetables, or tools), suggesting possible category-specific representations.

Despite these now well-known neurofunctional correlates of language processing, some clinicians still rely heavily upon the classic view of language, placing great emphasis on identifying and mapping Broca's and Wernicke's areas at the exclusion of the broader language network. When mapping language, the authors encourage clinicians to adopt a framework conceptualizing language as a distributed network of interacting linguistic systems in which damage to various regions within the larger network have the potential to cause impairment.

Sources of Localization

The first level of analysis in functional localization and surgical risk assessment comes from structural imaging, assessing proximity of tumor to potential eloquent areas based on known brain—behavior relationships. It is essential to integrate structural localization of the lesion with the presenting symptoms and neuropsychological findings, which often have localizing value and can indicate proximity to eloquent regions. For example, a patient presenting with transient aphasia that resolves with steroid treatment would suggest a lesion within proximity to language areas, though the transient nature of the deficit suggests that risk of surgically acquired language impairment may be avoided with careful surgical planning and intraoperative mapping. This contrasts with cases where more persistent deficits are present, in which predictive value is less clear.

Although basic structural scans and presenting neurocognitive signs and symptoms are informative, resecting brain tumors near eloquent cortex requires more exact determination of tumor proximity to the cortical gray and subcortical white matter involved in the particular neurocognitive process of concern. fMRI can be used to map the cortical surface and deep gray matter while DTI allows for detailed anatomic examination of potential functional white matter tracts. Combining functional gray matter localization from fMRI with white matter localization from DTI offers a more complete picture of functional anatomy and allows for more detailed determination of surgical approach and boundaries, while identifying possible targets for intraoperative DCS. DCS remains the gold standard for functional localization during surgery and can be used to test functional contributions of both gray and subcortical white matter. Each of the sources of localizing information described earlier are often insufficient for neurosurgical planning when used in isolation; however, when combined and the results converge, greater confidence is placed in the data for surgical decision-making.

Functional Magnetic Resonance Imaging

fMRI is a noninvasive technique for measuring brain activity by detecting changes in blood oxygenation and blood flow in response to neural activity. fMRI is increasingly used in the clinical setting to map motor, sensory, and higher level neurocognitive functions such as language. There are numerous language paradigms described in the literature with very little standardization or consensus on best practice for task selection. There are also very few normative studies on these different tasks, making it difficult to form expectations about typical activation patterns and complicating the detection and interpretation of deviant or absent patterns of activation. A detailed discussion of available language paradigms is beyond the scope of this chapter; however, the interested reader is directed to Black et al. (2017) for a review of some of the more commonly used tasks.

Support for the use of fMRI in neurosurgical planning comes primarily from studies examining concordance with invasive language mapping methods, such as the Wada test or intraoperative DCS, though the majority of studies involve epilepsy rather than brain tumor populations. In patients with epilepsy, fMRI has shown relatively strong concordance (85% agreement) with Wada testing in determining laterality of language dominance (Janecek et al., 2013). The concordance between fMRI and DCS language mapping has been more variable, with reported rates of 59%−100% sensitivity and 0%−97% specificity (Giussani et al., 2009). Factors explaining this variability are not well understood but may include differences between fMRI and DCS tasks, use of inappropriate fMRI control tasks that inadvertently subtract out essential language activation (e.g., rest as a control), differences in fMRI analysis methods, and potential effects of tumor on blood oxygen level−dependent (BOLD) signal detection. It should also be emphasized that fMRI reveals all brain regions activated during a particular task and does not differentiate between essential, supportive, or even unrelated areas of activation.

The ultimate validity test is whether fMRI can predict and mitigate neurocognitive outcome. Some data support the predictive utility of fMRI, including studies indicating that the degree of language lateralization to the surgical hemisphere predicts language (Bonelli et al., 2012; Sabsevitz et al., 2003) and verbal memory outcome in left temporal lobectomy patients (Binder, Sabsevitz, et al., 2008). Other studies have shown that fMRI language and memory mapping are superior to Wada testing for prediction of postoperative outcome in epilepsy patients (Binder et al., 2008; Dupont et al., 2010; Sidhu et al., 2015). It is important to note that it remains unclear whether these findings generalize to brain tumor patients or resections to other areas of the brain. In addition, while data suggest that lateralization of function via fMRI is predictive of outcome, to our knowledge there are no studies showing fMRI can predict outcome at a more precise level (e.g., voxel or cluster), which is desired for surgical planning purposes.

Various challenges specific to using fMRI for surgical mapping in patients with brain tumors should be considered. A critical limitation is the potential for neurovascular uncoupling (NVU). BOLD contrast fMRI assumes tight coupling between neural activity and cerebral blood flow changes; however, brain tumors can disrupt regional cerebral vasoactivity resulting in reduced or lost BOLD signal near the tumor, despite the presence of neuronally active tissue (Agarwal et al., 2016; Holodny et al., 2000; Pillai & Zacá, 2011; Schreiber, Hubbe, Ziyeh, & Hennig, 2000; Ulmer et al., 2003). NVU can have serious consequences in surgical planning, as false negative areas may be mistakenly labeled as "safe" and included in the surgical resection. Breath-hold cerebrovascular reactivity (CVR) mapping can be done to assess for NVU and is becoming a standard in clinical fMRI mapping (Zacá, Jovicich, Nadar, Voyvodic, & Pillai, 2014). During a CVR test, the patient induces hypercapnia by performing breath-holds, resulting in vasodilation and expected BOLD changes throughout the brain. Signal dropout indicating reduced CO_2 response suggests possible areas of NVU.

As alluded to above, another challenge of using fMRI in surgical mapping is that the technique identifies all brain regions that contribute to a task, including those essential to the process of interest, but also areas involved in supportive processes and even areas unrelated to the function of interest. Accordingly, the relevance of the activation is not always clear. Despite this, fMRI activation maps can be useful in the identification of potential targets for intraoperative DCS. Finally, some patients with brain tumors are unable to adequately perform the fMRI tasks due to preexisting problems (e.g., sensory impairment) or disease related deficits (e.g., aphasia) which are often elicited on preoperative neuropsychological testing. In such cases, considerable modification of tasks may be necessary (e.g., aural vs visual stimulus presentation, task simplification) and other sources of mapping should be considered (e.g., nTMS).

Direct Cortical Stimulation

DCS is an invasive technique used to localize functional tissue during awake brain surgery. Following craniotomy and exposure of the brain, the patient is awakened by the anesthesia team and DCS mapping is conducted through the coordinated efforts of the surgical team, including the neuropsychologist. The neurosurgeon utilizes a specialized probe (most commonly an Ojemann stimulator) to administer a focal electrical current to different areas, causing temporary excitation or a reversible lesion (i.e., transient disruption in neural activity). Motor, sensory, and higher level neurocognitive functions can be tested by the neuropsychologist, with stimulus presentation (e.g., line drawings of objects) coordinated with the administration of electrical stimulation to the exposed brain. Stimulation sites where a positive (e.g., motor/sensory activity) or negative (e.g., disruption of motor movement or speech) response is provoked are marked and avoided if possible during surgery. Unlike fMRI, DCS allows for testing of white matter tracts in addition to the cortical surface. fMRI and DTI can be used to identify targets for DCS, resulting in improved efficiency in the operating room.

A number of studies have supported the efficacy of DCS, with a large scale meta-analysis indicating that patients with glioma who underwent surgical resection with DCS mapping had greater extent of resection (75% vs 58% gross total resection) and lower rates of permanent severe neurological deficits (3.4% vs 8.2%) than those without DCS (De Witt Hamer, Robles, Zwinderman, Duffau, & Berger, 2012). Nonetheless, questions remain whether resecting a positive DCS site necessarily leads to postoperative deficit or whether sparing such sites protects against a deficit. Complicating this line of study is the lack of comprehensive, systematic neuropsychological testing at different postoperative time points. The timing of neuropsychological assessments in these studies is critical given the high rate of deficits seen immediately postoperatively with subsequent improvement within a few weeks.

Despite its usefulness in maximizing safe resection and mitigating surgical risk, DCS is not without limitations. It is an invasive technique requiring the patient to cooperate and maintain focus for potentially long periods of time, and some patients have difficulty tolerating the procedure. There is also risk of inducing seizures, which can limit how high the amperage of the stimulator can be set, potentially resulting in insufficient stimulation intensity (and false negatives) in some cases. After-discharges can also occur as an unwanted side effect of stimulation. After-discharges result in transient neuronal disruption and tend to be characterized by rhythmic discharges of spikes and sharp waves on electrocorticography. Accordingly, close coordination with a neurophysiologist is essential to differentiate the effects of after-discharges from stimulation effects. When selecting stimulation parameters, it must also be considered that spread of higher intensity current can disrupt more distant cortical sites, limiting confidence in the focality of positive hits. A further limitation is that DCS does not allow examination of larger neural networks or the ability to test multiple nodes simultaneously. It is possible that stimulation at one site might be insufficient to elicit a deficit, while simultaneous stimulation of 2 or more sites might produce sufficient disruption of the network needed to produce a deficit.

PRACTICAL APPLICATIONS

Preoperative Neuropsychological Assessment

Resection of brain tumors often occurs very shortly following the initial identification of the lesion via neuroimaging, particularly for cases involving suspected malignant glioma. Accordingly, neuromedical workup for surgical planning, including neuropsychological evaluation, is frequently conducted under very limited time constraints. That is, testing is typically performed within a week from the scheduled neurosurgery and sometimes even the day prior to resection. The testing appointment is also typically only one of a number of scheduled visits for the patient on a given day. Given these logistical constraints, flexibility in scheduling and a tailored assessment are often necessary. An efficient assessment also serves to improve the tolerability of testing, as many patients with brain tumors present with significant impairment and fatigability.

Despite the logistical considerations noted above, it is nonetheless essential for the neuropsychological assessment to broadly sample across neuro-cognitive and neurobehavioral domains, as patients with even relatively focal tumors may present with broad and nonlocalizing deficits. Core domains include attention, processing speed, visuospatial, language, executive functioning, memory, and motor functions. Given that patients are typically followed postoperatively and at regular intervals throughout the disease course, consideration of measures that are repeatable is of value. When possible, measures with alternate forms are preferred to minimize practice effects. Equally important is selection of measures with sound psychometric data and reliable change indices (RCIs), which assist in evaluating patients for intraindividual longitudinal change in neuropsychological status. While a thorough discussion of RCIs is beyond the scope of this chapter, we advocate for methods that account for practice effects, particularly given the often brief test—retest interval from presurgical to postoperative evaluations (for a comparison of RCI models, see Temkin, Heaton, Grant, & Dikmen, 1999). Table 9.1 displays an example of a relatively brief core battery of tests with good psychometric properties, many of which also have alternate forms available, and existing RCIs. Of course, it is often necessary to supplement any core battery with additional measures to answer specific referral questions or better determine functioning in a particular domain.

Assessment of mood, quality of life, and symptoms common to patients with brain tumors is also important for a comprehensive neuropsychological evaluation with this population. As previously described, in addition to baseline neurocognitive functioning, assessment of emotional symptoms conveys important prognostic information. In addition, mood and neurocognitive functioning contribute independently to patient quality of life. A myriad of other symptoms and comorbidities may also influence neurocognitive functioning, mood, and quality of life, such as fatigue, sleep disturbance, headaches, seizures, irritability, and sensory changes. Given that malignant glioma represent a progressive disease with poor prognosis, early identification of symptoms, alterations in quality of life, and changes in mood and neurocognitive functioning are paramount to facilitate early intervention and preserve patient well-being and daily functioning to the greatest extent possible. A few commonly utilized mood, symptom, and quality of life measures appropriate for brain tumor evaluations are also listed in Table 9.1.

When preoperative neuropsychological evaluations are conducted to determine the feasibility of brain mapping, additional measures may be needed. For instance, a formal evaluation of handedness (e.g., the Edinburgh Handedness Inventory) should be included when neuropsychological evaluation is utilized to inform presurgical fMRI language mapping. The degree of right or left handedness of a patient can aid interpretation of fMRI language

TABLE 9.1 A Core Battery for Neuropsychological Assessment in Patients With Brain Tumors

Domain	Test/Measure
Premorbid Functioning	Wechsler Test of Adult Reading
Memory	Hopkins Verbal Learning Test-Revised
	Brief Visuospatial Memory Test-Revised
Language	Boston Naming Test
	MAE: Controlled Oral Word Association
	MAE: Token Test
Attention/Processing Speed	WAIS-IV Digit Span
	WAIS-IV Arithmetic
	WAIS-IV Coding
	WAIS-IV Symbol Search
	Trail Making Test Part A
Executive Functioning	Trail Making Test Part B
	WAIS-IV Similarities
Visuospatial	WAIS-IV Block Design
Motor	Lafayette Grooved Pegboard
Mood/Symptom/Quality of Life	Beck Depression Inventory-II
	Beck Anxiety Inventory
	MD Anderson Symptom Inventory-Brain Tumor
	EORTC QLQ C30/BN20

mapping results in which localization of function does not conform to typical hemispheric expectations (Bookheimer, 2007). In addition, neuropsychological testing can determine the feasibility of the fMRI and intraoperative mapping paradigms for a given patient. For example, a patient with frank expressive aphasia may be unable to perform language tasks in the scanner or during intraoperative DCS. Accordingly, testing patients on tasks analogous to the fMRI and intraoperative paradigms can inform the selection of appropriate tasks for use during brain mapping.

Postoperative and Follow-Up Neuropsychological Assessment

Proper timing of postoperative neuropsychological assessment is critical. Patient performances during evaluations conducted shortly following neurosurgical intervention (e.g., days to a week after surgery) may be adversely affected by transient complications and side effects of surgery, such as resolving edema and the impact of pain medication. Indeed, patients tend to show rapid improvement in neurocognitive and physical functioning in the weeks following resection of intracranial lesions despite initial postoperative worsening (Duffau, Taillandier, Gatignol, & Capelle, 2006; Rostomily, Berger, Ojemann, & Lettich, 1991). However, patients with malignant glioma typically initiate further treatment, such as chemoradiation, in the weeks following neurosurgical resection. These therapies can have a detrimental effect upon neurocognitive functioning as noted earlier. Accordingly, the timing of neuropsychological evaluation in the postoperative period is dictated by the referral question and goals of assessment. In many cases, conducting neuropsychological evaluation within 3–5 weeks after resection allows for adequate spontaneous recovery as well as assessment of functioning prior to initiation of therapies with potential adverse impact on neuropsychological status. For those receiving radiation therapy, repeat assessment is indicated following completion of radiation to monitor for treatment related changes. Similarly, adjuvant chemotherapy and eventual tumor progression can both impact neurocognitive functioning and monitoring should continue at regular intervals throughout adjuvant treatment and the survivorship period.

Ultimately, the objective of well-planned and executed serial neuropsychological evaluations in a brain tumor setting is to provide information and resources to preserve or improve patient functioning and quality of life. As such, personalized and evidence-based recommendations at each point of evaluation are crucial. Patients who exhibit significant neurocognitive impairment can benefit from education and instruction regarding implementation of environmental modifications, external aids, and internal strategies for optimization of neurocognitive functioning (Ferguson et al., 2007; Gehring, Aaronson, Taphoorn, & Sitskoorn, 2010). For patients exhibiting more severe neurocognitive impairment, perhaps with accompanying daily functional limitations, neurorehabilitation is often recommended. Home health services may also be recommended for patients with neurocognitive impairment limiting autonomous self-care. A variety of pharmacotherapies (e.g., donepezil, memantine, methylphenidate, modafinil) to prevent and/or treat neurocognitive dysfunction have been studied with mixed evidence of success (Boele et al., 2013; Brown et al., 2013; Day et al., 2014).

Management of mood symptoms identified via neuropsychological assessment often involves referrals for individual psychotherapy and/or psychiatric consultation. As an adjunct, engagement in peer and professionally led support networks may be helpful. Complaints regarding fatigue and sleep disturbance are often elicited via interview or completion of inventories. When identified, education regarding sleep hygiene and energy conservation techniques may be provided, in addition to formal evaluation and treatment of sleep issues. Given the impact of brain tumor and treatment upon cognition, it is unsurprising that many patients experience disability. When demonstrably valid testing reveals neurocognitive impairment of a severity expected to interfere with work functioning, various workplace accommodations can be recommended. Importantly, feedback can be therapeutic in itself, as patient questions regarding neurocognitive functioning can be answered, and often times, anxieties allayed. Patients may also gain a sense of control over their situation through development and implementation of a targeted treatment plan.

Brain Mapping

Functional Magnetic Resonance Imaging

Before the patient even enters the scanner a number of steps should be considered to guide mapping efforts. Premapping preparation should include review of structural imaging and presenting history, neuropsychological testing results, and task-specific training. Reviewing structural imaging and medical records can help formulate hypotheses about what functions may be at risk based on anatomic proximity and presenting symptoms, which in turn can direct fMRI task selection to assess these at-risk functions. Neuropsychological testing can provide important information about the patient's functional level and help determine ability to participate in fMRI. Neuropsychological findings are also useful in identifying the neurocognitive processes disrupted by the tumor to further guide task selection, as such functions may be particularly vulnerable to surgical compromise. Task-based training familiarizes the patient with the fMRI paradigms prior to active scanning and allows the examiner to determine what tasks the patient can perform and whether accommodations need to be made (e.g., slowing down stimuli presentation rate, allowing an easier alternate response to items).

Most clinical fMRI scans use the subtraction method in which a control task is subtracted from an experimental task. The experimental condition is intended to engage or activate certain neurocognitive processes of interest (e.g., language), while the control task is intended to activate all the processes shared with the experimental condition (e.g., attentional systems, working memory, motor response) except for the process of interest. The choice of control task can have dramatic effects on activation patterns and is as important if not more so than selection of the activation paradigm. Many studies use passive fixation or rest as the control state, operating under the false assumption that brain is inactive or not engaged in the process of interest during rest. However, fMRI studies have shown that the same areas are active during semantic conceptual processing as those involved in the conscious resting state (Binder et al., 1999). Accordingly, when rest is used as a control, important language areas can be inadvertently subtracted out resulting in false negative findings (Binder, Swanson, Hammeke, & Sabsevitz, 2008). Passive control tasks, such as passive listening and passive viewing of visual stimuli, also run the risk of inadvertently subtracting out language activation due to incomplete disruption of ongoing language or conceptual processing from attentional drift. Sensory discrimination tasks (e.g., auditory tone or visual symbol discrimination) are preferable, as these can effectively control for basic auditory and visual processes, attention, and working memory without engaging language.

Another practical consideration is balancing tolerability and power of the fMRI design. fMRI signal contains a lot of noise and signal averaging is used to produce stable maps. Many clinical protocols emphasize short scanning time due to concerns about patient fatigue or employ multiple shorter paradigms to sample language. However, such choices appear to come with potential costs. Huettel and McCarthy (2001) used an event related visual stimulation paradigm and showed that only 50% of eventually activated voxels were activated at 50 trials, which is typical for clinical protocols, and it was not until around 150 trials that the number of activated voxels hit an asymptote. In addition, increasing number of trials resulted in greater spatial extent of activation. While it is unclear how these findings would generalize to language activation maps, it highlights the importance of power and the risk of an underpowered study not revealing the full extent of activation. Selection of appropriate control tasks may be even more challenging when attempting to map other neurocognitive domains (e.g., memory, executive functions) given the complexity of subprocesses involved in these tasks.

Neuropsychologists are increasingly called upon to not only aid in brain mapping task selection, but also administration and monitoring during the scan itself. Estimation or direct observation of task performance is essential, as performance in the scanner has been shown to effect extent of activation on language paradigms in both healthy individuals (Booth et al., 2003) and epilepsy patients (Weber et al., 2006). Unfortunately, many language paradigms used for clinical purposes do not include collection of measurable responses or allow for the monitoring of task performance in the scanner (e.g., covert word generation), limiting the examiner's ability to determine the adequacy of the data for analysis. This is particularly relevant when mapping clinical patients who may have significant neurocognitive impairment from their disease which can limit their in-scan performance. Post hoc testing outside the scanner can provide some information about the patient's ability to perform a given task, though fidelity to performance during actual scanning cannot be assumed. Accordingly, the authors advocate for selection of tasks in which live performance monitoring is possible.

Direct Cortical Stimulation

Following selection of tasks tapping the neurocognitive function of interest, it is essential to establish baseline level of performance prior to intraoperative DCS. Patients with brain tumors vary considerably with respect to their neurocognitive status and it is important to individualize the testing materials to their functional level. Baselining serves to both familiarize the patient with the testing procedures and determine the task items they can respond to reliably and accurately. Only items in which the patient consistently responds correctly should be included in the intraoperative DCS stimuli set. In this manner, the examiner has greater confidence that a disruption or deficit in performance during DCS is actually stimulation induced. In addition, electrical stimulation is generally limited to 5 seconds trains due to risk of inducing seizures. As such, only items that can be answered accurately within this time frame should be included.

DCS is often used to explore the cortical surface under the craniotomy exposure to assist in determining the best approach to a tumor. DCS can also be used to test areas bordering or approximating identified white matter tracts. However, there are periods during surgery where active resection is taking place without DCS. Testing of neurocognitive functions of interest during awake brain surgery is not necessarily restricted to active DCS. Recently, real-time neuropsychological testing (RTNT) has been proposed as an adjunct to DCS brain mapping (Skrap, Marin, Ius, Fabbro, & Tomasino, 2016). RTNT involves continuous neuropsychological testing throughout the surgical procedure to monitor for reversible or oscillating signs of neuropsychological dysfunction to identify an evolving deficit before it becomes more permanent. RTNT is associated with increased extent of resection and patient performance on RTNT during surgery closely resembled neuropsychological performance at 1-week postsurgery, suggesting positive prediction of outcome. We suggest a combination of targeted discrete DCS trials and RTNT for best outcome, though further validation of this combined approach is needed.

Much like fMRI, there is very little consensus on task selection in DCS. Automatic speech tasks, such as counting or reciting the Pledge of Allegiance, and visual object naming are commonly used during DCS, while more systematic, sophisticated, and multidimensional assessment is rarer. An individualized and functional anatomically informed approach to task selection is ideal. Specific functions deemed at risk via tumor proximity to known eloquent regions or the presence of deficits on preoperative neuropsychological evaluation should be considered candidates for mapping. For example, phonological retrieval would be at risk in a patient with a tumor in proximity of the posterior STG, arcuate fasciculus, and/or supramarginal gyrus, and inclusion of a speech repetition task should be considered. In this case, repetition of nonwords likely provides for more sensitive mapping than repetition of words or object naming (Sierpowska et al., 2016). Similarly, for left anterior temporal lobe tumors, an auditory descriptive naming paradigm might be included along with a visual object naming task, given report of more auditory than visual naming sites in this area (Hamberger, Goodman, Perrine, & Tamny, 2001). Auditory naming tasks are also useful for resections near semantic speech comprehension areas given the auditory comprehension demands of the task. Regardless, selection of tasks should be based upon careful inspection of tumor location, presenting neuropsychological functioning, and thorough understanding of functional network organization.

CASE EXAMPLE

A highly educated lawyer presented with a 5.5×3.5 cm^2 lobulated, centrally necrotic, ring enhancing mass in the left ventral temporo-occipital region, identified as GBM per postoperative histopathological study. Initial presenting symptoms were visual changes and problems with reading. Specifically, words no longer looked familiar to him and his reading slowed considerably. These difficulties resolved with steroid administration. Fig. 9.2 shows results of presurgical imaging. The mass was located in immediate proximity to the optic radiations on its superior medial border and ventral occipital fasciculus on the medial border, which has been implicated in reading. fMRI was performed using a panel of tasks, including rhyme decisions, silent word generation, listening comprehension, and reading comprehension. Results showed strong left hemisphere dominance for language with robust BOLD activation near the superior lateral margin of the tumor in the posterior aspect of left middle temporal gyrus. Also notable was activation from reading tasks near the anterior medial margin of the tumor in the left ventral temporo-occipital area, an area thought to be responsible for visual word recognition. Preoperative neuropsychological testing was notable for a partial right

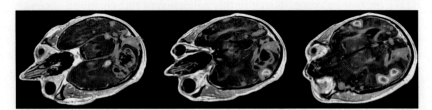

FIGURE 9.2 DTI was performed with a 3D isotropic technique on a 3 T scanner in 15 noncollinear directions. DTI fractional anisotropy maps are shown for three axial slices through the temporal lobes. BOLD fMRI language mapping using silent word generation, rhyming, listening comprehension, and reading comprehension tasks are also overlaid on the structural images. The optic radiations can be observed (green fibers) wrapping around the superior, medial, and posterior margins of the tumor as they course toward occipital visual cortex. fMRI language mapping showed left dominant activation observed for phonologic and semantic language tasks (green to red activation blobs lateral and superior to the tumor). Activation within the posterior aspect of left middle temporal gyrus, in the area thought to be responsible for semantic storage and retrieval, has immediate proximity to the superior lateral margin of the tumor. Activation in the left ventral temporo-occipital area, an area thought to be responsible for visual word recognition, is approximately 5 mm from the anterior medial margin of the tumor.

homonymous hemianopsia. No evidence of aphasia or other cognitive dysfunction was found on exam.

The patient was taken to the operating room and a modified version of an auditory descriptive naming task was administered. For this task, the patient reads a description of an object and is required to respond by stating the name of the object described (e.g., read "what a king wears on his head" and respond with "crown"). DCS of the anterior border of the tumor that corresponded to the area of BOLD activation during reading based tasks produced a transient pure alexia (i.e., inability to read the sentence), which was interpreted as resulting from disruption of the adjacent visual word form area or the associated white matter connections (i.e., ventral occipital fasciculus). DCS along the superior lateral border that corresponded to an area of language BOLD activation produced a few semantic paraphasic errors. The subsequent surgical approach mitigated damage to these areas essential to reading and descriptive naming.

Following recovery from surgery, neuropsychological testing revealed no changes or deficits in reading or language functions and the patient successfully returned to his law practice. In this case, appropriate task selection was essential. If a visual object naming task was used instead of the descriptive naming task, the patient would have performed without error during DCS near the visual word form area since reading is not involved in visual object naming. Further, if the visual word form area and its underlying white matter connections were compromised during surgery, a severe reading deficit would be probable. This case also exemplifies the importance of identifying DCS tasks that tap functions particularly relevant to a given patient (e.g., mapping math in an accountant, manual dexterity in a guitarist).

FUTURE DIRECTIONS

Traditional paper-and-pencil neuropsychological tests remain fundamental tools for both clinical practice and research involving patients with primary brain tumors. However, pressures exist to streamline and expedite clinical practice and to improve standardization to facilitate data sharing and interpretability of results. To this end, computerized neuropsychological assessment may be useful, and neuropsychologists should become aware of the potential benefits and pitfalls of such platforms. Essential to the development of these next generation tools will be establishing the adequacy of their reliability and validity with brain tumor populations.

Very little is known about the psychometric and normative properties of most brain mapping paradigms. Many paradigms remain unstandardized, with stimuli and task demands varying widely across institutions. Administration of tasks during DCS often involves presentation of stimuli on flashcards (e.g., visual object naming), which can be cumbersome and inefficient for use in the intraoperative environment. Frequently, timing of DCS is loosely tied to stimulus presentation, as it is common practice for neurosurgeons to begin stimulation when the patient initiates a prompt (e.g., stating "this is a..." in the case of a visual object naming task). Tablet-based platforms may represent an avenue for improved standardization of paradigms. Tablet-based administration might involve coordinated presentation of stimuli with an auditory cue for timing DCS, in addition to facilitating flexible task design, and enabling capture of more nuanced patient response data (e.g., vocal reaction times). Despite the standardization and normative limitations of current brain mapping paradigms noted earlier, numerous tasks exist for mapping language functions that may be easily adapted for tablet-based administration. In contrast, other domains remain lacking in terms of available paradigms, despite increasing interest in mapping broader neuro-cognitive functions. Future work is needed to develop and validate these paradigms for both fMRI and DCS mapping.

Resting state fMRI is an emerging technique with recent experimental application to brain mapping in patients with brain tumors. Extraction of various networks (including language and motor) appears possible from resting state scans, with early evidence supporting concordance with intraoperative DCS (Wang et al., 2015). This technique may be particularly useful in patients in which task-based fMRI paradigms are not feasible, such as those with sensory or severe neurocognitive impairment. Another increasingly common imaging modality is intraoperative MRI, allowing neuroimaging to be performed in the operating room during resection. This facilitates assessment of changes in lesion characteristics and potential cerebral shift as the operation proceeds, as well as ongoing determination of resection proximity to eloquent cortex and important white matter pathways. Potentially, intraoperative imaging may include fMRI to update brain mapping results throughout the surgery.

Ultrasonic surgical aspirators use high energy sound waves to remove tissue and are commonly used to surgically resect brain tumors. These aspirators may be useful not only for the dissection of tissue but potentially also for mapping purposes. That is, aspirators have been shown to transiently interfere with or inhibit motor (Carrabba et al., 2008) and possibly language (Sierpowska et al., 2016) function. As such, combining RTNT with ultrasonic mapping may offer an opportunity to continuously map sensorimotor and language functions without disrupting surgical work flow.

Neuropsychology is now recognized as essential to the integrated care of patients with brain tumors. However, the traditional focus of practice and research with this population centers largely on assessment more so than intervention. Efficacious interventions for the neurocognitive sequelae of patients with brain tumors are sorely needed and neuropsychologists are well positioned to lead this important line of inquiry and practice.

REFERENCES

Acquaye, A. A., Vera-Bolanos, E., Armstrong, T. S., Gilbert, M. R., & Lin, L. (2013). Mood disturbance in glioma patients. *Journal of Neuro-oncology*, *113*, 505−512.

Agarwal, S., Sair, H. I., Airan, R., Hua, J., Jones, C. K., Heo, H.-Y., . . . Pillai, J. J. (2016). Demonstration of brain tumor-induced neurovascular uncoupling in resting-state fMRI at ultrahigh field. *Brain Connectivity*, *6*(4), 267−272.

Anderson, S. W., Damasio, H., & Tranel, D. (1990). Neuropsychological impairments associated with lesions caused by tumor or stroke. *Archives of Neurology*, *47*, 397−405.

Anderson, S. W., & Ryken, T. C. (2008). Intracranial tumors. In J. E. Morgan, & J. H. Ricker (Eds.), *Textbook of clinical neuropsychology* (pp. 578−587). New York: Taylor & Francis.

Armstrong, T. S., Wefel, J. S., Wang, M., Gilbert, M. R., Won, M., Bottomley, A., . . . Mehta, M. (2013). Net clinical benefit analysis of Radiation Therapy Oncology Group 0525: A phase III trial comparing conventional adjuvant temozolomide with dose-intensive temozolomide in patients with newly diagnosed glioblastoma. *Journal of Clinical Oncology*, *31*(32), 4076−4084.

Arnold, S. D., Forman, L. M., Brigidi, B. D., Carter, K. E., Schweitzer, H. A., Quinn, H. E., . . . Raynor, R. H. (2008). Evaluation and characterization of generalized anxiety and depression in patients with primary brain tumors. *Neuro-Oncology*, *10*(2), 171−181.

Austermuehle, A., Cocjin, J., Reynolds, R., Agrawal, S., Sepeta, L., Gaillard, W. D., . . . Theodore, W. H. (2017). Language functional MRI and direct cortical stimulation in epilepsy preoperative planning. *Annals of Neurology*, *81*(4), 526−537.

Binder, J. R. (2011). fMRI of language systems: Methods and applications. In S. H. Faro, F. B. Mohamed, M. Law, & J. T. Ulmer (Eds.), *Functional neuroradiology: Principles and clinical applications* (pp. 393−417). Boston, MA: Springer.

Binder, J. R. (2015). The Wernicke area. *Neurology*, *85*(24), 2170.

Binder, J. R., Desai, R. H., Graves, W. W., & Conant, L. L. (2009). Where is the semantic system? A critical review and meta-analysis of 120 functional neuroimaging studies. *Cerebral Cortex*, *19*(12), 2767−2796.

Binder, J. R., Frost, J. A., Hammeke, T. A., Bellgowan, P. S. F., Rao, S. M., & Cox, R. W. (1999). Conceptual processing during the conscious resting state: A functional MRI study. *Journal of Cognitive Neuroscience*, *11*(1), 80−93.

Binder, J. R., Medler, D. A., Westbury, C. F., Liebenthal, E., & Buchanan, L. (2006). Tuning of the human left fusiform gyrus to sublexical orthographic structure. *NeuroImage*, *33*(2), 739−748.

Binder, J. R., Sabsevitz, D. S., Swanson, S. J., Hammeke, T. A., Raghavan, M., & Mueller, W. M. (2008). Use of preoperative functional MRI to predict verbal memory decline after temporal lobe epilepsy surgery. *Epilepsia*, *49*(8), 1377−1394.

Binder, J. R., Swanson, S. J., Hammeke, T. A., & Sabsevitz, D. S. (2008). A comparison of five fMRI protocols for mapping speech comprehension systems. *Epilepsia*, *49*(12), 1980−1997.

Black, D. F., Vachha, B., Mian, A., Faro, S. H., Maheshwari, M., Sair, H. I., . . . Welker, K. (2017). American Society of Functional Neuroradiology-recommended fMRI paradigm algorithms for presurgical language assessment. *American Journal of Neuroradiology*, *38*(10), E65−E73.

Boele, F. W., Douw, L., de Groot, M., van Thuijl, H. F., Cleijne, W., Heimans, J. J., . . . Klein, M. (2013). The effect of modafinil on fatigue, cognitive functioning, and mood in primary brain tumor patients: A multicenter randomized controlled trial. *Neuro Oncology*, *15*(10), 1420−1428.

Bonelli, S. B., Thompson, P. J., Yogarajah, M., Vollmar, C., Powell, R. H., Symms, M. R., ... Duncan, J. S. (2012). Imaging language networks before and after anterior temporal lobe resection: Results of a longitudinal fMRI study. *Epilepsia*, *53*(4), 639−650.

Bookheimer, S. (2007). Pre-surgical language mapping with functional magnetic resonance imaging. *Neuropsychology Review*, *17*(2), 145−155.

Booth, J. R., Burman, D. D., Meyer, J. R., Gitelman, D. R., Parrish, T. B., & Mesulam, M. M. (2003). Relation between brain activation and lexical performance. *Human Brain Mapping*, *19*(3), 155−169.

Brown, P. D., Pugh, S., Laack, N. N., Wefel, J. S., Khuntia, D., Meyers, C., ... Radiation Therapy Oncology Group (RTOG). (2013). Memantine for the prevention of cognitive dysfunction in patients receiving whole-brain radiotherapy: A randomized, double-blind, placebo-controlled trial. *Neuro Oncology*, *15*(10), 1429−1437.

Buchmann, N., Gempt, J., Stoffel, M., Foerschler, A., Meyer, B., & Ringel, F. (2011). Utility of diffusion tensor-imaged (DTI) motor fiber tracking for the resection of intracranial tumors near the corticospinal tract. *Acta Neurochirurgica*, *153*(1), 68−74.

Bulik, M., Jancalek, R., Vanicek, J., Skoch, A., & Mechl, M. (2013). Potential of MR spectroscopy for assessment of glioma grading. *Clinical Neurology and Neurosurgery*, *115*(2), 146−153.

Carrabba, G., Mandonnet, E., Fava, E., Capelle, L., Gaini, S. M., Duffau, H., & Bello, L. (2008). Transient inhibition of motor function induced by the Cavitron ultrasonic surgical aspirator during brain mapping. *Neurosurgery*, *63*, E178−E179.

Cohen, L., Lehéricy, S., Chochon, F., Lemer, C., Rivaud, S., & Dehaene, S. (2002). Language-specific tuning of visual cortex? Functional properties of the Visual Word Form Area. *Brain*, *125*(5), 1054−1069.

Day, J., Zienius, K., Gehring, K., Grosshans, D., Taphoorn, M., Grant, R., ... Brown, P. D. (2014). Interventions for preventing and ameliorating cognitive deficits in adults treated with cranial irradiation. *Cochrane Database Systematic Review*, *18*(12).

Dehaene, S., Jobert, A., Naccache, L., Ciuciu, P., Poline, J. B., Bihan, D. L., & Cohen, L. (2004). Letter binding and invariant recognition of masked words: Behavioral and neuroimaging evidence. *Psychological Science*, *15*(5), 307−313.

Dehaene, S., Naccache, L., Cohen, L., Bihan, D. L., Mangin, J.-F., Poline, J.-B., & Rivière, D. (2001). Cerebral mechanisms of word masking and unconscious repetition priming. *Nature Neuroscience*, *4*, 752.

De Witt, I., & Rauschecker, J. P. (2012). Phoneme and word recognition in the auditory ventral stream. *Proceedings of the National Academy of Sciences*, *109*(8), E505−E514.

De Witt Hamer, P. C., Robles, S. G., Zwinderman, A. H., Duffau, H., & Berger, M. S. (2012). Impact of intraoperative stimulation brain mapping on glioma surgery outcome: A meta-analysis. *Journal of Clinical Oncology*, *30*(20), 2559−2565.

Duffau, H. (2017). A two-level model of interindividual anatomo-functional variability of the brain and its implications for neurosurgery. *Cortex*, *86*, 303−313.

Duffau, H., & Mandonnet, E. (2013). The "onco-functional balance" in surgery for diffuse low-grade glioma: Integrating the extent of resection with quality of life. *Acta Neurochirurgica*, *155*(6), 951−957.

Duffau, H., Taillandier, L., Gatignol, P., & Capelle, L. (2006). The insular lobe and brain plasticity: Lessons from tumor surgery. *Clinical Neurology and Neurosurgery*, *108*(6), 543−548.

Dupont, S., Duron, E., Samson, S., Denos, M., Volle, E., Delmaire, C., ... Baulac, M. (2010). Functional MR imaging or Wada test: Which is the better predictor of individual postoperative memory outcome? *Radiology*, *255*(1), 128−134.

Ferguson, R. J., Ahles, T. A., Saykin, A. J., McDonald, B. C., Furstenberg, C. T., Cole, B. F., & Mott, L. A. (2007). Cognitive-behavioral management of chemotherapy-related cognitive change. *Psycho-Oncology, 16*, 772–777.

Gehring, K., Aaronson, N. K., Taphoorn, M. J., & Sitskoorn, M. M. (2010). Interventions for cognitive deficits in patients with a brain tumor: An update. *Expert Reviews in Anticancer Therapy, 10*(11), 1779–1795.

Gathinji, M., McGirt, M. J., Attenello, F. J., Chaichana, K. L., Than, K., Olivi, A., ... Quinones-Hinojosa, A. (2009). Association of preoperative depression and survival after resection of malignant brain astrocytoma. *Surgical Neurology, 71*(3), 299–303.

Gehring, K., Sawyer, A., Etzel, C., Lang, F., & Wefel, J. (2011a). Prediction of memory outcomes after resection of high-grade glioma. *Neuro-Oncology, 13*, iii73.

Gehring, K., Sawyer, A., Etzel, C., Lang, F., & Wefel, J. (2011b). Prediction of language outcomes after resection of high-grade glioma. *Neuro-Oncology, 13*, iii75.

Giussani, C., Roux, F.-E., Ojemann, J., Sganzerla, E., Pirillo, D., & Papagno, C. (2009). Is preoperative functional magnetic resonance imaging reliable for language areas mapping in brain tumor surgery? Review of language functional magnetic resonance imaging and direct cortical stimulation correlation studies. *Neurosurgery, 66*(1), 113–120.

Hamberger, M. J., Goodman, R. R., Perrine, K., & Tamny, T. (2001). Anatomic dissociation of auditory and visual naming in the lateral temporal cortex. *Neurology, 56*(1), 56.

Henriksson, R., Asklund, T., & Poulsen, H. S. (2011). Impact of therapy on quality of life, neurocognitive function and their correlates in glioblastoma multiforme: A review. *Journal of Neuro-Oncology, 104*(3), 639–646.

Holodny, A. I., Schulder, M., Liu, W.-C., Wolko, J., Maldjian, J. A., & Kalnin, A. J. (2000). The effect of brain tumors on BOLD functional MR imaging activation in the adjacent motor cortex: Implications for image-guided neurosurgery. *American Journal of Neuroradiology, 21*(8), 1415.

Huettel, S. A., & McCarthy, G. (2001). The effects of single-trial averaging upon the spatial extent of fMRI activation. *Neuroreport, 12*(11), 2411–2416.

Janecek, J. K., Swanson, S. J., Sabsevitz, D. S., Hammeke, T. A., Raghavan, M., Rozman, M. E., & Binder, J. R. (2013). Language lateralization by fMRI and Wada testing in 229 patients with epilepsy: Rates and predictors of discordance. *Epilepsia, 54*(2), 314–322.

Johnson, D. R., Sawyer, A. M., Meyers, C. A., O'neill, B. P., & Wefel, J. S. (2012). Early measures of cognitive function predict survival in patients with newly diagnosed glioblastoma. *Neuro-Oncology, 14*(6), 808–816.

Keime-Guibert, F., Napolitano, M., & Delattre, J. Y. (1998). Neurological complications of radiotherapy and chemotherapy. *Journal of Neurology, 245*(11), 695–708.

Kelm, A., Sollmann, N., Ille, S., Meyer, B., Ringel, F., & Krieg, S. M. (2017). Resection of gliomas with and without neuropsychological support during awake craniotomy—Effects on surgery and clinical outcome. *Frontiers in Oncology, 7*(176).

Kesler, S. R., Noll, K., Cahill, D. P., Rao, G., & Wefel, J. S. (2017). The effect of IDH1 mutation on the structural connectome in malignant astrocytoma. *Journal of Neuro-Oncology, 131*(3), 565–574.

Lacroix, M., Abi-Said, D., Fourney, D. R., Gokaslan, Z. L., Shi, W., DeMonte, F., ... Hess, K. (2001). A multivariate analysis of 416 patients with glioblastoma multiforme: Prognosis, extent of resection, and survival. *Journal of Neurosurgery, 95*(2), 190–198.

Litofsky, N. S., Farace, E., Anderson, F., Jr, Meyers, C. A., Huang, W., & Laws, E. R., Jr (2004). Depression in patients with high-grade glioma: Results of the Glioma Outcomes Project. *Neurosurgery, 54*(2), 358–367.

Louis, D. N., Ohgaki, H., Wiestler, O. D., & Cavenee, W. K. (2016). *World Health Organization histological classification of tumours of the central nervous system*. France: International Agency for Research on Cancer.

Mabray, M. C., Barajas, R. F., & Cha, S. (2015). Modern brain tumor imaging. *Brain Tumor Research and Treatment, 3*(1), 8−23.

Meyers, C. A., & Hess, K. R. (2003). Multifaceted end points in brain tumor clinical trials: Cognitive deterioration precedes MRI progression. *Neuro-Oncology, 5*(2), 89−95.

Meyers, C. A., Hess, K. R., Yung, W. A., & Levin, V. A. (2000). Cognitive function as a predictor of survival in patients with recurrent malignant glioma. *Journal of Clinical Oncology, 18* (3), 646−650.

Noll, K., Bradshaw, M., Weinberg, J., & Wefel, J. S. (2017a). Neurocognitive functioning is associated with functional independence in newly diagnosed patients with temporal lobe glioma. *Neuro-Oncology Practice*. Available from https://doi.org/10.1093/nop/npx028. (Advance online publication).

Noll, K. R., Bradshaw, M. E., Weinberg, J. S., & Wefel, J. S. (2017b). Relationships between neurocognitive functioning, mood, and quality of life in patients with temporal lobe glioma. *Psycho-Oncology, 26*(5), 617−624.

Noll, K. R., Garbarino, A., Turner, C., Verhaak, A. M. S., & Wefel, J. S. (2014). Depression and executive functioning in relation to survival among patients with glioblastoma, Abstract *Neuro-Oncology, 16*(Suppl. 5), v136.

Noll, K. R., Weinberg, J. S., Ziu, M., Benveniste, R. J., Suki, D., & Wefel, J. S. (2015). Neurocognitive changes associated with surgical resection of left and right temporal lobe glioma. *Neurosurgery, 77*(5), 777−785.

Ostrom, Q. T., Gittleman, H., Liao, P., Rouse, C., Chen, Y., Dowling, J., ... Barnholtz-Sloan, J. (2014). CBTRUS statistical report: Primary brain and central nervous system tumors diagnosed in the United States in 2007−2011. *Neuro-Oncology, 16*(Suppl. 4), iv1−iv63.

Papanicolaou, A. C., Rezaie, R., Narayana, S., Choudhri, A. F., Boop, F. A., & Wheless, J. W. (2017). On the relative merits of invasive and non-invasive pre-surgical brain mapping: New tools in ablative epilepsy surgery. *Epilepsy Research*. Available from https://doi.org/ 10.1016/j.eplepsyres.2017.07.002. (Advance online publication).

Pelletier, G., Verhoef, M. J., Khatri, N., & Hagen, N. (2002). Quality of life in brain tumor patients: The relative contributions of depression, fatigue, emotional distress, and existential issues. *Journal of Neuro-Oncology, 57*(1), 41−49.

Pillai, J. J., & Zacá, D. (2011). Clinical utility of cerebrovascular reactivity mapping in patients with low grade gliomas. *World Journal of Clinical Oncology, 2*(12), 397−403.

Polk, T. A., & Farah, M. J. (2002). Functional MRI evidence for an abstract, not perceptual, word-form area. *Journal of Experimental Psychology: General, 131*(1), 65−72. Available from https://doi.org/10.1037/0096-3445.131.1.65.

Rostomily, R. C., Berger, M. S., Ojemann, G. A., & Lettich, E. (1991). Postoperative deficits and functional recovery following removal of tumors involving the dominant hemisphere supplementary motor area. *Journal of Neurosurgery, 75*(1), 62−68.

Sabsevitz, D. S., Swanson, S. J., Hammeke, T. A., Spanaki, M. V., Possing, E. T., Morris, G. L., ... Binder, J. R. (2003). Use of preoperative functional neuroimaging to predict language deficits from epilepsy surgery. *Neurology, 60*(11), 1788.

Satoer, D., Vork, J., Visch-Brink, E., Smits, M., Dirven, C., & Vincent, A. (2012). Cognitive functioning early after surgery of gliomas in eloquent areas. *Journal of Neurosurgery, 117* (5), 831−838.

Schreiber, A., Hubbe, U., Ziyeh, S., & Hennig, J. (2000). The influence of gliomas and nonglial space-occupying lesions on blood-oxygen-level-dependent contrast enhancement. *American Journal of Neuroradiology, 21*(6), 1055.

Sidhu, M. K., Stretton, J., Winston, G. P., Symms, M., Thompson, P. J., Koepp, M. J., & Duncan, J. S. (2015). Memory fMRI predicts verbal memory decline after anterior temporal lobe resection. *Neurology, 84*(15), 1512.

Sierpowska, J., Gabarrós, A., Fernandez-Coello, A., Camins, À., Castañer, S., Juncadella, M., ... Rodríguez-Fornells, A. (2016). Words are not enough: Nonword repetition as an indicator of arcuate fasciculus integrity during brain tumor resection. *Journal of Neurosurgery, 126*(2), 435−445. Available from https://doi.org/10.3171/2016.2.JNS151592.

Skrap, M., Marin, D., Ius, T., Fabbro, F., & Tomasino, B. (2016). Brain mapping: A novel intraoperative neuropsychological approach. *Journal of Neurosurgery, 125*(4), 877−887. Available from https://doi.org/10.3171/2015.10.JNS15740.

Taphoorn, M. J. B., & Klein, M. (2004). Cognitive deficits in adult patients with brain tumors. *Lancet Neurology, 3*, 159−168.

Temkin, N. R., Heaton, R. K., Grant, I., & Dikmen, S. S. (1999). Detecting significant change in neuropsychological test performance: A comparison of four models. *Journal of the International Neuropsychological Society, 5*(4), 357−369.

Tran, B., & Rosenthal, M. A. (2010). Survival comparison between glioblastoma multiforme and other incurable cancers. *Journal of Clinical Neuroscience, 17*(4), 417−421.

Tremblay, P., & Dick, A. S. (2016). Broca and Wernicke are dead, or moving past the classic model of language neurobiology. *Brain and Language, 162*, 60−71.

Tucha, O., Smely, C., Preier, M., & Lange, K. W. (2000). Cognitive deficits before treatment among patients with brain tumors. *Neurosurgery, 47*, 324−334.

Ulmer, J. L., Krouwer, H. G., Mueller, W. M., Ugurel, M. S., Kocak, M., & Mark, L. P. (2003). Pseudo-reorganization of language cortical function at fMR imaging: A consequence of tumor-induced neurovascular uncoupling. *American Journal of Neuroradiology, 24*(2), 213.

Wang, D., Buckner, R. L., Fox, M. D., Holt, D. J., Holmes, A. J., Stoecklein, S., ... Baker, J. T. (2015). Parcellating cortical functional networks in individuals. *Nature Neuroscience, 18*(12), 1853−1860.

Weber, B., Wellmer, J., Schür, S., Dinkelacker, V., Ruhlmann, J., Mormann, F., ... Fernández, G. (2006). Presurgical Language fMRI in patients with drug-resistant epilepsy: Effects of task performance. *Epilepsia, 47*(5), 880−886.

Wefel, J. S., Noll, K. R., Rao, G., & Cahill, D. P. (2016). Neurocognitive function varies by IDH1 genetic mutation status in patients with malignant glioma prior to surgical resection. *Neuro-Oncology, 18*(12), 1656−1663.

Wefel, J. S., Noll, K. R., & Scheurer, M. E. (2016). Neurocognitive functioning and genetic variation in patients with primary brain tumours. *Lancet Oncology, 17*(3), e97−e108.

Wefel, J. S., & Schagen, S. B. (2012). Chemotherapy-related cognitive dysfunction. *Current Neurology and Neuroscience Reports, 12*(3), 267−275.

Wu, A. S., Witgert, M. E., Lang, F. F., Xiao, L., Bekele, B. N., Meyers, C. A., ... Wefel, J. S. (2011). Neurocognitive function before and after surgery for insular gliomas. *Journal of Neurosurgery, 115*(6), 1115−1125.

Zacà, D., Jovicich, J., Nadar, S. R., Voyvodic, J. T., & Pillai, J. J. (2014). Cerebrovascular reactivity mapping in patients with low grade gliomas undergoing presurgical sensorimotor mapping with BOLD fMRI. *Journal of Magnetic Resonance Imaging, 40*(2), 383−390.

Chapter 10

The Role of the Neuropsychologist in Deep Brain Stimulation

Alexander I. Tröster

Department of Clinical Neuropsychology and Center for Neuromodulation, Barrow Neurological Institute, Phoenix, AZ, United States

BRIEF HISTORY

Modern deep brain stimulation (DBS) is used primarily to treat movement disorders and neurosurgical interventions for movement disorders have a long history. Surgery targeting the pyramidal system, specifically precentral cortical excisions to relieve athetotic movements, date to 1908 (Horsley, 1909), but the extrapyramidal system was not targeted to treat parkinsonian movement disorders until Meyers carried out bilateral caudate lesions in a woman with postenecephalitic parkinsonism in 1939 (Abel, Walch, & Howard, 2016) and several other cases by 1942 (Meyers, 1942). Surgical treatments for movement disorders evolved, probably as a function of increased knowledge of basal ganglia function and physiology and advances in technology. These treatments, especially popular in the 1950s and 1960s, included thalamotomy (Hassler & Riechert, 1954) and pallidotomy (Cooper, 1954; Narabayashi, Okuma, & Shikiba, 1956), and later, lesions of the subthalamic region including the fields of Forel, zona incerta, and posterior subthalamic area (Andy, Jurko, & Sias, 1963; Houdart, Mamo, Dondey, & Cophignon, 1965). Lesions of the subthalamic nucleus (STN) proper (Alvarez et al., 2001; Patel et al., 2003) were avoided for a long time. Ablative surgeries for movement disorders subsided with the advent of levodopa and newer dopaminomimetic agents until limitations of pharmacotherapy were realized and pallidotomy was "rediscovered" (Laitinen, Bergenheim, & Hariz, 1992). Although focused ultrasound lesioning for movement disorders had been used by Meyers and colleagues (Meyers et al., 1959), this method (albeit with much advanced technology) has recently been resurrected for thalamotomy in essential tremor (ET) (Elias et al., 2013;

Neurosurgical Neuropsychology. DOI: https://doi.org/10.1016/B978-0-12-809961-2.00011-4

Lipsman et al., 2013a) and in thalamotomy (Bond et al., 2017) and subthalamotomy for Parkinson's disease (PD) (Martinez-Fernandez et al., 2018).

Neuropsychological studies of early ablative surgeries for movement disorders have been reviewed in detail (Wilkinson & Tröster, 1998). These studies tended to be few in number and often circumscribed, and typically evaluated verbal skills and memory after thalamotomy (Almgren, Andersson, & Kullberg, 1969; Blumetti & Modesti, 1980; Jurko & Andy, 1973; Kocher, Siegfried, & Perret, 1982; Niebuhr, 1962; Van Buren, Li, Shapiro, Henderson, & Sadowsky, 1973; Vilkki & Laitinen, 1974), but very few examined the neuropsychological effects of early pallidotomy (Laitinen, 2000) or subthalamotomy (Meier & Story, 1967). The probably most extensive body of neuropsychological studies of early thalamotomy and pallidotomy was published by coworkers of Irving Cooper and a summary of this work (Riklan & Levita, 1969) is essential reading for anyone contemplating a course of studies of the neurobehavioral outcomes of stereotactic and functional neurosurgery for movement disorders. Of course modern pallidotomy underwent much more rigorous neuropsychological scrutiny than early interventions (Alegret et al., 2000; Green et al., 2002; Kubu, Grace, & Parrent, 2000; Obwegeser et al., 2000; Rettig et al., 2000; Tröster, Woods, & Fields, 2003; York, Levin, Grossman, & Hamilton, 1999) and detailed neuropsychological outcomes of modern subthalamotomy have also been reported (Bickel et al., 2010; McCarter, Walton, Rowan, Gill, & Palomo, 2000; Obeso et al., 2017; Patel et al., 2003).

Given irreversibility of lesions and the risk of cognitive and behavioral morbidity with bilateral lesioning procedures, many advocated for use of DBS. The history of "early DBS" has been reviewed elsewhere (Blomstedt & Hariz, 2010). Stimulation via externalized electrodes for movement disorders such as PD was used in the 1960s (Bechtereva, Bondartchuk, Smirnov, Meliutcheva, & Shandurina, 1975; Sem-Jacobsen, 1965). The stimulation protocol was usually designed to predict satisfactory response to lesioning and required the patient to return frequently for intermittent stimulation over days or weeks. Bechtereva and colleagues, however, were the first to use stimulation for purely therapeutic purposes in movement disorders (including PD, dystonia, and Wilson's disease), and they intermittently stimulated patients with externalized electrodes for as much as 18 months.

Today, DBS devices consist of an implantable pulse generator or generators (akin to a pacemaker), typically implanted below the collarbone, that is connected via an extension running beneath the skin of the neck to the head where it connects to the leads which have multiple electrodes (contacts) and are implanted in the target structure (Fig. 10.1). Two possible electrical field shapes are illustrated in Fig. 10.2. Some of the newer devices allow shaping of the electrical field and current steering so that more circumscribed areas of the target can be stimulated, thereby presumably enhancing desired effects while minimizing side effects.

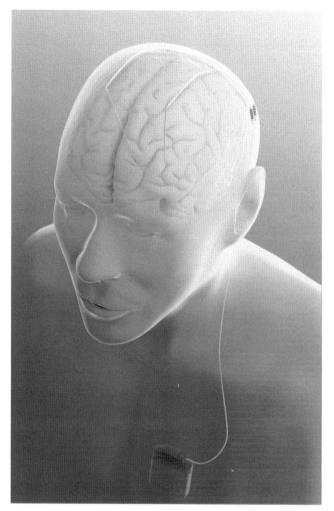

FIGURE 10.1 Connection of deep brain simulation electrode and implantable pulse generator. *Reprinted with permission from Medtronic, Inc., © 2017.*

The first report of a chronically, implanted system to treat intention tremor in multiple sclerosis via contingent stimulation in the 1970s was published in 1980 (Brice & McLellan, 1980). Benabid and colleagues in France studied thalamic stimulation for PD tremor in the 1980s and published outcomes of a DBS trial in the early 1990s (Benabid et al., 1991), while the first large scale North American study reported its results in 1997 (Koller et al., 1997). In the United States, DBS was first approved by the Food and Drug Administration (FDA) to treat tremor in PD and ET in 1997, to treat the symptoms of advanced PD in 2002, and to treat persons with PD who have

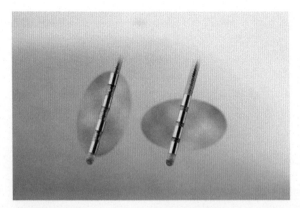

FIGURE 10.2   Examples of four-contact deep brain stimulation leads with different spacing and electrical field shapes and sizes. *Reprinted with permission from Medtronic, Inc., © 2017.*

motor complications early in the disease (after a minimum of 4 years) in 2016. Two additional DBS systems for ET and/or PD were FDA-approved in 2015 and 2017. DBS was approved in the United States as a humanitarian use device for dystonia in 2003 and for obsessive−compulsive disorder in 2009.

Although they are not DBS devices per se, two related devices have recently been approved by the FDA. One is an magnetic resonance imaging−guided focused ultrasound lesioning device (FDA-approved for ET in 2016) and the other is a *cortical* responsive neurostimulation (closed loop) device approved by the FDA as an adjunctive treatment for partial seizures in 2013. A DBS device for anterior thalamic stimulation for epilepsy (as an adjunct treatment) received the CE mark in Europe in 2010 and was approved by the FDA in the United States in 2018.

Unlike early lesioning procedures, modern DBS has been subjected to detailed and extensive neuropsychological scrutiny already since the 1990s. Although the sections that follow summarize recent outcomes research, from a historical point it is worth noting how soon after approval of DBS neuropsychological outcomes were reported (and indeed, neuropsychological data were often submitted with FDA approval applications). Early neuropsychological studies in the modern DBS era examined the outcomes of thalamic DBS for PD (Caparros-Lefebvre, Blond, Pécheux, Pasquier, & Petit, 1992; Tröster, Fields, Wilkinson, Busenbark, et al., 1997) and ET (Tröster et al., 1999), pallidal (GPi) DBS for PD (Ardouin et al., 1999; Burchiel, Anderson, Favre, & Hammerstad, 1999; Fields, Tröster, Wilkinson, Pahwa, & Koller, 1999; Tröster, Fields, Wilkinson, Pahwa, et al., 1997; Vingerhoets et al., 1999), and STN DBS for PD (Alegret et al., 2001; Saint-Cyr, Trépanier, Kumar, Lozano, & Lang, 2000; Woods, Fields, & Tröster, 2002). In addition, neuropsychological outcomes were published early on for STN (Jahanshahi,

Czernecki, & Zurowski, 2011; Kleiner-Fisman et al., 2007) and pallidal DBS for dystonia (Hälbig et al., 2005; Pillon et al., 2006). Neuropsychological data for obsessive−compulsive disorder treatment with DBS have been slower to emerge (Denys et al., 2010; Gabriels, Cosyns, Nuttin, Demeulemeester, & Gybels, 2003; Greenberg et al., 2010; Schoenberg et al., 2015), perhaps because of the limited number of patients treated and the variety of stimulation targets used. Finally, it is to be noted that neuromodulation is an evolving therapy and several investigational uses of DBS are occurring and new indications for DBS are being identified. Some of the indications for DBS being investigated are presented in Table 10.1, but we will not within the confines of this chapter review neuropsychological aspects of investigational DBS.

CURRENT KNOWLEDGE BASE AND PRACTICE

Parkinson's Disease

PD is probably the most frequently treated movement disorder in the DBS arena. The disease is characterized by four cardinal motor signs: Tremor, bradykinesia, rigidity and postural abnormality Accompanying nonmotor (including cognitive and emotional) features are well recognized. PD rarely occurs before age 50 years and onset is most often in the mid-60s. Reported overall prevalence rates of PD range from 167 to 5703 per 100,000 persons per year (Wirdefeldt, Adami, Cole, Trichopoulos, & Mandel, 2011). Mild cognitive impairment (PD-MCI) prevalence is about 30% (Goldman et al., 2013). The most accepted prevalence rates of dementia in PD are 20% to 40% (Mohr, Mendis, & Grimes, 1995) with the best quality studies disclosing a prevalence of 31% (Aarsland, Perry, Brown, Larsen, & Ballard, 2005). Clinically relevant depressive symptoms are observed in about 35% of persons with PD (Aarsland, Taylor, & Weintraub, 2014).

Given the neurobehavioral features of PD, and the fact that DBS is not indicated for dementia, it is probably almost universal practice that patients undergo neuropsychological evaluation, or at least cognitive screening, prior to DBS. This practice is consistent with consensus statements (Bronstein et al., 2011; Lang et al., 2006). Indeed, although the exact nature and severity of neurobehavioral deficits that renders DBS less or inappropriate for a given person remains controversial (Okun, Fernandez, Rodriguez, & Foote, 2007), almost half of those persons excluded from DBS on screening are excluded for neuropsychological reasons (Lopiano et al., 2002), and 10%−15% may be excluded on the basis of neuropsychological evaluation even after screening (Abboud et al., 2014; Coleman, Kotagal, Patil, & Chou, 2014). Many centers also almost routinely evaluate patients after surgery so as to facilitate determination of the etiology of potential cognitive changes and treatment decision-making.

TABLE 10.1 Some Established and Current and Past Investigational Indications for Deep Brain Stimulation, Targets (With Sample References), and Availability of Neuropsychological Outcome Data

Condition	Target	Established or Investigational	Availability of Published Neuropsychological Data
Parkinson's disease	STN & GPi (Weaver et al., 2009; Weaver et al., 2012; Rothlind et al., 2015), VIM (Oertel et al., 2017; Cury et al., 2017), PPN (Khan et al., 2012, Welter et al., 2015, Schrader et al., 2013, Liu et al., 2015), zona incerta (Kan et al., 2012), motor cortex (Moro et al., 2011, Bentivoglio et al., 2012, De Rose et al., 2012)	STN and GPi established	Good for STN and GPi
Tremor (ET, MS, Holmes, post-traumatic)	Motor cortex (Moro et al., 2011), VIM (Cury et al., 2017, Torres et al., 2010, Chen et al., 2016, Eisinger et al., 2018), zona incerta/posterior subthalamic area (Eisinger et al., 2018, Sandvik et al., 2011, Hägglund et al., 2016, Fytagoridis et al., 2016)	VIM established	Limited for ET; very limited or absent for others
Dystonia	GPi (Volkmann et al., 2012, Gruber et al., 2010), VIM (Cury et al., 2017, Gruber et al., 2010), STN (Ostrem et al., 2014)	GPi and VIM established	Limited
Tardive dystonia/dyskinesia	GPi (Capelle et al., 2010, Shaikh et al., 2015), STN (Deng et al., 2017)	GPi established	Limited
Alzheimer's disease	Fornix (Laxton et al., 2010, Lozano et al., 2016), nucleus basalis of Meynert (Kuhn et al., 2015)	Investigational	Limited
Huntington's disease	GPi (Gonzalez et al., 2014, Rasche et al., 2016)	Investigational	Limited
Multiple system atrophy (MSA)	STN (Zhu et al., 2014, Meissner et al., 2016, Thavanesan et al., 2014)	Investigational	Absent
Neuroacanthocytosis	GPi (Nakano et al., 2015, Lee et al., 2015, Lim et al., 2013), thalamus (Nakano et al., 2015)	Investigational	Absent

Epilepsy	STN (Wille et al., 2011), anterior thalamic nucleus (Fisher et al., 2010, Lehtimäki et al., 2016), centromedian thalamic nucleus (Valentín et al., 2013, Son et al., 2016), hippocampus (Tellez-Zenteno et al., 2006, McLachlan et al., 2010, Wiebe et al., 2013), nucleus accumbens (Kowski et al., 2015), caudate (Chkhenkeli et al., 2004), cerebellum (Velasco et al., 2005), mammillary bodies and mammillothalamic tract (Duprez et al., 2005)	Investigational	Limited for hippocampus; very limited or absent for others
Pain	periventricular gray, ventrocaudal (sensory) thalamus (Boccard et al., 2012, Pereira et al., 2013), anterior cingulate cortex (Boccard et al., 2014)	Investigational	Very limited or absent
Cluster headaches and neuralgiform headaches	hypothalamus (Fontaine et al., 2010, Chabardès et al., 2016, Piacentino et al., 2014), ventral tegmental area (Akram et al., 2016, Miller et al., 2016)	Investigational	Absent
Depression	subcallosal cingulate white matter (Holtzheimer et al., 2012), medial forebrain bundle (Schlaepfer et al., 2013, Bewernick et al., 2017), internal capsule (Bergfeld et al., 2016, Dougherty et al., 2015), thalamic peduncle (Raymaekers et al., 2017), ventral striatum (Dougherty et al., 2015)	Investigational	Limited
Obsessive-compulsive disorder	internal capsule (Greenberg et al., 2010), nucleus accumbens (Figee et al., 2013, Denys et al., 2010), subthalamic nucleus (Mallet et al., 2008), inferior thalamic peduncle (Jiménez-Ponce et al., 2009)	Investigational	Very limited or absent
Tourette's syndrome	GPi (Cannon et al., 2012), centromedian thalamic nucleus (Maling et al., 2012, Okun et al., 2013)	Investigational	Very limited or absent
"Impulsive and violent behavior"	Posterior hypothalamus (Torres et al., 2013, Franzini et al., 2013)	Investigational	Absent
Obesity	Lateral hypothalamus (Whiting et al., 2013), ventromedial hypothalamus (Torres et al., 2012), nucleus accumbens (Mantione et al., 2010)	Investigational	Absent

(Continued)

TABLE 10.1 (Continued)

Condition	Target	Established or Investigational	Availability of Published Neuropsychological Data
Anorexia/bulimia	Subcallosal cingulate (Lipsman et al., 2013), nucleus accumbens (Zhang et al., 2013)	Investigational	Absent
Drug abuse	Nucleus accumbens (Kuhn et al., 2014, Müller et al., 2016)	Investigational	Absent
Tinnitus	Locus of caudate neurons (Kuhn et al., 2011, Larson et al., 2013), VIM (Shi et al., 2009)	Investigational	Absent

ET, essential tremor; *MS*, multiple sclerosis; *GPI*, globus pallidus internus; *STN*, subthalamic nucleus; *VIM*, ventral intermediate nucleus of the thalamus; *PPN*, pedunculopontine nucleus.

Neurobehavioral Adverse Events After Subthalamic Nucleus and Global Pallidus Deep Brain Stimulation for Parkinson's Disease

Although thalamic DBS might still be rarely undertaken for PD, the predominant targets in PD are the GPi and STN. Consequently, the focus in this paper is on outcomes using these two targets. A recent review of the state of the knowledge regarding neuropsychological outcomes and their prediction highlights that DBS in PD is relatively safe and that although serious adverse events (AEs) are rare, mild cognitive or emotional deficits are not uncommon within 3−12 months of surgery (Tröster, 2017). AEs identified by clinician or patient report show that serious cognitive AEs are rare, especially over the first 3−6 months of treatment with STN or GPi DBS. The most commonly reported serious cognitive AE is confusion or delirium (occurring in less than 3% of persons), while nonserious confusion may occur in up to 11% of patients. The higher rates of confusion (20%−24%) reported at 24 months might reflect disease progression, medication, and other medical conditions. The most common and consistently reported psychiatric AE in larger and/or controlled trials is depression (about 5% serious and 5% nonserious depression within 6 months of surgery but as high as 37% over 24 months). The higher rates of depression over time are not surprising given the natural history of the disease and estimated depression prevalence. Other complications may include mania, psychosis, apathy, impulsive and compulsive behaviors, and suicide. Unfortunately, long-term data comparing outcomes in surgically and medically treated groups in regard to these complications are lacking.

Neuropsychological Outcomes of Subthalamic Nucleus and Global Pallidus Deep Brain Stimulation for Parkinson's Disease

Generally, studies using neuropsychological instruments and behavioral rating scales report higher rates of cognitive decline after DBS than do studies relying on patient/clinician report. There are now many studies examining neurobehavioral outcomes after DBS and thus it becomes instructive to review recent meta-analyses of findings The properties and characteristics of recent meta-analyses are presented in Table 10.2, and effect sizes concerning outcomes either in cognitive and behavioral domains or on specific neuropsychological tests are presented in Tables 10.3 and 10.4, respectively. It is emphasized that while meta-analyses are useful in detecting treatment effects when the effects are small or when studies yield inconsistent findings and have small samples, their value is determined by the quality of studies included in the analysis. The tables also reveal that the majority of effect sizes refer to patient groups' changes relative to their own preoperative baseline, some meta-analyses combine outcomes from different surgical targets or uni- and bilateral interventions and some report effect sizes relative to change in the DBS group compared to a control group or DBS of another target.

TABLE 10.2 Properties and Characteristics of Meta-Analyses of the Neuropsychological Effects of Deep Brain Stimulation

Reference	Number of Studies Included	Sample Size	Comments	Cognitive Domain (D) or Test (T) Effect Sizes/Mean Standardized Differences (MSD) Provided	Fixed or Random Effects Model(s)	STN SMD	GPi SMD	STN Versus GPi Direct Comparison SMD	Unilateral Versus Bilateral Comparison SMD
Parsons et al. (2006)	28	612		D	Random	Yes	No	No	No
Combs et al. (2015)	38	1622	Includes mix of unilateral and bilateral DBS	D	Random	Yes	Yes	No	No
Wang et al. (2016)	7	521	Includes RCTs only; seven studies from four trials; mix of unilateral and bilateral	D	Both depending on heterogeneity	No	No	Yes	No
Xie et al. (2016)	10	797	RCTs and nonrandomized, controlled trials only	T	Both depending on heterogeneity	Yes	No	No	No

Elgebaly et al. (2017)	4	345	Four RCT in meta-analysis, seven studies for qualitative analysis	T	Both depending on heterogeneity	No	No	Yes	Yes (subgroup analysis)
Martínez-Martínez et al. (2017)	50	Total not given; 69–246 per test	Executive function tests only; includes one GPi study	T	Both depending on heterogeneity	Yes (but includes 1 GPi)	No	No	No

GPi, Globus pallidus internus; *RCT*, randomized controlled trial; *SMD*, standardized mean difference (effect size); *STN*, subthalamic nucleus.
Source: From Tröster (2017).

TABLE 10.3 Standardized or Weighted Mean Differences (Effect Sizes) Reported by Meta-Analyses of Effects of Deep Brain Stimulation on Cognition and Emotion Domains and Quality of Life

Cognitive/ Emotion Domain	Parsons et al. (2006) STN (Change Within Group)	Combs et al. (2015) STN (Change Within Group)	Combs et al. (2015) GPi (Change Within Group)	Wang et al. (2016) STN Versus GPi Difference (Change Between Groups)
Global/cognitive screening	0.04	− 0.24	0.23	− 4.30[a,b]
Attention/ concentration	0.02	− 0.12	− 0.19	− 0.21[a]
Executive function	− 0.08	− 0.13	0.00	− 0.12
Psychomotor speed	0.22	− 0.16	− 0.16	
Verbal functions (memory)	− 0.21			
Verbal fluency	− 0.64	− 0.40	− 0.22	− 0.24[a]
Phonemic fluency	− 0.51	− 0.36	− 0.19	− 2.93[a]
Semantic fluency	− 0.73	− 0.48	− 0.24	− 1.55
Visuoperceptual functions and visual memory	0.06			
Learning and memory		− 0.12	− 0.09	− 0.16[a]
Language		0.04	0.01	0.05
Visuoperceptual/ spatial skills		− 0.02	0.12	
Depression symptoms				1.37
Anxiety symptoms				− 0.02
Quality of life				− 0.15

[a]*STN greater decline than GPi.*
[b]*Based on Mattis Dementia Rating Scale (DRS) so high likelihood the result is heavily influenced by semantic fluency decline; GPi, globus pallidus internus; STN, subthalamic nucleus; − represents decline from preoperative baseline or worse performance by STN than GPi.*
Source: From Tröster (2017).

TABLE 10.4 Standardized or Weighted Mean Differences (Effect Sizes) Reported by Meta-Analyses of Effects of Deep Brain Stimulation on Specific Neuropsychological Tests

Neuropsychological Test	Xie et al. (2016) STN (vs Control) (Change Between Groups)	Martínez-Martínez et al. (2017) STN (But Includes 1 GPi Study) (Change Within Group)	Elgebaly et al. (2017) STN Versus GPi (Direct Comparison) (Change Between Groups)	Elgebaly et al. (2017) Bilateral and Unilateral (Indirect Comparison)
Mini mental state exam	0.06			
Mattis Dementia Rating Scale	− 0.21[a]			
Digit span total		− 0.05		
Digit span forward			0.08	
Digit span backward	− 0.14		0.19	
WAIS-R digit symbol			− 0.16	
WAIS-R arithmetic			0.02	
Stroop word reading			− 0.21	
Stroop color naming			− 0.31[c]	
Stroop color-word	− 0.20[a]	− 0.21	− 0.16	
Trailmaking Part A	0.03	− 0.02	− 0.05	
Trailmaking Part B	− 0.39	0.05	− 0.14	
Trailmaking A−B		− 0.04		
Raven's matrices	− 0.15	0.06		

(Continued)

TABLE 10.4 (Continued)

Neuropsychological Test	Xie et al. (2016) STN (vs Control) (Change Between Groups)	Martínez-Martínez et al. (2017) STN (But Includes 1 GPi Study) (Change Within Group)	Elgebaly et al. (2017) STN Versus GPi (Direct Comparison) (Change Between Groups)	Elgebaly et al. (2017) Bilateral and Unilateral (Indirect Comparison)
Wisconsin card sorting		0.06[b]		
Verbal fluency overall		−0.27		
Verbal fluency—phonemic	−0.49[a]			−0.04/−0.05
Verbal fluency—semantic	−0.39[a]			−0.09/−0.29
Boston naming test	0.02		−0.11	
Rey auditory verbal learning test—immediate recall	−2.06[a]			
Rey auditory verbal learning test—delayed recall	−1.41[a]			
Paired associate learning	−0.69			
Beck depression inventory				0.15/0.36

[a]Significantly greater decline in STN DBS versus control.
[b]Different versions of WCST considering categories, errors, and perseverations.
[c]Favors GPi versus STN DBS.
Source: From Tröster (2017).

The data presented in Tables 10.3 and 10.4 reveal that effect sizes associated with changes within cognitive domains or on specific tests are usually small (occasionally moderate) after both GPi and STN DBS. The most consistent and largest effect sizes (pre- to postoperative change) are reported for verbal fluency (semantic marginally greater than phonemic). This finding is consistent with prior reports that mild to moderate verbal fluency reductions occur in 25%−50% of persons after bilateral DBS (Fields & Tröster, 2000; Funkiewiez et al., 2004). Clinically it is worth noting that the fluency changes usually persist and, probably due to the natural course of the disease, might even worsen between 5 and 8 years after surgery (Fasano et al., 2010). Importantly, the immediate postoperative verbal fluency decrements appear not to foretell more rapid or extensive cognitive decline (Borden et al., 2014).

Cognitive changes after DBS are not, however, limited to verbal fluency. One meta-analysis including controlled (albeit not necessarily randomized) studies, reported greater memory decrements in STN DBS groups than controls (Xie, Meng, Xiao, Zhang, & Zhang, 2016). This finding parallels those from earlier controlled studies of STN DBS showing declines in attention and memory (Smeding et al., 2006; York et al., 2008). STN DBS may be associated with small declines in executive function (Parsons, Rogers, Braaten, Woods, & Tröster, 2006), but interestingly, patients may subjectively perceive improved executive functions after DBS (Pham et al., 2015). Overall, recent meta-analyses' findings are consistent with large randomized, controlled trails showing relatively small declines on a narrow range of tests, and the most robust effects of DBS on verbal fluency.

Psychiatric issues after DBS were addressed in an early meta-analysis (Appleby, Duggan, Regenberg, & Rabins, 2007) but this analysis did not report effect sizes and combined studies of DBS for a variety of conditions and anatomical targets. A very useful, more recent study reviewed rates of various psychiatric complications in trials of medical therapy and STN and GPi DBS using various ascertainment methods (e.g., rating scales, self-report, etc.) (Castrioto, Lhommee, Moro, & Krack, 2014). For depression, based on clinical symptoms, rates at 3−6 months ranged from 0% for BMT to 0%−5% for STN DBS (no estimates at 6 months for GPi DBS). Rates of change were not reported for studies using depression scales, but STN, unlike best medical therapy (BMT), yielded group-wise reductions in symptom severity regardless of scale used. Suicidal ideation rates over 6−24 months ranged from 0.7% to 1.5% for STN DBS and 0%−0.7% for GPi DBS. Completed suicide rates between 6 and 24 months after surgery ranged from 0% to 0.8% for BMT, 0% to 1.3% for STN DBS, and 0.7% for GPi DBS (at 24 months). Apathy was estimated to occur in 1.3%−5% of persons between 3 and 6 months after STN DBS. Psychosis, interestingly, may occur at higher rates in BMT (9%) compared to DBS (2.2%−6%).

Two of the largest and best-designed comparisons of STN and GPi DBS (VA CSP-468 and The Netherlands SubThalamic and Pallidal Stimulation (NSTAPS) study) (Odekerken et al., 2013; Odekerken et al., 2015; Rothlind et al., 2015) and three other controlled studies of STN DBS published detailed neuropsychological follow-up analyses (Foki et al., 2017; Tramontana et al., 2015; Tröster, Jankovic, Tagliati, Peichel, & Okun, 2017). These studies are worth mentioning for their quality and because they were not all included in the meta-analyses. Both of the STN and GPi comparisons are especially useful because they determined reliable changes in neuropsychological functioning after DBS. The follow-up of the CSP-468 study (Rothlind et al., 2015) examined neuropsychological data from 117 BMT, 80 GPi DBS, and 84 STN DBS patients before and 6 months after GPi or STN DBS surgery. Because there were minimal meaningful differences between STN and GPi DBS outcomes (STN performed better on one measure of learning and memory while GPi performed better on one measure of processing speed), the groups were combined for further analyses. Factor analysis disclosed that the neuropsychological tests administered covered five domains: Processing speed, working memory, language, memory, and executive functions. Cognitive change at 6 months was determined on the basis of reliable change indices (RCI) derived from mean change in the BMT control group, and a decline in a domain was defined as reliable change having occurred on at least one-third of the measures used to evaluate the domain (the number of measures/scores per domain varies). Declines in multiple domains were seen in 3% of BMT and 11% of DBS patients 6 months after DBS. Importantly, although those showing and not showing multiple-domain cognitive declines both had significantly improved self-reported functioning and quality of life (QoL), the gain was attenuated in the group with cognitive declines.

The report of the NSTAPS 12-month neurobehavioral outcomes among 62 GPi (3 of 65 subjects withdrew after randomization) and 63 STN DBS (Odekerken et al., 2013) used a broad composite outcome based on loss of important relationships, loss of professional activity, decline per RCI on three or more neuropsychological tests, and diagnosis of anxiety, depression or psychosis for 3 months or more. Among the GPi DBS group, 58% had negative neurobehavioral composite scores (at least one of the listed adverse behavioral events), while 56% of the STN DBS group had at least one of the four negative neurobehavioral indicators within 12 months of surgery. Cognitive decline (per RCI on at least three measures) occurred in 27% of GPi and 35% of STN DBS (a nonsignificant difference). More specific 12-month neuropsychological test data (available for 58 GPi and 56 STN DBS patients) were published in a follow-up paper (Odekerken et al., 2015). Despite having administered multiple tests, many with multiple scores, only Stroop task word reading and color naming (but not interference) and Trailmaking Part B declined more in the STN than GPi DBS group. Using the composite of changes on at least 3 of 12 tests per RCI, 29% of GPi, and 39% of STN experienced declines 12 months after surgery (no significant differences between GPi and STN DBS). Cognitive decline was associated with older age but not, unlike in the VA cooperative study, with QoL changes.

Several other trials again have shown circumscribed cognitive declines after STN DBS. One study of 18 STN DBS patients compared cognitive changes on the Neuropsychological Test Battery Vienna short-form over 12 months in that group against changes in 25 PD undergoing BMT, 24 MCI (non-PD), and 12 healthy control subjects (Foki et al., 2017). Using RCI, 11% of the DBS group, but none of the BMT group showed phonemic verbal fluency declines. In a comparison of neuropsychological changes in 101 PD patients 3 months after STN DBS, and in 35 patients who had STN electrodes implanted but stimulation had not yet been activated, both groups showed decrements in semantic and switching verbal fluency (Tröster et al., 2017). Only the stimulation group evidenced a decline in phonemic verbal fluency and on the Stroop task (suggesting these declines may be related to either stimulation or a combination lesion and stimulation effect). The stimulation group, however, evidenced more frequent improvements in depressive symptomatology than the delayed activation control group.

One recent report is of note because it documented neuropsychological outcomes of bilateral STN DBS in 15 patients *early* in the disease (i.e., these patients had been treated with antiparkinsonian medication for 6 months to 4 years). The outcomes after DBS were compared to those in a group of 15 early PD patients who underwent BMT (Tramontana et al., 2015). The STN DBS group showed greater declines at 12 months in phonemic fluency, digit span, and on the Stroop task, and lesser gains on the Wisconsin Card Sorting Test (WCST perseverative errors) and some trials of the Paced Auditory Serial Addition Test (PASAT). Overall, then, cognitive changes after DBS appear to be qualitatively similar in early and advanced PD, with the most obvious changes seen in fluency and executive functions.

Some Important Neuropsychological Issues in Deep Brain Stimulation

Several questions arise regarding the neurobehavioral outcomes of DBS. Perhaps the most important is whether good and poor outcomes can be confidently predicted from preoperative evaluation. To date the quest to identify neuropsychological outcome predictors has proved difficult. Probably the main reason for this is that persons for whom DBS is considered inappropriate for neuropsychological reasons are excluded from DBS, meaning outcome data are never obtained to verify that the nature and severity of preoperative neuropsychological deficits indeed foretell a poor outcome. One is thus confronted with identifying risk among a relatively homogeneous population of patients that rarely experience severe neuropsychological deficits after DBS. Nonetheless, there are isolated reports of dementia or significant functional declines after DBS (Hariz et al., 2000; Rektorova, Hummelova, & Balaz, 2011). Although neuropsychological and other predictors of DBS outcome have rarely been reliably identified and replicated, factors often considered in preoperative evaluation and as being related to outcomes (e.g., cognitive, length of hospitalization) are presented in Table 10.5.

Another question that has arisen recently is whether neuropsychological outcomes differ depending upon whether DBS surgery is done with the patient awake using electrophysiological guidance or under general anesthesia using direct anatomical targeting. Preliminary findings suggest that cognitive morbidity after "asleep" surgery is minimal (Sidiropoulos et al., 2016) and that asleep DBS may even be associated with better QoL and speech outcomes than awake surgery (Brodsky et al., 2017). However, these findings require replication.

A common point of debate is whether STN and or GPi DBS might be superior regarding neuropsychological outcomes. While some suggest that GPi DBS might be safer, the two largest, randomized trials (Odekerken et al., 2015; Rothlind et al., 2015) and relevant meta-analyses (see Tables 10.3 and 10.4) reviewed earlier have failed to identify any clinically meaningful differences in neuropsychological outcomes (although minor differences tend to favor GPi) (Tröster, 2017).

Although DBS is rarely performed in persons with dementia, mild cognitive impairment in PD (MCI-PD) is typically not exclusionary. Nonetheless, very little research has addressed the changes in the occurrence of PD-MCI before and after DBS, and whether PD-MCI prior to surgery heightens risk for poor neuropsychological outcomes. One study found that frequency of PD-MCI increased from 47% prior to surgery to 63% an average of 9 months after surgery among 30 STN DBS patients (Yaguez et al., 2014). However, two studies suggest that preoperative MCI per se may have less predictive utility than specific cognitive deficits, for example, in attention. One study of 130 patients included 60% that had multiple-domain MCI and 21% with single domain MCI prior to surgery. PD-MCI was not predictive of length of hospital stay or postoperative confusion after bilateral STN DBS patients, but persons with preoperative attention impairments were significantly more likely to have hospital stays of 3 or more days (39% vs 12%), and tended to have postoperative confusion more frequently (11% vs 3%) (Abboud et al., 2015). In a study of 103 bilateral STN DBS patients, 63% of whom had MCI at baseline (Kim et al., 2014) the annual rate of decline over 7 years on the Mini Mental State Examination (MMSE) was small (0.4 ± 1.7), and not associated with MCI diagnosis. 10 of the 103 patients developed dementia and the probability of developing dementia was significantly greater among those with than without baseline PD-MCI. However, given similar rates of

TABLE 10.5 Neuropsychological, Patient Demographic, Disease, and Anatomical Factors Impacting Deep Brain Stimulation Outcome and Possibly Considered During Patient Selection

Neuropsychological	Demographic	Disease	Anatomical
Overall level of cognitive impairment (Witt et al., 2011)	Age (Hrabovsky et al., 2017; Smeding, Speelman, Huizenga, Schuurman, & Schmand, 2011)	Motor symptom severity (Merola et al., 2014)	White matter lesion burden (Blume et al., 2017)
Attention (Abboud et al., 2015; Smeding et al., 2011)		Axial symptom severity (Daniels et al., 2010; Fukaya et al., 2017)	Thalamic and hippocampal volumes (Aybek et al., 2009; Geevarghese et al., 2016)
Executive dysfunction (Kim et al., 2014; Pilitsis et al., 2005)		Age at disease onset (Fukaya et al., 2017)	Greater intermammillary distance (third ventricular size; indirect measure of atrophy) (Hrabovsky et al., 2017)
Intelligence (Yaguez et al., 2014)		Baseline dopaminergic medication dosage (Daniels et al., 2010; Kim et al., 2014)	
List learning (Yaguez et al., 2014)		Baseline response to dopaminergic medication (Smeding et al., 2011)	
Apathy (Drapier et al., 2005)			
Hallucinations (Blume et al., 2017)			
Anxiety (Schoenberg, Mash, Bharucha, Francel, & Scott, 2008)			
Visuospatial impairment (Abboud et al., 2015)			

MMSE decline in those with and without PD-MCI at baseline, the higher probability of dementia in the PD-MCI-diagnosed group probably reflects disease progression and the fact that PD-MCI is a risk factor for dementia in PD (Hoogland et al., 2017). Such a conclusion was drawn by the authors of a study of 184 patients, 23% of whom had PD-MCI prior to surgery. After surgery, among those developing dementia, dementia affected those with PD-MCI sooner (median 6 years) than those without PD-MCI (median 11 years). Because no cases of dementia were observed early after DBS the authors concluded that the more precocious development of dementia in the PD-MCI group might reflect the natural history of the disease (Merola et al., 2014).

It is increasingly recognized that it is critical to address patient expectations regarding DBS outcomes prior to surgery (Maier et al., 2013). Specifically those persons dissatisfied with outcomes, despite motor improvement, had unrealistic expectations and greater preoperative depression and apathy symptoms (Maier et al., 2013). Expectations regarding magnitude and durability of DBS effect on the most troubling symptoms as well as AEs should be reviewed. In one study patients were asked to rate their QoL and to indicate on the same scale (Parkinson's Disease Questionnaire (PDQ-39)) where they expected to see themselves after surgery. QoL improved significantly 6 months after surgery but there was a marked discrepancy between expected and actual (much smaller) change (Hasegawa, Samuel, Douiri, & Ashkan, 2014). Nonetheless most patients rated themselves as satisfied and having their expectations fulfilled on visual analog scales. Satisfaction with outcome was related to fulfillment of expectations but not with QoL changes, suggesting that satisfaction may be more related to having expectations fulfilled than QoL improved. Another study, in contrast, found that satisfaction with outcome was related to QoL change: Those with mixed and satisfactory outcomes showed QoL gains whereas those who were dissatisfied with DBS outcome failed to show QoL improvement (Maier et al., 2016).

Essential Tremor

ET is the most common movement disorder. The hallmark of ET is a kinetic tremor, usually of the upper extremities and more rarely affecting the head and lower limbs. Patients may also show postural tremor, resting tremor (related to basal ganglia dysfunction), and intention tremor (due to cerebellar involvement). Although initially called familial or benign tremor, the tremor can actually be quite functionally limiting with significant detrimental impact on QoL. Prevalence of ET among those older than 60 years ranges from about 1.3% to 5% (Louis & Ferreira, 2010).

Outcome of Deep Brain Stimulation for Essential Tremor

The most common surgical target in ET is the ventral intermediate nucleus of the thalamus (Vim). DBS of the subthalamic region has also been undertaken (Elble, Shih, & Cozzens, 2018; Sandvik, Koskinen, Lundquist, & Blomstedt, 2012), although ataxia after thalamic DBS has been hypothesized to be related to current spread to the ventrocaudal subthalamic area (Reich et al., 2016). Neuropsychological studies of DBS for ET are few in number and generally limited to small samples (Tröster & Tucker, 2005), with the largest studies documenting outcomes in 40−49 patients (Tröster et al., 1999; Woods, Fields, Lyons, Pahwa, & Tröster, 2003). Generally, thalamic DBS is deemed relatively safe from a cognitive standpoint, and subtle improvements, perhaps related to practice effects (related to repeated test exposure), may be seen on visuoconstruction/perceptual and memory tasks. Improvements are also seen in dexterity (Tröster et al., 1999). These gains tend to persist over 12 months of follow-up (Fields et al., 2003). Isolated decrements are observed in verbal fluency, and these may be more likely to occur in persons with preoperative verbal fluency deficits (Fields et al., 2003). When mild cognitive declines occur, they are more likely to occur among persons with tremor onset after age 37 years and when pulse widths of 120 μs or greater are used in stimulation (Woods et al., 2003). One study, albeit with a small sample (n = 9), is helpful in evaluating neuropsychological impact of DBS in ET because patients were followed up at 1 and 6 years and the study compared effects with the stimulator tuned on and off (Heber et al., 2013). That study reported no cognitive declines but improvement in reaction time.

Neurobehavioral AE rates in clinical trials are difficult to interpret, because studies differ in how systematically and by which ascertainment methods the data were collected (if at all). A recent review including 17 studies and 430 patients who underwent DBS for ET reported an overall postsurgical depression rate of about 1.2%, word finding problems in about 0.5%, and attention/mild cognitive decrements in about 0.9% (Flora, Perera, Cameron, & Maddern, 2010). These estimates are likely conservative because not all studies may have collected data about all types of neurobehavioral AEs. At any rate, neuropsychological evaluation has been recommended as part of the minimum standard of the pre-DBS workup in ET (Shah, Leventhal, Persad, Patil, & Chou, 2016).

Dystonia

Dystonia is a hyperkinetic movement disorder involving sustained or intermittent muscle contractions causing abnormal, often repetitive, movements and/or postures. Dystonic movements are often patterned, twisting, and sometimes tremulous. Dystonia is the third most common movement disorder. Prevalence of primary dystonia has been estimated at 15.2 per 100,000 (Epidemiological Study of Dystonia in Europe Collaborative, 2000). Traditionally, dystonia has been classified on the basis of etiology (primary vs secondary), distribution (generalized, segmental, multifocal vs focal) and age at onset (early vs late). A more recent formulation overlaps with the original system but classifies the condition along two axes: Etiology and a host of clinical features (Albanese et al., 2013).

Outcome of Deep Brain Stimulation for Dystonia

The globus pallidus is the most common surgical target for dystonia, but other targets have also been used (e.g., STN). Very few neuropsychological studies have been conducted and these were expertly reviewed recently (Jahanshahi, 2017). Jahanshahi noted that five studies of bilateral GPi or STN DBS in idiopathic or DYT1 dystonia used detailed neuropsychological assessment (Dinkelbach et al., 2015; Halbig et al., 2005; Jahanshahi et al., 2014; Owen, Gimeno, Selway, & Lin, 2015; Pillon et al., 2006) while a handful of other studies mentioned cognitive outcomes in less detail. On the basis of these studies Jahanshahi concluded that DBS treatment of idiopathic or DYT1 dystonia is not associated with change in major domains of cognition but that declines after DBS may be observed on isolated tests of sustained attention (Jahanshahi et al., 2014) and alternating category verbal fluency (Dinkelbach et al., 2015). Isolated and usually small improvements were deemed by her to be difficult to interpret in the absence of controlled studies and potential tests practice effects.

PRACTICAL APPLICATION AND ISSUES

Issues often discussed by members of DBS treatment teams include the fit of the neuropsychological evaluation within the overall patient evaluation process, the purpose of neuropsychological evaluation, the instruments to be used, and potential adaptations of tests for use with persons with movement disorders patients.

Neuropsychology Within the Overall Evaluation Process of the Deep Brain Stimulation Candidate

Several disciplines and steps are involved in the evaluation of DBS candidates. Evaluative studies include neurological, neurosurgical, and neuropsychological evaluation, neuroradiological and other laboratory investigations, preoperative education, and multidisciplinary team discussion. At our center (Barrow Neurological Institute) there is some flexibility to the process depending upon where the person enters the evaluation process and whether certain steps in the process need to be repeated. For example, when a patient is initially markedly depressed and inadequately treated they may be

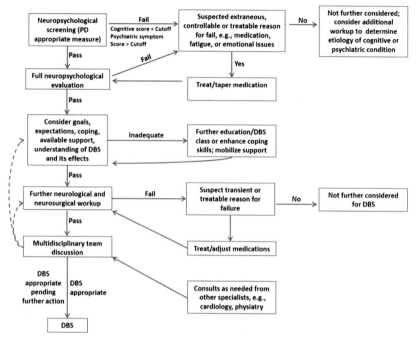

FIGURE 10.3 Flowchart of deep brain stimulation candidacy evaluation process for persons with Parkinson's disease. A very similar process is followed for persons with other movement disorders. *From Tröster et al. (2018).*

reconsidered after further treatment of depression but they may not need to undergo *all* of the evaluation steps again. The typical screening and evaluation process at Barrow Neurological Institute, and its flexibility, is highlighted in Fig. 10.3 which shows the relationship between neuropsychological screening and evaluation outcomes and progress through subsequent outcome-dependent evaluation and treatment steps.

Purposes of the Neuropsychological Evaluation

Neuropsychological evaluation prior to surgery has at least four general goals:

- Ascertaining that the pattern of a person's cognitive and emotional assets and liabilities is generally consistent with the movement disorder in question rather than an atypical parkinsonism or other disorder such as bipolar disorder or Alzheimer's disease.
- To determine whether the individual's cognitive and emotional functioning and coping resources allow them to:

- understand the DBS process and anticipated effects (including potential AEs);
- express their therapeutic goals and preferences;
- provide informed consent (including suggested reading level of educational and consent documents); and
- cooperate and comply with the evaluation and peri- and postoperative demands.
- To determine whether the nature and severity of cognitive and emotional liabilities constitute significant concerns or relative contraindications to DBS and to recommend potential treatment and reevaluation (e.g., if the person has medication-related cognitive deficits, marked depression, misconceptions about DBS, or expectations inconsistent with known treatment benefits and AEs).
- To provide information relevant to the individual's and treatment team's decision-making regarding surgical target.

Evaluation after surgery is usually carried out to:

- document quantitatively possible adverse neurobehavioral events;
- adjudicate whether neuropsychological complications are related to microlesion effects, stimulation, disease progression, medication effects or other comorbidity;
- identify potential rehabilitative and compensatory strategies for neuropsychological deficits; and
- facilitate placement decisions in persons with severe complications.

Content of the Neuropsychological Evaluation

Despite the evaluation of similar domains of cognition and emotion, the test content of DBS neuropsychological evaluations is variable among US centers (Burandt, Lebowitz, Tröster, & O'Connor, 2008). This variability is likely to be greater across international boundaries. In addition, some centers may rely on computerized or brief screening tests, and the advantages and limitations of those approaches have been discussed (Marras, Tröster, Kulisevsky, & Stebbins, 2014; Tröster, 2017). Table 10.6 presents the current domains evaluated (and tests used) at Barrow Neurological Institute. In addition to the tests, which are usually completed in 2.5 hours or less, persons being evaluated for DBS undergo a neuropsychological interview lasting 30−60 minutes, depending upon complexity of the person's medical, surgical, psychiatric, and psychosocial history. Medical records are reviewed prior to appointment and potential motor and sensory limitations (and patient's motor fluctuations and dyskinesias) are addressed with the patient and caregiver (and physician when needed) when the appointment is made. This allows for planning of an efficient examination and of necessary test modifications. In addition to covering the traditional areas of inquiry, the interview

TABLE 10.6 Neuropsychological Domains Evaluated in Deep Brain Stimulation Candidates and Some Tests and Scales Used to Evaluate These Domains at the Barrow Neurological Institute (BNI)

Domain	Test or Scale
Estimate of premorbid intelligence and current word reading level	Wide Range Achievement Test, 5th edition (WRAT-5) or Wechsler Test of Adult Reading (WTAR)
Overall level of cognitive functioning or cognitive screening	Mattis Dementia Rating Scale, 2nd edition (DRS-2) and/or Montreal Cognitive Assessment (MOCA)
Intelligence estimate	Wechsler Abbreviated Scale of Intelligence, 2nd edition (WASI-II) (two- or four-subtest version)
Attention/working memory	Trailmaking Test Part A; Digit Span; Stroop Neuropsychological Screening Test (SNST)
Executive function	Trailmaking Test Part B; Wisconsin Card Sorting Test (WCST-64); Phonemic (Letter) Verbal Fluency
Language	Semantic Verbal Fluency; Boston Naming Test (BNT; 60-item) or Neuropsychological Assessment Battery (NAB) Naming Test
Visuoperceptual and visuospatial	Judgment of Line Orientation (JLO); Hooper Visual Organization Test (VOT)
Learning and memory	Wechsler Memory Scale, 4th edition (WMS-IV) Logical Memory subtest; Hopkins Verbal Learning Test, revised (HVLT-R); Brief Visuospatial Memory Test, revised (BVMT-R); Test of Memory and Learning—Senior Edition (TOMAL-SE): Location memory subtest
Activities of daily living	Lawton and Brody Instrumental ADL scale
Quality of life	Parkinson's Disease Questionnaire (PDQ-39); Quality of Life in Essential Tremor (QUEST)
Emotion	Beck Depression Inventory, 2nd edition (BDI-II); Beck Anxiety Inventory (BDI-II); Apathy Evaluation Scale (Starkstein version); Center for Neurologic Studies—Lability Scale (CNS-LS) for Pseudobulbar Affect (PBA)
Impulsivity	Questionnaire for Impulsive-Compulsive Disorders in Parkinson's Disease-Rating Scale (QUIP-RS)
Occasionally used other tests (most common)	Personality Assessment Inventory (PAI); Clock Drawing; Grooved Pegboard; Finger Tapping; Coping Responses Inventory, Adult Version (CRI-A); NAB List Learning, Story Learning, and Medication Instructions subtests

Source: Adapted from Tröster et al. (2018).

specifically addresses the individual's expectations and goals for DBS, their understanding of the procedure and its potential effects, and their insight into current motor, cognitive, and emotional deficits. The evaluation also addresses social support, family dynamics, and caregiver expectations. Behavioral observations regarding the individual's interaction with health care providers and family are also used to infer whether there are barriers to care provision. The ability to cope with the stress of evaluation is considered as an index of a person's likely ability to tolerate lengthy investigations, surgery, and demanding postoperative visits.

Possible Test Modifications and Accommodations for Persons With Movement Disorders

When evaluating patients with movement disorders, awareness of the potential impact of various features of movement disorders (e.g., motor fluctuations, sleep disturbance and fatigability, choreiform and dystonic dyskinesias, dysarthria, hypersalivation) on test performance needs to be considered. Testing methods may need to be modified. If slurred speech is evident, patients may be asked to repeat responses although this is frustrating to some patients, perhaps necessitating testing over multiple brief sessions. Hypophonia may be addressed with an amplification device. Tests requiring pointing rather than oral responses may be more appropriate for patients with speech impairment.

A patient with tremor, dyskinesia, dystonia, or apraxia may require help from the examiner when completing tests or questionnaires requiring writing, circling of alternatives, or filling in of multiple choice blanks. Thus, such scales might be administered orally, with the examiner making the necessary written notation. On some tasks, such as card sorting or tower tests, the examiner may need to hold and move the cards or blocks/beads as instructed by the patient (standard timing cannot be used in such cases). In general, tests with significant motor demands are better avoided with patients who have movement disorders. Though nonmotor tasks might be administered when patients have dyskinesias, the patient may still be distracted by these movements, and this needs to be considered in interpreting the test results. Persons with dystonia may have significant pain and require shorter testing sessions. Twisted posture may necessitate that the examiner hold visual stimuli so that they can be seen and motoric tasks may be challenging. For ET patients it is helpful to observe postural and kinetic tremor and to perhaps have the patient write their name and draw an Archimedes spiral. This may reveal whether tasks with significant motor components can be validly administered.

In person who have somnolence, fatigue, severe motor "off" periods, or frequent fluctuations, frequent breaks and rest periods may be required. Unless there is need to compare performances "on" and "off" medications before surgery, or on and off stimulation after surgery, it is recommended that patients are tested while on medications (though anticholinergics are best tapered prior to evaluation) and on stimulation. Testing during the off state is unnecessarily challenging to patient and examiner, and the patient may experience dysphoria and anxiety that complicate testing and test interpretation.

CASE

An issue frequently confronted by neuropsychologists evaluating DBS candidates is whether uncovered cognitive deficits are related to the movement disorder or medication. This is of importance because medication-related cognitive deficits are often reversible, meaning that someone for whom DBS is seemingly inappropriate may be deemed to be a good candidate after medication is adjusted or eliminated and the patient has been reevaluated and shown to demonstrate cognitive improvement.

The profound impact that anticholinergics, even in seemingly low doses, can have on memory in PD is illustrated by the memory test results shown in Table 10.7. This man had notable memory deficits on initial evaluation (and, given apparent inappropriateness of DBS, only a brief test battery was administered). However, memory deficits were atypical in quality for PD and an effect of anticholinergics used to treat tremor was suspected. He was tapered off anticholinergic medication (trihexyphenidyl, 2 mg, four times daily) and reevaluated 6 weeks later. Memory scores improved well beyond the extent expected by practice effects alone.

FUTURE DIRECTIONS

Neuromodulation is an evolving therapy: Technological advances are frequently made and new indications for DBS are being pursued based on plausible theoretical rationales and preliminary data although there is still much to understand about the effects of DBS for currently accepted indications (Budman et al., 2018; Tröster, 2018). Questions remain unanswered about the relative neuropsychological safety of DBS of various targets and a crucial issue remains the reliable prediction of adverse outcomes. This is only likely to be answered by large multicenter studies and establishment of collaborative databases may be fruitful in this regard. What is clear is that neuropsychology's role is rapidly moving from demonstrating safety of treatment. Recently, studies have begun to address the impact of DBS in cognitive impairment, specifically dementia associated with Alzheimer's disease (using fornix DBS) (Lozano et al., 2016; Ponce et al., 2016) and PD (using DBS of the nucleus basalis of Meynert) (Gratwicke et al., 2017). As cognitive and behavioral outcomes become primary rather than secondary clinical trial endpoints, neuropsychology will be centrally involved in patient selection and in identifying underlying cognitive and neural mechanisms of DBS effects.

TABLE 10.7 Cognitive Screening, Language, and Memory Test Results in a Patient With Parkinson's Disease Before and About 6 Weeks Later, After Discontinuation of Anticholinergic Medication (Trihexyphenidyl)

Test	Baseline		Postmedication Change	
	Raw Score	T-Score/ Percentile	Raw Score	T-Score/ Percentile
WASI				
Full scale IQ	89	23rd percentile		
Dementia Rating Scale—2				
Attention	36	53	35	50
Initiation/perseveration	35	43	36	50
Construction	4	33	6	50
Conceptualization	31	37	30	37
Memory	20	33	23	50
Total score	126	33	130	40
Letter fluency	32	49	–	
Animal naming	20	59	–	
Boston naming test	55		–	
Wechsler Memory Scale—III				
Working memory index	88		99	
Logical memory I immediate recall	17	4	34	10
Logical memory II delayed recall	8	5	20	11
Logical memory % retained	57	8	95	14
Facial recognition immediate recall	28	7	–	
Facial recognition delayed recall	29	8		
Facial recognition % retained	100	12		
California Verbal Learning Test—II				
Total trials 1–5	27	38	32	45
Short delay free recall	4	40	5	40
Long delay free recall	3	35	6	40
Recognition hits	10	25	9	25
Recognition discriminability	0.9	30	2	45

Source: From Tröster and Garrett (2018).

REFERENCES

Aarsland, D., Perry, R., Brown, A., Larsen, J. P., & Ballard, C. (2005). Neuropathology of dementia in Parkinson's disease: A prospective, community-based study. *Annals of Neurology*, *58*(5), 773−776.

Aarsland, D., Taylor, J. P., & Weintraub, D. (2014). Psychiatric issues in cognitive impairment. *Movement Disorders*, *29*(5), 651−662. Available from https://doi.org/10.1002/mds.25873.

Abboud, H., Floden, D., Thompson, N. R., Genc, G., Oravivattanakul, S., Alsallom, F., ... Fernandez, H. H. (2015). Impact of mild cognitive impairment on outcome following deep brain stimulation surgery for Parkinson's disease. *Parkinsonism & Related Disorders*, *21*(3), 249−253. Available from https://doi.org/10.1016/j.parkreldis.2014.12.018.

Abboud, H., Mehanna, R., Machado, A., Ahmed, A., Gostkowski, M., Cooper, S., ... Fernandez, H. H. (2014). Comprehensive, multidisciplinary deep brain stimulation screening for Parkinson's patients: No room for "short cuts". *Movement Disorders Clinical Practice*, *1*(4), 336−341. Available from https://doi.org/10.1002/mdc3.12090.

Abel, T. J., Walch, T., & Howard, M. A., 3rd (2016). Russell Meyers (1905−1999): Pioneer of functional and ultrasonic neurosurgery. *Journal of Neurosurgery*, *125*(6), 1589−1595. Available from https://doi.org/10.3171/2015.9.JNS142811.

Akram, H., Miller, S., Lagrata, S., Hyam, J., Jahanshahi, M., Hariz, M., ... Zrinzo, L. (2016). Ventral tegmental area deep brain stimulation for refractory chronic cluster headache. *Neurology*, *86*(18), 1676−1682.

Albanese, A., Bhatia, K., Bressman, S. B., Delong, M. R., Fahn, S., Fung, V. S., ... Teller, J. K. (2013). Phenomenology and classification of dystonia: A consensus update. *Movement Disorders*, *28*(7), 863−873. Available from https://doi.org/10.1002/mds.25475.

Alegret, M., Junqué, C., Valldeoriola, F., Vendrell, P., Pilleri, M., Rumià, J., ... Tolosa, E. (2001). Effects of bilateral subthalamic stimulation on cognitive function in Parkinson's disease. *Archives of Neurology*, *58*(8), 1223−1227.

Alegret, M., Vendrell, P., Junqué, C., Valldeoriola, F., Nobbe, F. A., Rumià, J., & Tolosa, E. (2000). Effects of unilateral posteroventral pallidotomy on 'on−off' cognitive fluctuations in Parkinson's disease. *Neuropsychologia*, *38*(5), 628−633.

Almgren, P. E., Andersson, A. L., & Kullberg, G. (1969). Differences in verbally expressed cognition following left and right ventrolateral thalamotomy. *Scandinavian Journal of Psychology*, *10*(4), 243−249.

Alvarez, L., Macias, R., Guridi, J., Lopez, G., Alvarez, E., Maragoto, C., ... Obeso, J. A. (2001). Dorsal subthalamotomy for Parkinson's disease. *Movement Disorders*, *16*(1), 72−78.

Andy, O. J., Jurko, M. F., & Sias, F. R., Jr. (1963). Subthalamotomy in treatment of parkinsonian tremor. *Journal of Neurosurgery*, *20*, 861−871.

Appleby, B. S., Duggan, P. S., Regenberg, A., & Rabins, P. V. (2007). Psychiatric and neuropsychiatric adverse events associated with deep brain stimulation: A meta-analysis of ten years' experience. *Movement Disorders*, *22*(12), 1722−1728. Available from https://doi.org/10.1002/mds.21551.

Ardouin, C., Pillon, B., Peiffer, E., Bejjani, P., Limousin, P., Damier, P., ... Pollak, P. (1999). Bilateral subthalamic or pallidal stimulation for Parkinson's disease affects neither memory nor executive functions: A consecutive series of 62 patients. *Annals of Neurology*, *46*(2), 217−223.

Aybek, S., Lazeyras, F., Gronchi-Perrin, A., Burkhard, P. R., Villemure, J. G., & Vingerhoets, F. J. (2009). Hippocampal atrophy predicts conversion to dementia after STN-DBS in Parkinson's disease. *Parkinsonism & Related Disorders*, *15*(7), 521−524. Available from https://doi.org/10.1016/j.parkreldis.2009.01.003.

Bechtereva, N. P., Bondartchuk, A. N., Smirnov, V. M., Meliutcheva, L. A., & Shandurina, A. N. (1975). Method of electrostimulation of the deep brain structures in treatment of some chronic diseases. *Confinia Neurologica*, *37*(1−3), 136−140.

Benabid, A. L., Pollak, P., Gervason, C., Hoffmann, D., Gao, D. M., Hommel, M., . . . de Rougemont, J. (1991). Long-term suppression of tremor by chronic stimulation of the ventral intermediate thalamic nucleus. *Lancet*, *337*(8738), 403−406.

Bentivoglio, A. R., Fasano, A., Piano, C., Soleti, F., Daniele, A., Zinno, M., . . . Meglio, M. (2012). Unilateral extradural motor cortex stimulation is safe and improves Parkinson disease at 1 year. *Neurosurgery*, *71*(4), 815−825.

Bergfeld, I. O., Mantione, M., Hoogendoorn, M. L., Ruhé, H. G., Notten, P., van Laarhoven, J., . . . Schene, A. H. (2016). Deep brain stimulation of the ventral anterior limb of the internal capsule for treatment-resistant depression: A randomized clinical trial. *JAMA Psychiatry*, *73* (5), 456−464.

Bewernick, B. H., Kayser, S., Gippert, S. M., Switala, C., Coenen, V. A., & Schlaepfer, T. E. (2017). Deep brain stimulation to the medial forebrain bundle for depression-long-term outcomes and a novel data analysis strategy. *Brain Stimulation*, *10*(3), 664−671.

Bickel, S., Alvarez, L., Macias, R., Pavon, N., Leon, M., Fernandez, C., . . . Litvan, I. (2010). Cognitive and neuropsychiatric effects of subthalamotomy for Parkinson's disease. *Parkinsonism & Related Disorders*, *16*(8), 535−539. Available from https://doi.org/10.1016/j.parkreldis.2010.06.008.

Blomstedt, P., & Hariz, M. I. (2010). Deep brain stimulation for movement disorders before DBS for movement disorders. *Parkinsonism & Related Disorders*, *16*(7), 429−433. Available from https://doi.org/10.1016/j.parkreldis.2010.04.005.

Blume, J., Lange, M., Rothenfusser, E., Doenitz, C., Bogdahn, U., Brawanski, A., & Schlaier, J. (2017). The impact of white matter lesions on the cognitive outcome of subthalamic nucleus deep brain stimulation in Parkinson's disease. *Clinical Neurology and Neurosurgery*, *159*, 87−92. Available from https://doi.org/10.1016/j.clineuro.2017.05.023.

Blumetti, A. E., & Modesti, L. M. (1980). Long term cognitive effects of stereotactic thalamotomy on non-parkinsonian dyskinetic patients. *Applied Neurophysiology*, *43*(3−5), 259−262.

Boccard, S. G., Pereira, E. A., Moir, L., Aziz, T. Z., . . . Green, A. L. (2012). Long-term outcomes of deep brain stimulation for neuropathic pain. *Neurosurgery*, *72*(2), 221−231.

Boccard, S. G., Fitzgerald, J. J., Pereira, E. A., Moir, L., Van Hartevelt, T. J., Kringelbach, M. L., . . . Aziz, T. Z. (2014). Targeting the affective component of chronic pain: a case series of deep brain stimulation of the anterior cingulate cortex. *Neurosurgery*, *74*(6), 628−637.

Bond, A. E., Shah, B. B., Huss, D. S., Dallapiazza, R. F., Warren, A., Harrison, M. B., . . . Elias, W. J. (2017). Safety and efficacy of focused ultrasound thalamotomy for patients with medication-refractory, tremor-dominant Parkinson's disease: A randomized clinical trial. *JAMA Neurology*, *74*(12), 1412−1418. Available from https://doi.org/10.1001/jamaneurol.2017.3098.

Borden, A., Wallon, D., Lefaucheur, R., Derrey, S., Fetter, D., Verin, M., & Maltete, D. (2014). Does early verbal fluency decline after STN implantation predict long-term cognitive outcome after STN-DBS in Parkinson's disease? *Journal of the Neurological Sciences*, *346*(1-2), 299−302. Available from https://doi.org/10.1016/j.jns.2014.07.063.

Brice, J., & McLellan, L. (1980). Suppression of intention tremor by contingent deep-brain stimulation. *Lancet*, *1*(8180), 1221−1222.

Brodsky, M. A., Anderson, S., Murchison, C., Seier, M., Wilhelm, J., Vederman, A., & Burchiel, K. J. (2017). Clinical outcomes of asleep vs awake deep brain stimulation for Parkinson's disease. *Neurology*, *89*, 1944−1950. Available from https://doi.org/10.1212/WNL.0000000000004630.

Bronstein, J. M., Tagliati, M., Alterman, R. L., Lozano, A. M., Volkmann, J., Stefani, A., ... DeLong, M. R. (2011). Deep brain stimulation for Parkinson's disease: An expert consensus and review of key issues. *Archives of Neurology*, *68*(2), 165. Available from https://doi.org/ 10.1001/archneurol.2010.260.

Budman, E., Deeb, W., Martinez-Ramirez, D., Pilitsis, J. G., Peng-Chen, Z., Okun, M. S., & Ramirez-Zamora, A. (2018). Potential indications for deep brain stimulation in neurological disorders: An evolving field. *European Journal of Neurology*, *25*(3). Available from https:// doi.org/10.1111/ene.13548, 434-e430.

Burandt, C. A., Lebowitz, B. K., Tröster, A. I., & O'Connor, M. G. (2008). The role of neuro-psychology in the evaluation of surgical candidates for deep brain stimulation in Parkinson's disease: A survey study. *The Clinical Neuropsychologist*, *22*, 390.

Burchiel, K. J., Anderson, V. C., Favre, J., & Hammerstad, J. P. (1999). Comparison of pallidal and subthalamic nucleus deep brain stimulation for advanced Parkinson's disease: Results of a randomized, blinded pilot study. *Neurosurgery*, *45*(6), 1375−1382. (discussion 1382-1374).

Cannon, E., Silburn, P., Coyne, T., O'Maley, K., Crawford, J. D., & Sachdev, P. S. (2012). Deep brain stimulation of anteromedial globus pallidus interna for severe Tourette's syndrome. *American Journal of Psychiatry*, *169*(8), 860−866.

Capelle, H. H., Blahak, C., Schrader, C., Baezner, H., Kinfe, T. M., Herzog, J., ... Krauss, J. K. (2010). Chronic deep brain stimulation in patients with tardive dystonia without a history of major psychosis. *Movement Disorders*, *25*(10), 1477−1481.

Caparros-Lefebvre, D., Blond, S., Pécheux, N., Pasquier, F., & Petit, H. (1992). Evaluation neu-ropsychologique avant et après stimulation thalamique chez 9 parkinsoniens. *Revue Neurologique*, *148*(2), 117−122.

Castrioto, A., Lhommee, E., Moro, E., & Krack, P. (2014). Mood and behavioural effects of sub-thalamic stimulation in Parkinson's disease. *Lancet Neurology*, *13*(3), 287−305. Available from https://doi.org/10.1016/S1474-4422(13)70294-1.

Chabardès, S., Carron, R., Seigneuret, E., Torres, N., Goetz, L., Krainik, A., ... Benabid, A. L. (2016). Endoventricular deep brain stimulation of the third ventricle: Proof of concept and application to cluster headache. *Neurosurgery*, *79*(6), 806−815.

Chen, T., Mirzadeh, Z., Chapple, K., Lambert, M., Dhall, R., & Ponce, F. A. (2016). "Asleep" deep brain stimulation for essential tremor. *Journal of Neurosurgery*, *124*(6), 1842−1849.

Chkhenkeli, S. A., Šramka, M., Lortkipanidze, G. S., Rakviashvili, T. N., Bregvadze, E. S., Magalashvili, G. E., ... Chkhenkeli, I. S. (2004). Electrophysiological effects and clinical results of direct brain stimulation for intractable epilepsy. *Clinical Neurology and Neurosurgery*, *106*(4), 318−329.

Coleman, R. R., Kotagal, V., Patil, P. G., & Chou, K. L. (2014). Validity and efficacy of screen-ing algorithms for assessing deep brain stimulation candidacy in Parkinson's disease. *Movement Disorders Clinical Practice*, *1*(4), 342−347. Available from https://doi.org/ 10.1002/mdc3.12103.

Combs, H. L., Folley, B. S., Berry, D. T., Segerstrom, S. C., Han, D. Y., Anderson-Mooney, A. J., ... van Horne, C. (2015). Cognition and depression following deep brain stimulation of the subthalamic nucleus and globus pallidus pars internus in Parkinson's disease: A meta-analysis. *Neuropsychology Review*, *25*, 439−454. Available from https://doi.org/10.1007/ s11065-015-9302-0.

Cooper, I. S. (1954). Intracerebral injection of procaine into the globus pallidus in hyperkinetic disorders. *Science*, *119*(3091), 417−418.

Cury, R. G., Fraix, V., Castrioto, A., Fernández, M. A. P., Krack, P., Chabardes, S., . . . Moro, E. (2017). Thalamic deep brain stimulation for tremor in Parkinson disease, essential tremor, and dystonia. *Neurology*, *89*(13), 1416−1423.

Daniels, C., Krack, P., Volkmann, J., Pinsker, M. O., Krause, M., Tronnier, V., . . . Witt, K. (2010). Risk factors for executive dysfunction after subthalamic nucleus stimulation in Parkinson's disease. *Movement Disorders*, *25*(11), 1583−1589. Available from https://doi. org/10.1002/mds.23078.

Deng, Z. D., Li, D. Y., Zhang, C. C., Pan, Y. X., Zhang, J., Jin, H., . . . Sun, B. M. (2017). Long-term follow-up of bilateral subthalamic deep brain stimulation for refractory tardive dystonia. *Parkinsonism & Related Disorders*, *41*, 58−65.

Denys, D., Mantione, M., Figee, M., van den Munckhof, P., Koerselman, F., Westenberg, H., . . . Schuurman, R. (2010). Deep brain stimulation of the nucleus accumbens for treatment-refractory obsessive-compulsive disorder. *Archives of General Psychiatry*, *67*(10), 1061−1068.

De Rose, M., Guzzi, G., Bosco, D., Romano, M., Lavano, S. M., Plastino, M., . . . Lavano, A. (2012). Motor cortex stimulation in Parkinson's disease. *Neurology Research International*, 2012. Available from https://doi.org/10.1155/2012/502096.

Dinkelbach, L., Mueller, J., Poewe, W., Delazer, M., Elben, S., Wolters, A., . . . Sudmeyer, M. (2015). Cognitive outcome of pallidal deep brain stimulation for primary cervical dystonia: One year follow up results of a prospective multicenter trial. *Parkinsonism & Related Disorders*, *21*(8), 976−980. Available from https://doi.org/10.1016/j.parkreldis.2015.06.002.

Dougherty, D. D., Rezai, A. R., Carpenter, L. L., Howland, R. H., Bhati, M. T., O'Reardon, J. P., . . . Cusin, C. (2015). A randomized sham-controlled trial of deep brain stimulation of the ventral capsule/ventral striatum for chronic treatment-resistant depression. *Biological Psychiatry*, *78*(4), 240−248.

Drapier, S., Raoul, S., Drapier, D., Leray, E., Lallement, F., Rivier, I., . . . Verin, M. (2005). Only physical aspects of quality of life are significantly improved by bilateral subthalamic stimulation in Parkinson's disease. *Journal of Neurology*, *252*(5), 583−588.

Duprez, T. P., Serieh, B. A., & Raftopoulos, C. (2005). Absence of memory dysfunction after bilateral mammillary body and mammillothalamic tract electrode implantation: Preliminary experience in three patients. *American Journal of Neuroradiology*, *26*(1), 195−198.

Elble, R. J., Shih, L., & Cozzens, J. W. (2018). Surgical treatments for essential tremor. *Expert Review of Neurotherapeutics*, 1−19. Available from https://doi.org/10.1080/14737175.2018.1445526.

Eisinger, R. S., Wong, J., Almeida, L., Ramirez-Zamora, A., Cagle, J. N., Giugni, J. C., . . . Hess, C. W. (2017). Ventral Intermediate Nucleus Versus Zona Incerta Region Deep Brain Stimulation In Essential Tremor. *Movement Disorders Clinical Practice*.

Elias, W. J., Huss, D., Voss, T., Loomba, J., Khaled, M., Zadicario, E., . . . Wintermark, M. (2013). A pilot study of focused ultrasound thalamotomy for essential tremor. *New England Journal of Medicine*, *369*(7), 640−648. Available from https://doi.org/10.1056/NEJMoa1300962.

Elgebaly, A., Elfil, M., Attia, A., Magdy, M., & Negida, A. (2017). Neuropsychological performance changes following subthalamic versus pallidal deep brain stimulation in Parkinson's disease: a systematic review and metaanalysis. *CNS Spectrum*, 1−14. Available from https://doi.org/10.1017/S1092852917000062.

Epidemiological Study of Dystonia in Europe Collaborative, G. (2000). A prevalence study of primary dystonia in eight European countries. *Journal of Neurology*, *247*(10), 787−792.

Fasano, A., Romito, L. M., Daniele, A., Piano, C., Zinno, M., Bentivoglio, A. R., & Albanese, A. (2010). Motor and cognitive outcome in patients with Parkinson's disease 8 years after subthalamic implants. *Brain*, *133*(9), 2664−2676. Available from https://doi.org/10.1093/brain/awq221.

Fields, J. A., & Tröster, A. I. (2000). Cognitive outcomes after deep brain stimulation for Parkinson's disease: A review of initial studies and recommendations for future research. *Brain and Cognition*, *42*(2), 268−293.

Fields, J. A., Tröster, A. I., Wilkinson, S. B., Pahwa, R., & Koller, W. C. (1999). Cognitive outcome following staged bilateral pallidal stimulation for the treatment of Parkinson's disease. *Clinical Neurology and Neurosurgery*, *101*(3), 182−188.

Fields, J. A., Tröster, A. I., Woods, S. P., Higginson, C. I., Wilkinson, S. B., Lyons, K. E., . . . Pahwa, R. (2003). Neuropsychological and quality of life outcomes 12 months after unilateral thalamic stimulation for essential tremor. *Journal of Neurology, Neurosurgery, and Psychiatry*, *74*(3), 305−311.

Figee, M., Luigjes, J., Smolders, R., Valencia-Alfonso, C. E., Van Wingen, G., De Kwaasteniet, B., . . . Levar, N. (2013). Deep brain stimulation restores frontostriatal network activity in obsessive-compulsive disorder. *Nature Neuroscience*, *16*(4), 386−387.

Fisher, R., Salanova, V., Witt, T., Worth, R., Henry, T., Gross, R., . . . Kaplitt, M. (2010). Electrical stimulation of the anterior nucleus of thalamus for treatment of refractory epilepsy. *Epilepsia*, *51*(5), 899−908.

Flora, E. D., Perera, C. L., Cameron, A. L., & Maddern, G. J. (2010). Deep brain stimulation for essential tremor: A systematic review. *Movement Disorders*, *25*(11), 1550−1559. Available from https://doi.org/10.1002/mds.23195.

Foki, T., Hitzl, D., Pirker, W., Novak, K., Pusswald, G., Auff, E., & Lehrner, J. (2017). Assessment of individual cognitive changes after deep brain stimulation surgery in Parkinson's disease using the Neuropsychological Test Battery Vienna short version. *Wiener Klinische Wochenschrift*. Available from https://doi.org/10.1007/s00508-017-1169-z.

Fontaine, D., Lazorthes, Y., Mertens, P., Blond, S., Géraud, G., Fabre, N., . . . Paquis, P. (2010). Safety and efficacy of deep brain stimulation in refractory cluster headache: a randomized placebo-controlled doubleblind trial followed by a 1-year open extension. *The Journal of Headache and Pain*, *11*(1), 23−31.

Franzini, A., Broggi, G., Cordella, R., Dones, I., & Messina, G. (2013). Deep-brain stimulation for aggressive and disruptive behavior. *World Neurosurgery*, *80*(3), S29-e11.

Fukaya, C., Watanabe, M., Kobayashi, K., Oshima, H., Yoshino, A., & Yamamoto, T. (2017). Predictive factors for long-term outcome of subthalamic nucleus deep brain stimulation for Parkinson's disease. *Neurologia Medico-Chirurgica*, *57*(4), 166−171. Available from https://doi.org/10.2176/nmc.oa.2016-0114.

Funkiewiez, A., Ardouin, C., Caputo, E., Krack, P., Fraix, V., Klinger, H., . . . Pollak, P. (2004). Long term effects of bilateral subthalamic nucleus stimulation on cognitive function, mood, and behaviour in Parkinson's disease. *Journal of Neurology, Neurosurgery and Psychiatry*, *75*(6), 834−839.

Fytagoridis, A., Åström, M., Samuelsson, J., & Blomstedt, P. (2016). Deep brain stimulation of the caudal zona incerta: Tremor control in relation to the location of stimulation fields. *Stereotactic and Functional Neurosurgery*, *94*(6), 363−370.

Gabriels, L., Cosyns, P., Nuttin, B., Demeulemeester, H., & Gybels, J. (2003). Deep brain stimulation for treatment-refractory obsessive-compulsive disorder: Psychopathological and neuropsychological outcome in three cases. *Acta Psychiatrica Scandinavica*, *107*(4), 275−282.

Geevarghese, R., Lumsden, D. E., Costello, A., Hulse, N., Ayis, S., Samuel, M., & Ashkan, K. (2016). Verbal memory decline following DBS for Parkinson's disease: Structural volumetric MRI relationships. *PLoS One*, *11*(8), e0160583. Available from https://doi.org/10.1371/journal.pone.0160583.

Gonzalez, V., Cif, L., Biolsi, B., Garcia-Ptacek, S., Seychelles, A., Sanrey, E., ... James, S. (2014). Deep brain stimulation for Huntington's disease: Long-term results of a prospective open-label study. *Journal of Neurosurgery*, *121*(1), 114−122.

Goldman, J. G., Holden, S., Bernard, B., Ouyang, B., Goetz, C. G., & Stebbins, G. T. (2013). Defining optimal cutoff scores for cognitive impairment using Movement Disorder Society Task Force criteria for mild cognitive impairment in Parkinson's disease. *Movement Disorders*, *28*(14), 1972−1979. Available from https://doi.org/10.1002/mds.25655.

Gratwicke, J., Zrinzo, L., Kahan, J., Peters, A., Beigi, M., Akram, H., ... Foltynie, T. (2017). Bilateral deep brain stimulation of the nucleus basalis of Meynert for Parkinson's disease dementia: A randomized clinical trial. *JAMA Neurology*. Available from https://doi.org/10.1001/jamaneurol.2017.3762.

Green, J., McDonald, W. M., Vitek, J. L., Haber, M., Barnhart, H., Bakay, R. A., ... DeLong, M. R. (2002). Neuropsychological and psychiatric sequelae of pallidotomy for PD: Clinical trial findings. *Neurology*, *58*(6), 858−865.

Greenberg, B. D., Gabriels, L. A., Malone, D. A., Rezai, A. R., Friehs, G. M., Okun, M. S., ... Malloy, P. F. (2010). Deep brain stimulation of the ventral internal capsule/ventral striatum for obsessive-compulsive disorder: worldwide experience. *Molecular Psychiatry*, *15*(1), 64−79.

Gruber, D., Kühn, A. A., Schoenecker, T., Kivi, A., Trottenberg, T., Hoffmann, K. T., ... Asmus, F. (2010). Pallidal and thalamic deep brain stimulation in myoclonus-dystonia. *Movement Disorders*, *25*(11), 1733−1743.

Halbig, T. D., Gruber, D., Kopp, U. A., Schneider, G. H., Trottenberg, T., & Kupsch, A. (2005). Pallidal stimulation in dystonia: Effects on cognition, mood, and quality of life. *Journal of Neurology, Neurosurgery and Psychiatry*, *76*(12), 1713−1716.

Hariz, M. I., Johansson, F., Shamsgovara, P., Johansson, E., Hariz, G. M., & Fagerlund, M. (2000). Bilateral subthalamic nucleus stimulation in a parkinsonian patient with preoperative deficits in speech and cognition: Persistent improvement in mobility but increased dependency: A case study. *Movement Disorders*, *15*(1), 136−139.

Hasegawa, H., Samuel, M., Douiri, A., & Ashkan, K. (2014). Patients' expectations in subthalamic nucleus deep brain stimulation surgery for Parkinson's disease. *World Neurosurgery*, *82* (6). Available from https://doi.org/10.1016/j.wneu.2014.02.001, 1295-1299 e1292.

Hassler, R., & Riechert, T. (1954). Indications and localization of stereotactic brain operations. *Nervenarzt*, *25*(11), 441−447.

Hägglund, P., Sandström, L., Blomstedt, P., & Karlsson, F. (2016). Voice tremor in patients with essential tremor: Effects of deep brain stimulation of caudal zona incerta. *Journal of Voice*, *30*(2), 228−233.

Heber, I. A., Coenen, V. A., Reetz, K., Schulz, J. B., Hoellig, A., Fimm, B., & Kronenbuerger, M. (2013). Cognitive effects of deep brain stimulation for essential tremor: Evaluation at 1 and 6 years. *Journal of Neural Transmission (Vienna)*, *120*(11), 1569−1577. Available from https://doi.org/10.1007/s00702-013-1030-0.

Holtzheimer, P. E., Kelley, M. E., Gross, R. E., Filkowski, M. M., Garlow, S. J., Barrocas, A., ... Moreines, J. L. (2012). Subcallosal cingulate deep brain stimulation for treatment-resistant unipolar and bipolar depression. *Archives of General Psychiatry*, *69*(2), 150−158.

Hoogland, J., Boel, J. A., de Bie, R. M. A., Geskus, R. B., Schmand, B. A., Dalrymple-Alford, J. C., ... MDS Study Group "Validation of Mild Cognitive Impairment in Parkinson's Disease". (2017). Mild cognitive impairment as a risk factor for Parkinson's disease dementia. *Movement Disorders, 32*(7), 1056−1065. Available from https://doi.org/10.1002/mds.27002.

Horsley, V. (1909). The Linacre lecture on the function of the so-called motor area of the brain: Delivered to the Master and Fellows of St. John's College, Cambridge, May 6th, 1909. *British Medical Journal, 2*(2533), 121−132.

Houdart, R., Mamo, H., Dondey, M., & Cophignon, J. (1965). Results of subthalamic coagulations in Parkinson's disease (apropos of 50 cases). *Revue Neurologique, 112*(6), 521−529.

Hrabovsky, D., Balaz, M., Rab, M., Feitova, V., Hummelova, Z., Novak, Z., & Chrastina, J. (2017). Factors responsible for early postoperative mental alterations after bilateral implantation of subthalamic electrodes. *British Journal of Neurosurgery, 31*(2), 212−216. Available from https://doi.org/10.1080/02688697.2016.1226256.

Jahanshahi, M. (2017). Neuropsychological and neuropsychiatric features of idiopathic and DYT1 dystonia and the impact of medical and surgical treatment. *Archives of Clinical Neuropsychology, 32*(7), 888−905. Available from https://doi.org/10.1093/arclin/acx095.

Jahanshahi, M., Czernecki, V., & Zurowski, A. M. (2011). Neuropsychological, neuropsychiatric, and quality of life issues in DBS for dystonia. *Movement Disorders, 26*(Suppl. 1), S63−S78. Available from https://doi.org/10.1002/mds.23511.

Jahanshahi, M., Torkamani, M., Beigi, M., Wilkinson, L., Page, D., Madeley, L., ... Tisch, S. (2014). Pallidal stimulation for primary generalised dystonia: Effect on cognition, mood and quality of life. *Journal of Neurology, 261*(1), 164−173. Available from https://doi.org/10.1007/s00415-013-7161-2.

Jiménez-Ponce, F., Velasco-Campos, F., Castro-Farfán, G., Nicolini, H., Velasco, A. L., ... Salín-Pascual, R., ... Criales, J. L. (2009). Preliminary study in patients with obsessive-compulsive disorder treated with electrical stimulation in the inferior thalamic peduncle. *Operative Neurosurgery, 65*, 203−209.

Jurko, M. F., & Andy, O. J. (1973). Psychological changes correlated with thalamotomy site. *Journal of Neurology, Neurosurgery, and Psychiatry, 36*(5), 846−852.

Khan, S., Gill, S. S., Mooney, L., White, P., Whone, A., Brooks, D. J., & Pavese, N. (2012). Combined pedunculopontine-subthalamic stimulation in Parkinson disease. *Neurology, 78*(14), 1090−1095.

KhanWeaver, F. M., Follett, K., Stern, M., Hur, K., Harris, C., Marks, W. J., ... Pahwa, R. (2009). Bilateral deep brain stimulation vs best medical therapy for patients with advanced Parkinson disease: A randomized controlled trial. *JAMA, 301*(1), 63−73.

Kowski, A. B., Voges, J., Heinze, H. J., Oltmanns, F., Holtkamp, M., & Schmitt, F. C. (2015). Nucleus accumbens stimulation in partial epilepsy—a randomized controlled case series. *Epilepsia, 56*, 6.

Kim, H. J., Jeon, B. S., Paek, S. H., Lee, K. M., Kim, J. Y., Lee, J. Y., ... Ehm, G. (2014). Long-term cognitive outcome of bilateral subthalamic deep brain stimulation in Parkinson's disease. *Journal of Neurology, 261*(6), 1090−1096. Available from https://doi.org/10.1007/s00415-014-7321-z.

Kleiner-Fisman, G., Liang, G. S., Moberg, P. J., Ruocco, A. C., Hurtig, H. I., Baltuch, G. H., ... Stern, M. B. (2007). Subthalamic nucleus deep brain stimulation for severe idiopathic dystonia: Impact on severity, neuropsychological status, and quality of life. *Journal of Neurosurgery, 107*(1), 29−36.

Kocher, U., Siegfried, J., & Perret, E. (1982). Verbal and nonverbal learning ability of Parkinson patients before and after unilateral ventrolateral thalamotomy. *Applied Neurophysiology*, *45* (3), 311−316.

Koller, W., Pahwa, R., Busenbark, K., Hubble, J., Wilkinson, S., Lang, A., … Olanow, C. W. (1997). High-frequency unilateral thalamic stimulation in the treatment of essential and parkinsonian tremor. *Annals of Neurology*, *42*(3), 292−299.

Kubu, C. S., Grace, G. M., & Parrent, A. G. (2000). Cognitive outcome following pallidotomy: The influence of side of surgery and age of patient at disease onset. *Journal of Neurosurgery*, *92*(3), 384−389.

Kuhn, J., Gründler, T. O., Bauer, R., Huff, W., Fischer, A. G., Lenartz, D., … Sturm, V. (2011). Successful deep brain stimulation of the nucleus accumbens in severe alcohol dependence is associated with changed performance monitoring. *Addiction Biology*, *16*(4), 620−623.

Kuhn, J., Moller, M., Treppmann, J. F., Bartsch, C., Lenartz, D., Gründler, T. O., … Sturm, V. (2014). Deep brain stimulation of the nucleus accumbens and its usefulness in severe opioid addiction. *Molecular Psychiatry*, *19*(2), 145−147.

Kuhn, J., Hardenacke, K., Lenartz, D., Gruendler, T., Ullsperger, M., Bartsch, C., … Schulz, R. J. (2015). Deep brain stimulation of the nucleus basalis of Meynert in Alzheimer's dementia. *Molecular Psychiatry*, *20*(3), 353−360.

Laitinen, L. V. (2000). Behavioral complications of early pallidotomy. *Brain and Cognition*, *42* (3), 313−323.

Laitinen, L. V., Bergenheim, A. T., & Hariz, M. I. (1992). Leksell's posteroventral pallidotomy in the treatment of Parkinson's disease [see comments]. *Journal of Neurosurgery*, *76*(1), 53−61.

Lang, A. E., Houeto, J. L., Krack, P., Kubu, C., Lyons, K. E., Moro, E., … Voon, V. (2006). Deep brain stimulation: preoperative issues. *Movement Disorders*, *21*(Suppl. 14), S171−S196.

Larson, P. S., & Cheung, S. W. (2013). A stroke of silence: tinnitus suppression following placement of a deep brain stimulation electrode with infarction in area LC: Case report. *Journal of Neurosurgery*, *118*(1), 192−194.

Laxton, A. W., Tang-Wai, D. F., McAndrews, M. P., Zumsteg, D., Wennberg, R., Keren, R., … Lozano, A. M. (2010). A phase I trial of deep brain stimulation of memory circuits in Alzheimer's disease. *Annals of Neurology*, *68*(4), 521−534.

Lee, J. H., Cho, W. H., Cha, S. H., & Kang, D. W. (2015). Globus pallidus interna deep brain stimulation for chorea- acanthocytosis. *Journal of Korean Neurosurgical Society*, *57*(2), 143−146.

Lehtimäki, K., Möttönen, T., Järventausta, K., Katisko, J., Tähtinen, T., Haapasalo, J., … Peltola, J. (2016). Outcome based definition of the anterior thalamic deep brain stimulation target in refractory epilepsy. *Brain Stimulation*, *9*(2), 268−275.

Lim, T. T., Fernandez, H. H., Cooper, S., Wilson, K. M. K., & Machado, A. G. (2013). Successful deep brain stimulation surgery with intraoperative magnetic resonance imaging on a difficult neuroacanthocytosis case: Case report. *Neurosurgery*, *73*(1), E184−E188.

Lipsman, N., Woodside, D. B., Giacobbe, P., Hamani, C., Carter, J. C., Norwood, S. J., … Smith, G. S. (2013). Subcallosal cingulate deep brain stimulation for treatment-refractory anorexia nervosa: a phase 1 pilot trial. *The Lancet*, *381*(9875), 1361−1370.

Lipsman, N., Schwartz, M. L., Huang, Y., Lee, L., Sankar, T., Chapman, M., … Lozano, A. M. (2013a). MR-guided focused ultrasound thalamotomy for essential tremor: A proof-of-concept study. *Lancet Neurology*, *12*(5), 462−468. Available from https://doi.org/10.1016/S1474-4422(13)70048-6.

Liu, H. G., Zhang, K., Yang, A. C., & Zhang, J. G. (2015). Deep brain stimulation of the subthalamic and pedunculopontine nucleus in a patient with Parkinson's disease. *Journal of Korean Neurosurgical Society, 57*(4), 303−306.

Lopiano, L., Rizzone, M., Bergamasco, B., Tavella, A., Torre, E., Perozzo, P., & Lanotte, M. (2002). Deep brain stimulation of the subthalamic nucleus in PD: An analysis of the exclusion causes. *Journal of the Neurological Sciences, 195*(2), 167−170.

Louis, E. D., & Ferreira, J. J. (2010). How common is the most common adult movement disorder? Update on the worldwide prevalence of essential tremor. *Movement Disorders, 25*(5), 534−541. Available from https://doi.org/10.1002/mds.22838.

Lozano, A. M., Fosdick, L., Chakravarty, M. M., Leoutsakos, J. M., Munro, C., Oh, E., ... Tang-Wai, D. F. (2016). A phase II study of fornix deep brain stimulation in mild Alzheimer's disease. *Journal of Alzheimer's Disease, 54*(2), 777−787.

Lozano, A. M., Fosdick, L., Chakravarty, M. M., Leoutsakos, J. M., Munro, C., Oh, E., ... Smith, G. S. (2016). A phase II study of Fornix deep brain stimulation in mild Alzheimer's disease. *Journal of Alzheimer's Disease, 54*(2), 777−787. Available from https://doi.org/10.3233/JAD-160017.

Maier, F., Lewis, C. J., Horstkoetter, N., Eggers, C., Dembek, T. A., Visser-Vandewalle, V., ... Timmermann, L. (2016). Subjective perceived outcome of subthalamic deep brain stimulation in Parkinson's disease one year after surgery. *Parkinsonism & Related Disorders, 24*, 41−47. Available from https://doi.org/10.1016/j.parkreldis.2016.01.019.

Maier, F., Lewis, C. J., Horstkoetter, N., Eggers, C., Kalbe, E., Maarouf, M., ... Timmermann, L. (2013). Patients' expectations of deep brain stimulation, and subjective perceived outcome related to clinical measures in Parkinson's disease: A mixed-method approach. *Journal of Neurology, Neurosurgery and Psychiatry, 84*(11), 1273−1281. Available from https://doi.org/10.1136/jnnp-2012-303670.

Maling, N., Hashemiyoon, R., Foote, K. D., Okun, M. S., & Sanchez, J. C. (2012). Increased thalamic gamma band activity correlates with symptom relief following deep brain stimulation in humans with Tourette's syndrome. *PloS ONE, 7*(9), e44215.

Mallet, L., Polosan, M., Jaafari, N., Baup, N., Welter, M. L., Fontaine, D., ... Raoul, S. (2008). Subthalamic nucleus stimulation in severe obsessive−compulsive disorder. *New England Journal of Medicine, 359*(20), 2121−2134.

Mantione, M., van de Brink, W., Schuurman, P. R., & Denys, D. (2010). Smoking cessation and weight loss after chronic deep brain stimulation of the nucleus accumbens: therapeutic and research implications: case report. *Neurosurgery, 66*(1), E218-E218.

Marras, C., Tröster, A. I., Kulisevsky, J., & Stebbins, G. T. (2014). The tools of the trade: A state of the art "How to Assess Cognition" in the patient with Parkinson's disease. *Movement Disorders, 29*(5), 584−596. Available from https://doi.org/10.1002/mds.25874.

Martinez-Fernandez, R., Rodriguez-Rojas, R., Del Alamo, M., Hernandez-Fernandez, F., Pineda-Pardo, J. A., Dileone, M., ... Obeso, J. A. (2018). Focused ultrasound subthalamotomy in patients with asymmetric Parkinson's disease: A pilot study. *Lancet Neurology, 17*(1), 54−63. Available from https://doi.org/10.1016/S1474-4422(17)30403-9.

Martinez-Martinez, A. M., Aguilar, O. M., & Acevedo-Triana, C. A. (2017). Meta-Analysis of the relationship between deep brain stimulation in patients with Parkinson's disease and performance in evaluation tests for executive brain functions. *Parkinson's Disease, 2017*, 9641392. Available from https://doi.org/10.1155/2017/9641392.

McLachlan, R. S., Pigott, S., Tellez-Zenteno, J. F., Wiebe, S., & Parrent, A. (2010). Bilateral hippocampal stimulation for intractable temporal lobe epilepsy: Impact on seizures and memory. *Epilepsia, 51*(2), 304−307.

McCarter, R. J., Walton, N. H., Rowan, A. F., Gill, S. S., & Palomo, M. (2000). Cognitive functioning after subthalamic nucleotomy for refractory Parkinson's disease. *Journal of Neurology, Neurosurgery, and Psychiatry, 69*(1), 60−66.

Meier, M. J., & Story, J. L. (1967). Selective impairment of Porteus Maze Test performance after right subthalamotomy. *Neuropsychologia, 5,* 181−189.

Merola, A., Rizzi, L., Artusi, C. A., Zibetti, M., Rizzone, M. G., Romagnolo, A., . . . Lopiano, L. (2014). Subthalamic deep brain stimulation: Clinical and neuropsychological outcomes in mild cognitive impaired parkinsonian patients. *Journal of Neurology, 261*(9), 1745−1751. Available from https://doi.org/10.1007/s00415-014-7414-8.

Meissner, W. G., Laurencin, C., Tranchant, C., Witjas, T., Viallet, F., Guehl, D., . . . Vital, A. (2016). Outcome of deep brain stimulation in slowly progressive multiple system atrophy: A clinico-pathological series and review of the literature. *Parkinsonism & Related Disorders, 24,* 69−75.

Meyers, R. (1942). The modification of alternating tremors, rigidity, and festination by surgery of the basal ganglia. *Research Publications—Association for Research in Nervous and Mental Disease, 21,* 602−665.

Meyers, R., Fry, W. J., Fry, F. J., Dreyer, L. L., Schultz, D. F., & Noyes, R. F. (1959). Early experiences with ultrasonic irradiation of the pallidofugal and nigral complexes in hyperkinetic and hypertonic disorders. *Journal of Neurosurgery, 16*(1), 32−54. Available from https://doi.org/10.3171/jns.1959.16.1.0032.

Miller, S., Akram, H., Lagrata, S., Hariz, M., Zrinzo, L., & Matharu, M. (2016). Ventral tegmental area deep brain stimulation in refractory short-lasting unilateral neuralgiform headache attacks. *Brain, 139*(10), 2631−2640.

Mohr, E., Mendis, T., & Grimes, J. D. (1995). Late cognitive changes in Parkinson's disease with an emphasis on dementia. *Advances in Neurology, 65,* 97−113.

Moro, E., Schwalb, J. M., Piboolnurak, P., Poon, Y. Y. W., Hamani, C., Hung, S. W., . . . Lozano, A. M. (2011). Unilateral subdural motor cortex stimulation improves essential tremor but not Parkinson's disease. *Brain, 134*(7), 2096−2105.

Müller, U. J., Sturm, V., Voges, J., Heinze, H. J., Galazky, I., Büntjen, L., . . . Bogerts, B. (2016). Nucleus Accumbens Deep Brain Stimulation for Alcohol Addiction−Safety and Clinical Long-term Results of a Pilot Trial. *Pharmacopsychiatry, 49*(04), 170−173.

Nakano, N., Miyauchi, M., Nakanishi, K., Saigoh, K., Mitsui, Y., & Kato, A. (2015). Successful combination of pallidal and thalamic stimulation for intractable involuntary movements in patients with neuroacanthocytosis. *World Neurosurgery, 84*(4), 1177-e1.

Narabayashi, H., Okuma, T., & Shikiba, S. (1956). Procaine oil blocking of the globus pallidus. *A.M.A. Archives of Neurology and Psychiatry, 75*(1), 36−48.

Niebuhr, H., Jr. (1962). Some psychological aspects of patients with Parkinson's disease before and after ventro-lateral thalamotomy. In E. A. Spiegel, & H. T. Wycis (Eds.), *Stereoencephalotomy. Part II Clinical and physiological applications* (Vol. II, pp. 349−357). New York: Grune and Stratton.

Obeso, I., Casabona, E., Rodriguez-Rojas, R., Bringas, M. L., Macias, R., Pavon, N., . . . Jahanshahi, M. (2017). Unilateral subthalamotomy in Parkinson's disease: Cognitive, psychiatric and neuroimaging changes. *Cortex, 94,* 39−48. Available from https://doi.org/10.1016/j.cortex.2017.06.006.

Obwegeser, A. A., Uitti, R. J., Lucas, J. A., Witte, R. J., Turk, M. F., & Wharen, R. E., Jr. (2000). Predictors of neuropsychological outcome in patients following microelectrode-guided pallidotomy for Parkinson's disease. *Journal of Neurosurgery, 93*(3), 410−420.

Odekerken, V. J., Boel, J. A., Geurtsen, G. J., Schmand, B. A., Dekker, I. P., de Haan, R. J., ... Group, N.S. (2015). Neuropsychological outcome after deep brain stimulation for Parkinson disease. *Neurology*, *84*(13), 1355–1361. Available from https://doi.org/10.1212/WNL.0000000000001419.

Oertel, M. F., Schüpbach, W. M. M., Ghika, J. A., Stieglitz, L. H., Fiechter, M., Kaelin-Lang, A., ... Pollo, C. (2017). Combined thalamic and subthalamic deep brain stimulation for tremor-dominant Parkinson's disease. *Acta Neurochirurgica*, *159*(2), 265–269.

Odekerken, V. J., van Laar, T., Staal, M. J., Mosch, A., Hoffmann, C. F., Nijssen, P. C., ... de Bie, R. M. (2013). Subthalamic nucleus versus globus pallidus bilateral deep brain stimulation for advanced Parkinson's disease (NSTAPS study): A randomised controlled trial. *Lancet Neurology*, *12*(1), 37–44. Available from https://doi.org/10.1016/S1474-4422(12)70264-8.

Okun, M. S., Fernandez, H. H., Rodriguez, R. L., & Foote, K. D. (2007). Identifying candidates for deep brain stimulation in Parkinson's disease: The role of the primary care physician. *Geriatrics*, *62*(5), 18–24.

Okun, M. S., Foote, K. D., Wu, S. S., Ward, H. E., Bowers, D., Rodriguez, R. L., ... Mink, J. W. (2013). A trial of scheduled deep brain stimulation for Tourette syndrome: moving away from continuous deep brain stimulation paradigms. *JAMA Neurology*, *70*(1), 85–94.

Ostrem, J. L., Markun, L. C., Glass, G. A., Racine, C. A., Volz, M. M., Heath, S. L., ... Starr, P. A. (2014). Effect of frequency on subthalamic nucleus deep brain stimulation in primary dystonia. *Parkinsonism & Related Disorders*, *20*(4), 432–438.

Owen, T., Gimeno, H., Selway, R., & Lin, J. P. (2015). Cognitive function in children with primary dystonia before and after deep brain stimulation. *European Journal of Paediatric Neurology*, *19*(1), 48–55. Available from https://doi.org/10.1016/j.ejpn.2014.09.004.

Parsons, T. D., Rogers, S. A., Braaten, A. J., Woods, S. P., & Tröster, A. I. (2006). Cognitive sequelae of subthalamic nucleus deep brain stimulation in Parkinson's disease: A meta-analysis. *Lancet Neurology*, *5*(7), 578–588.

Patel, N. K., Heywood, P., O'Sullivan, K., McCarter, R., Love, S., & Gill, S. S. (2003). Unilateral subthalamotomy in the treatment of Parkinson's disease. *Brain*, *126*(Pt 5), 1136–1145.

Pereira, E. A., Boccard, S. G., Linhares, P., Chamadoira, C., Rosas, M. J., Abreu, P., ... Aziz, T. Z. (2013). Thalamic deep brain stimulation for neuropathic pain after amputation or brachial plexus avulsion. *Neurosurgical Focus*, *35*(3), E7.

Pham, U. H., Andersson, S., Toft, M., Pripp, A. H., Konglund, A. E., Dietrichs, E., ... Solbakk, A. K. (2015). Self-reported executive functioning in everyday life in Parkinson's disease after three months of subthalamic deep brain stimulation. *Parkinson's Disease*, *2015*, 461453. Available from https://doi.org/10.1155/2015/461453.

Pilitsis, J. G., Rezai, A. R., Boulis, N. M., Henderson, J. M., Busch, R. M., & Kubu, C. S. (2005). A preliminary study of transient confusional states following bilateral subthalamic stimulation for Parkinson's disease. *Stereotactic and Functional Neurosurgery*, *83*(2–3), 67–70.

Piacentino, M., D'Andrea, G., Perini, F., & Volpin, L. (2014). Drug-resistant cluster headache: long-term evaluation of pain control by posterior hypothalamic deep-brain stimulation. *World Neurosurgery*, *81*(2), 442-e11.

Pillon, B., Ardouin, C., Dujardin, K., Vittini, P., Pelissolo, A., Cottencin, O., ... Vidailhet, M. (2006). Preservation of cognitive function in dystonia treated by pallidal stimulation. *Neurology*, *66*(10), 1556–1558.

Ponce, F. A., Asaad, W. F., Foote, K. D., Anderson, W. S., Rees Cosgrove, G., Baltuch, G. H., ... Smith, G. S. (2016). Bilateral deep brain stimulation of the fornix for Alzheimer's disease: Surgical safety in the ADvance trial. *Journal of Neurosurgery, 125*(1), 75−84.

Rasche, D., Zittel, S., Tadic, V., Moll, C., Fellbrich, A., Brüggemann, N., ... Münchau, A. (2016). EP 39. Clinical experience with deep brain stimulation in Huntington's disease. *Clinical Neurophysiology, 127*(9), e192.

Raymaekers, S., Luyten, L., Bervoets, C., Gabriëls, L., & Nuttin, B. (2017). Deep brain stimulation for treatmentresistant major depressive disorder: A comparison of two targets and long-term follow-up. *Translational Psychiatry, 7*(10), e1251.

Reich, M. M., Brumberg, J., Pozzi, N. G., Marotta, G., Roothans, J., Astrom, M., ... Isaias, I. U. (2016). Progressive gait ataxia following deep brain stimulation for essential tremor: Adverse effect or lack of efficacy? *Brain, 139*(11), 2948−2956. Available from https://doi.org/10.1093/brain/aww223.

Rektorova, I., Hummelova, Z., & Balaz, M. (2011). Dementia after DBS surgery: A case report and literature review. *Parkinson's Disease, 2011*, 679283. Available from https://doi.org/10.4061/2011/679283.

Rettig, G. M., York, M. K., Lai, E. C., Jankovic, J., Krauss, J. K., Grossman, R. G., & Levin, H. S. (2000). Neuropsychological outcome after unilateral pallidotomy for the treatment of Parkinson's disease. *Journal of Neurology, Neurosurgery, and Psychiatry, 69*(3), 326−336.

Riklan, M., & Levita, E. (1969). *Subcortical correlates of human behavior: A psychological study of thalamic and basal ganglia surgery*. Baltimore, MD: Williams & Wilkins.

Rothlind, J. C., York, M. K., Carlson, K., Luo, P., Marks, W. J., Weaver, F. M., ... Reda, D. (2015). Neuropsychological changes following deep brain stimulation surgery for Parkinson's disease: Comparisons of treatment at pallidal and subthalamic targets versus best medical therapy. *Journal of Neurology Neurosurgery Psychiatry, 86*(6), 622−629.

Rothlind, J. C., York, M. K., Carlson, K., Luo, P., Marks, W. J., Jr., ... Group, C.S.P.S. (2015). Neuropsychological changes following deep brain stimulation surgery for Parkinson's disease: Comparisons of treatment at pallidal and subthalamic targets versus best medical therapy. *Journal of Neurology, Neurosurgery and Psychiatry, 86*(6), 622−629. Available from https://doi.org/10.1136/jnnp-2014-308119.

Saint-Cyr, J. A., Trépanier, L. L., Kumar, R., Lozano, A. M., & Lang, A. E. (2000). Neuropsychological consequences of chronic bilateral stimulation of the subthalamic nucleus in Parkinson's disease. *Brain, 123*(Pt 10), 2091−2108.

Sandvik, U., Koskinen, L. O., Lundquist, A., & Blomstedt, P. (2011). Thalamic and subthalamic deep brain stimulation for essential tremor: Where is the optimal target? *Neurosurgery, 70*(4), 840−846.

Schlaepfer, T. E., Bewernick, B. H., Kayser, S., Mädler, B., & Coenen, V. A. (2013). Rapid effects of deep brain stimulation for treatment-resistant major depression. *Biological Psychiatry, 73*(12), 1204−1212.

Schoenberg, M. R., Maddux, B. N., Riley, D. E., Whitney, C. M., Ogrocki, P. K., Gould, D., & Maciunas, R. J. (2015). Five-months-postoperative neuropsychological outcome from a pilot prospective randomized clinical trial of thalamic deep brain stimulation for Tourette syndrome. *Neuromodulation, 18*(2), 97−104. Available from https://doi.org/10.1111/ner.12233.

Schrader, C., Seehaus, F., Capelle, H. H., Windha Checn gen, A., Windhagen, H., & Krauss, J. K. (2013). Effects of pedunculopontine area and pallidal DBS on gait ignition in Parkinson's disease. *Brain Stimulation, 6*(6), 856−859, 2017.

Schoenberg, M. R., Mash, K. M., Bharucha, K. J., Francel, P. C., & Scott, J. G. (2008). Deep brain stimulation parameters associated with neuropsychological changes in subthalamic nucleus stimulation for refractory Parkinson's disease. *Stereotactic and Functional Neurosurgery*, *86*(6), 337−344. Available from https://doi.org/10.1159/000163554, 000163554 [pii].

Sem-Jacobsen, C. W. (1965). Depth electrographic stimulation and treatment of patients with Parkinson's disease including neurosurgical technique. *Acta Neurologica Scandinavica. Supplementum*, *13*(Pt 1), 365−377.

Shah, N., Leventhal, D., Persad, C., Patil, P. G., & Chou, K. L. (2016). A suggested minimum standard deep brain stimulation evaluation for essential tremor. *Journal of the Neurological Sciences*, *362*, 165−168. Available from https://doi.org/10.1016/j.jns.2016.01.041.

Shaikh, A. G., Mewes, K., DeLong, M. R., Gross, R. E., Triche, S. D., Jinnah, H. A., ... Aia, P. (2015). Temporal profile of improvement of tardive dystonia after globus pallidus deep brain stimulation. *Parkinsonism & Related Disorders*, *21*(2), 116−119.

Shi, Y., Burchiel, K. J., Anderson, V. C., & Martin, W. H. (2009). Deep brain stimulation effects in patients with tinnitus. *Otolaryngology—Head and Neck Surgery*, *141*(2), 285−287.

Sidiropoulos, C., Rammo, R., Merker, B., Mahajan, A., LeWitt, P., Kaminski, P., ... Schwalb, J. M. (2016). Intraoperative MRI for deep brain stimulation lead placement in Parkinson's disease: 1 year motor and neuropsychological outcomes. *Journal of Neurology*, *263*(6), 1226−1231. Available from https://doi.org/10.1007/s00415-016-8125-0.

Smeding, H. M., Speelman, J. D., Huizenga, H. M., Schuurman, P. R., & Schmand, B. (2011). Predictors of cognitive and psychosocial outcome after STN DBS in Parkinson's disease. *Journal of Neurology, Neurosurgery, and Psychiatry*, *82*(7), 754−760. Available from https://doi.org/10.1136/jnnp.2007.140012.

Smeding, H. M., Speelman, J. D., Koning-Haanstra, M., Schuurman, P. R., Nijssen, P., van Laar, T., & Schmand, B. (2006). Neuropsychological effects of bilateral STN stimulation in Parkinson disease: a controlled study. *Neurology*, *66*(12), 1830−1836.

Son, B. C., Shon, Y. M., Choi, J. G., Kim, J., Ha, S. W., Kim, S. H., & Lee, S. H. (2016). Clinical outcome of patients with deep brain stimulation of the centromedian thalamic nucleus for refractory epilepsy and location of the active contacts. *Stereotactic and Functional Neurosurgery*, *94*(3), 187−197.

Tellez-Zenteno, J. F., McLachlan, R. S., Parrent, A., Kubu, C. S., & Wiebe, S. (2006). Hippocampal electrical stimulation in mesial temporal lobe epilepsy. *Neurology*, *66*(10), 1490−1494.

Thavanesan, N., Gillies, M., Farrell, M., Green, A. L., & Aziz, T. (2014). Deep brain stimulation in multiple system atrophy mimicking idiopathic Parkinson's disease. *Case Reports in Neurology*, *6*(3), 232−237.

Torres, C. V., Moro, E., Lopez-Rios, A. L., Hodaie, M., Chen, R., Laxton, A. W., ... Lozano, A. M. (2010). Deep brain stimulation of the ventral intermediate nucleus of the thalamus for tremor in patients with multiple sclerosis. *Neurosurgery*, *67*(3), 646−651.

Torres, C. V., Sola, R. G., Pastor, J., Pedrosa, M., Navas, M., García-Navarrete, E., ... García-Camba, E. (2013). Long-term results of posteromedial hypothalamic deep brain stimulation for patients with resistant aggressiveness. *Journal of Neurosurgery*, *119*(2), 277−287.

Torres, N., Chabardes, S., Piallat, B., Devergnas, A., & Benabid, A. L. (2012). Body fat and body weight reduction following hypothalamic deep brain stimulation in monkeys: an intra-ventricular approach. *International Journal of Obesity*, *36*(12), 1537−1544.

Tramontana, M. G., Molinari, A. L., Konrad, P. E., Davis, T. L., Wylie, S. A., Neimat, J. S., ... Charles, D. (2015). Neuropsychological effects of deep brain stimulation in subjects with early stage Parkinson's disease in a randomized clinical trial. *Journal of Parkinson's Disease*, *5*(1), 151−163. Available from https://doi.org/10.3233/JPD-140448.

Tröster, A. I. (2017). Some clinically useful information that neuropsychology provides patients, carepartners, neurologists, and neurosurgeons about deep brain stimulation for Parkinson's disease. *Archives of Clinical Neuropsychology, 32*(7), 810−828. Available from https://doi. org/10.1093/arclin/acx090.

Tröster, A. I. (2018). Successes and optimism in deep brain stimulation for neurological disorders: Ripe for a surgical time out? *European Journal of Neurology, 25,* 705−706. Available from https://doi.org/10.1111/ene.13593.

Tröster, A. I., Fields, J. A., Pahwa, R., Wilkinson, S. B., Straits-Tröster, K. A., Lyons, K., ... Koller, W. C. (1999). Neuropsychological and quality of life outcome after thalamic stimulation for essential tremor. *Neurology, 53*(8), 1774−1780.

Tröster, A. I., Fields, J. A., Wilkinson, S. B., Busenbark, K., Miyawaki, E., Overman, J., ... Koller, W. C. (1997). Neuropsychological functioning before and after unilateral thalamic stimulating electrode implantation in Parkinson's disease [electronic manuscript]. *Neurosurgical Focus, 2*(3), 1−6. (Article 9).

Tröster, A. I., Fields, J. A., Wilkinson, S. B., Pahwa, R., Miyawaki, E., Lyons, K. E., & Koller, W. C. (1997). Unilateral pallidal stimulation for Parkinson's disease: neurobehavioral functioning before and 3 months after electrode implantation. *Neurology, 49*(4), 1078−1083.

Tröster, A. I., Jankovic, J., Tagliati, M., Peichel, D., & Okun, M. S. (2017). Neuropsychological outcomes from constant current deep brain stimulation for Parkinson's disease. *Movement Disorders, 32*(3), 433−440. Available from https://doi.org/10.1002/mds.26827.

Tröster, A. I., & Tucker, K. A. (2005). Impact of essential tremor and its medical and surgical treatment on neuropsychological functioning, activities of daily living, and quality of life. In K. E. Lyons, & R. Pahwa (Eds.), *Handbook of Essential Tremor and Other Tremor Disorders* (pp. 117−131). Boca Raton, FL: Taylor and Francis.

Tröster, A. I., Woods, S. P., & Fields, J. A. (2003). Verbal fluency declines after pallidotomy: An interaction between task and lesion laterality. *Applied Neuropsychology, 10*(2), 69−75.

Tröster, A. I., & Garrett, R. (2018). Parkinson's disease and other movement disorders. In J. Morgan, & J. Ricker (Eds.), *Textbook of Clinical Neuropsychology* (2nd. Ed., pp. 507−559). New York: Routledge.

Tröster, A. I., Ponce, F. A., & Moguel-Cobos, G. (2018). Deep brain stimulation for Parkinson's disease: Current perspectives on patient selection with an emphasis on neuropsychology. *Journal of Parkinsonism and Restless Legs Syndrome, 8,* 1−16.

Van Buren, J. M., Li, C. L., Shapiro, D. Y., Henderson, W. G., & Sadowsky, D. A. (1973). A qualitative and quantitative evaluation of parkinsonians three to six years following thalamotomy. *Confinia Neurologica, 35*(4), 202−235.

Valentín, A., García Navarrete, E., Chelvarajah, R., Torres, C., Navas, M., Vico, L., ... Alarcon, G. (2013). Deep brain stimulation of the centromedian thalamic nucleus for the treatment of generalized and frontal epilepsies. *Epilepsia, 54*(10), 1823−1833.

Velasco, F., Carrillo-Ruiz, J. D., Brito, F., Velasco, M., Velasco, A. L., Marquez, I., & Davis, R. (2005). Double-blind, randomized controlled pilot study of bilateral cerebellar stimulation for treatment of intractable motor seizures. *Epilepsia, 46*(7), 1071−1081.

Vilkki, J., & Laitinen, L. V. (1974). Differential effects of left and right ventrolateral thalamotomy on receptive and expressive verbal performances and face-matching. *Neuropsychologia, 12*(1), 11−19.

Vingerhoets, G., Lannoo, E., van der Linden, C., Caemaert, J., Vandewalle, V., van den Abbeele, D., & Wolters, M. (1999). Changes in quality of life following unilateral pallidal stimulation in Parkinson's disease. *Journal of Psychosomatic Research, 46*(3), 247−255.

Volkmann, J., Wolters, A., Kupsch, A., Müller, J., Kühn, A. A., Schneider, G. H., . . . Deuschl, G. (2012). Pallidal deep brain stimulation in patients with primary generalised or segmental dystonia: 5-year follow-up of a randomised trial. *The Lancet Neurology*, *11*(12), 1029–1038.

Wang, J. W., Zhang, Y. Q., Zhang, X. H., Wang, Y. P., Li, J. P., & Li, Y. J. (2016). Cognitive and psychiatric effects of STN versus GPi deep brain stimulation in Parkinson's disease: A meta-analysis of randomized controlled trials. *PloS One*, *11*, e0156721. Available from https://doi.org/10.1371/journal.pone.0156721.

Weaver, F. M., Follett, K. A., Stern, M., Luo, P., Harris, C. L., Hur, K., . . . Pahwa, R. (2012). Randomized trial of deep brain stimulation for Parkinson disease thirty-six-month outcomes. *Neurology*, *79*(1), 55–65.

Welter, M. L., Demain, A., Ewenczyk, C., Czernecki, V., Lau, B., El Helou, A., . . . Grabli, D. (2015). PPNa-DBS for gait and balance disorders in Parkinson's disease: a double-blind, randomised study. *Journal of Neurology*, *262*(6), 1515–1525.

Whiting, D. M., Tomycz, N. D., Bailes, J., deJonge, L., Lecoultre, V., Wilent, B., . . . Cantella, D. (2013). Lateral hypothalamic area deep brain stimulation for refractory obesity: A pilot study with preliminary data on safety, body weight, and energy metabolism. *Journal of Neurosurgery*, *119*(1), 56–63.

Wiebe, S., Kiss, Z., & Ahmed, N. (2013). Medical vs electrical therapy for mesial temporal lobe epilepsy: A multicenter randomized trial. In American Epilepsy Society 2012 Annual Meeting.

Wilkinson, S. B., & Tröster, A. I. (1998). Surgical interventions in neurodegenerative disease: impact on memory and cognition. In A. I. Tröster (Ed.), *Memory in neurodegenerative disease: Biological, cognitive, and clinical perspectives* (pp. 362–376). Cambridge, UK: Cambridge University Press.

Wille, C., Steinhoff, B. J., Altenmüller, D. M., Staack, A. M., Bilic, S., Nikkhah, G., & Vesper, J. (2011). Chronic high-frequency deep-brain stimulation in progressive myoclonic epilepsy in adulthood—Report of five cases. *Epilepsia*, *52*(3), 489–496.

Wirdefeldt, K., Adami, H. O., Cole, P., Trichopoulos, D., & Mandel, J. (2011). Epidemiology and etiology of Parkinson's disease: A review of the evidence. *European Journal of Epidemiology*, *26*(Suppl. 1), S1–S58. Available from https://doi.org/10.1007/s10654-011-9581-6.

Witt, K., Daniels, C., Krack, P., Volkmann, J., Pinsker, M. O., Kloss, M., . . . Deuschl, G. (2011). Negative impact of borderline global cognitive scores on quality of life after subthalamic nucleus stimulation in Parkinson's disease. *Journal of the Neurological Sciences*, *310* (1–2), 261–266. Available from https://doi.org/10.1016/j.jns.2011.06.028.

Woods, S. P., Fields, J. A., Lyons, K. E., Pahwa, R., & Tröster, A. I. (2003). Pulse width is associated with cognitive decline after thalamic stimulation for essential tremor. *Parkinsonism and Related Disorders*, *9*, 295–300.

Woods, S. P., Fields, J. A., & Tröster, A. I. (2002). Neuropsychological sequelae of subthalamic nucleus deep brain stimulation in Parkinson's disease: A critical review. *Neuropsychology Review*, *12*(2), 111–126.

Xie, Y., Meng, X., Xiao, J., Zhang, J., & Zhang, J. (2016). Cognitive changes following bilateral deep brain stimulation of subthalamic nucleus in Parkinson's disease: A meta-analysis. *BioMed Research International*, *2016*, 3596415. Available from https://doi.org/10.1155/2016/3596415.

Yaguez, L., Costello, A., Moriarty, J., Hulse, N., Selway, R., Clough, C., . . . Ashkan, K. (2014). Cognitive predictors of cognitive change following bilateral subthalamic nucleus deep brain stimulation in Parkinson's disease. *Journal of Clinical Neuroscience, 21*(3), 445–450. Available from https://doi.org/10.1016/j.jocn.2013.06.005.

York, M. K., Dulay, M., Macias, A., Levin, H. S., Grossman, R., Simpson, R., & Jankovic, J. (2008). Cognitive declines following bilateral subthalamic nucleus deep brain stimulation for the treatment of Parkinson's disease. *Journal of Neurology, Neurosurgery and Psychiatry, 79*(7), 789–795. Available from https://doi.org/10.1136/jnnp.2007.118786.

York, M. K., Levin, H. S., Grossman, R. G., & Hamilton, W. J. (1999). Neuropsychological outcome following unilateral pallidotomy. *Brain, 122*(Pt 12), 2209–2220.

Zhang, H. W., Li, D. Y., Zhao, J., Guan, Y. H., Sun, B. M., & Zuo, C. T. (2013). Metabolic imaging of deep brain stimulation in anorexia nervosa: a 18F-FDG PET/CT study. *Clinical Nuclear Medicine, 38*(12), 943–948.

Zhu, X. Y., Pan, T. H., Ondo, W. G., Jimenez-Shahed, J., & Wu, Y. C. (2014). Effects of deep brain stimulation in relatively young-onset multiple system atrophy Parkinsonism. *Journal of the Neurological Sciences, 342*(1), 42–44.

Chapter 11

Future Directions in Deep Brain Stimulation

Tsinsue Chen and Francisco A. Ponce
Department of Neurosurgery, Barrow Neurological Institute, St. Joseph's Hospital and Medical Center, Phoenix, AZ, United States

BRIEF HISTORY

The first experimental use of electrical stimulation in the human brain was reported in 1874, with the author noting that stimulation produced contralateral muscle contractions in a woman undergoing an operation for a brain abscess (Bartholow, 1874). Before this time, the stimulation of animal brains had been performed to observe subsequent effects on behavior. In 1952, Delgado (1952) documented the first use of chronically implanted electrodes for experimental studies, and that report was followed by the first reports about the use of chronic stimulation for therapeutic purposes in patients with psychiatric pathologies (Heath, 1954; Pool, 1954). Early stimulation procedures often involved the intermittent delivery of stimulation, with electrodes sometimes being explanted after several weeks or months, and stimulators or pulse generators being external to the patient. In some cases, electrical stimulation was used to avoid the negative neurobehavioral side effects observed with repeated lesioning of Parkinson's disease (PD) patients (Sem-Jacobsen, 1965).

In 1980, Brice and McLellan (1980) reported the first case of a fully implanted neurostimulator system for chronic stimulation in a movement disorders patient with tremor and multiple sclerosis. Benabid and colleagues (Benabid et al., 1991, 1994, 1987) subsequently popularized high-frequency deep brain stimulation (DBS) therapy for the treatment of movement disorders in the 1980s and 1990s. The ability to mimic and reverse the effects that occurred after focal destruction of lesions revitalized the field of functional neurosurgery for movement disorders treatment. In the United States, DBS for essential tremor (ET) was approved by the US Food and Drug Administration (FDA) in 1997, and DBS for PD was approved by the FDA in 2002 (Medtronic, 2013). Since then, several level-1 evidence studies have demonstrated the superiority of DBS to best medical therapy alone for

Neurosurgical Neuropsychology. DOI: https://doi.org/10.1016/B978-0-12-809961-2.00012-6

treatment of PD (Deuschl et al., 2006; Schuepbach et al., 2013; Weaver et al., 2009; Williams et al., 2010), and it has become a mainstream therapy for treatment of movement disorders.

The incidence of neurobehavioral side effects increases after ablative and bilateral neurosurgical procedures for movement disorders. It is therefore important to assess the neuropsychological and cognitive function of potential DBS candidates to evaluate the potential risk for postoperative neurocognitive decline. A neuropsychological evaluation typically involves a medical records review, patient interview, family interview (if available), and behavioral observation. The domains assessed include general cognitive functioning, intelligence, executive functioning capacity, working attention and memory, language skills, motor function, memory, mood, spatial and visuoperceptual function, and quality of life. The administration of standardized neuropsychological tests allows the clinician to assess cognition, behavior, and emotion. Although general guidelines have been established regarding preoperative evaluation of PD patients, prognostic predictions for risk of postoperative cognitive decline are typically made in terms of probability and likelihood statements. In addition, there is no overarching consensus on a definition of a tolerable level of cognitive impairment and no definitive inclusion or exclusion neuropsychological criteria for patients undergoing DBS (Troster, 2009).

Nonetheless, a thorough preoperative evaluation does allow for a baseline measure of neuropsychological status against which to compare any postoperative changes in cognition or behavior. The capacity to make an informed decision about various treatment options and to give informed surgical consent can also be evaluated. Preoperative counseling of patients and caregivers regarding the likelihood of temporary cognitive side effects from surgery can mitigate concerns on the part of the caregivers postoperatively if such a decline occurs. Furthermore, assessment of mood, ability to cope, and general cognitive capacity can inform potential challenges that may arise with postoperative recovery and compliance with long-term therapy maintenance (i.e., follow-up programming sessions). Finally, in the event of postoperative complications, a neuropsychological evaluation after surgery can help elucidate whether postoperative behavior and cognitive changes may be related to the operation, medications, degree of stimulation, disease progression, or external psychosocial factors (Troster, 2009).

CURRENT PRACTICE

Three primary DBS targets are used to treat the motor symptoms of PD: the subthalamic nucleus (STN), the globus pallidus internus (GPi), and the ventral intermediate nucleus of the thalamus (VIM). The STN and GPi targets are used most commonly, and our understanding of their roles in the development of PD motor symptoms has evolved over time through animal studies

(Benazzouz et al., 2000; DeLong, 1990; Hamani & Lozano, 2003), imaging investigations (Eidelberg et al., 1997; Thobois, Guillouet, & Broussolle, 2001), and experience with lesioning procedures (Baron et al., 1996; Laitinen, 2000; Lombardi et al., 2000). STN targeting tends to be used more frequently than GPi on the basis of the belief that a greater degree of medication reduction and tremor improvement can be seen with STN than with GPi. In contemporary practice, the VIM is the least commonly selected target for PD. This is because the stimulation of the VIM treats only the tremor component of PD, whereas stimulation of the GPi and the STN also relieves other dopa-responsive motor symptoms, such as bradykinesia and rigidity. Thus, VIM stimulation in PD is typically reserved for patients with tremor-dominant symptoms and significant preoperative concerns about neurocognitive impairment.

Thalamic Stimulation

None of the three studies that have thoroughly examined neuropsychological outcomes after thalamic DBS for PD have demonstrated evidence of significant variations in postoperative cognitive changes (Caparros-Lefebvre et al., 1992; Loher et al., 2003; Troster et al., 1997). Unlike thalamotomy for PD, VIM DBS did not result in decreased verbal memory or fluency. In an evaluation of cognition of a patient both on and off stimulation, as well as on and off medication, Troster et al. (1998) noted improved verbal fluency associated with thalamic VIM stimulation. In a 12-month follow-up study of five patients, Woods et al. (2001) found sustained improvements in verbal fluency and memory after DBS. Although few studies have evaluated mood effects after VIM DBS for PD, Caparros-Lefebvre et al. (1992) did so and found a decline in depressive symptoms 4–10 days postoperatively.

Pallidal Stimulation

On the basis of results from a few studies with small samples, unilateral GPi DBS appears to be safe from a neurocognitive standpoint. Troster et al. (1997) found that none of the nine patients who underwent unilateral pallidal DBS experienced clinically significant neurocognitive changes at 3-month follow-up. Vingerhoets et al. (1999) also found no significant decline in cognitive function after unilateral GPi DBS, and noted that only 6 of 20 patients demonstrated any degree of decline in cognitive function. A small number of studies have evaluated the effects of bilateral GPi DBS. Pillon et al. (2000) found no cognitive decline in 56 patients at 12-month follow-up, and Ghika et al. (1998) found no changes in neuropsychological scores in six patients at 3-month follow-up. The Toronto group found selected cognitive impairments in their series. For example, four patients experienced a considerable decline in backward digit span scores (Trepanier et al., 2000). Although some studies evaluating self-reported mood state, as measured by the Beck Depression Inventory, did not find clinically significant improvements in depressive symptoms (Ghika et al., 1998; Vingerhoets et al., 1999). Fields et al. (1999) observed a reduction in anxiety. Higginson, Fields, and Troster (2001) noted improvements in neurophysiologic and subjective anxiety symptoms in patients who underwent either unilateral GPi ablation or DBS. For rare reported side effects of mania and hypersexuality after unilateral and bilateral GPi DBS, it is unclear whether they are related to the stimulation or to interactions between the stimulation and medication (Krause et al., 2001; Roane et al., 2002).

Subthalamic Nucleus Stimulation

Neuropsychological outcomes after STN DBS in PD patients have been studied more frequently than other targets, but reported incidences vary by the specific clinical evaluations used to assess neurocognitive status. Although many studies with formal neuropsychological testing have small sample sizes and lack control groups, several series have found mild cognitive changes, most frequently in regard to verbal fluency. Most clinical studies suggest that dramatically wide fluctuations in cognition are likely to be rare (Troster, 2009). Rodriguez-Oroz et al. (2005) reported severe, incapacitating deficits in 1%−2% of patients. Moderate deficits (mild functional consequences) and mild deficits (no functional consequences) occurred more frequently, affecting approximately 20% of patients.

There is no consensus on the clinical relevance of cognitive changes among those studies that have demonstrated larger degrees of cognitive decline. One systematic review reported a cognitive decline in 41% of STN DBS patients but provided few details about the nature and severity of the deficits (Temel et al., 2006). The controlled studies available do show that STN DBS is associated with a higher incidence of decline in verbal fluency. The first controlled series with formal neuropsychological evaluation compared eight bilateral STN DBS patients with eight unilateral pallidotomy patients and eight PD patients who received only medical therapy (Gironell et al., 2003). Both groups of surgical patients were evaluated in the "on-medication" state 1 month before and 6 months after the procedure, and the control group on medical therapy was tested 6 months after baseline evaluation. Results showed a decrease in semantic verbal fluency in the DBS group. In 2003, Moretti et al. (2003) reported that nine STN DBS patients (compared to nine medically treated PD patients) had found a decrease at 1, 6, and 12 months in verbal fluency and task performance on the Stroop Color and Word Test. Another study compared 99 STN DBS patients to 36 control PD patients and found a trend toward decreased cognitive functioning, verbal fluency, delayed recall, and visual attention in the surgical group (Smeding et al., 2006).

Deuschl et al. (2006) evaluated 6-month outcomes in patients randomized to bilateral STN or best medical therapy and found no significant changes from baseline in scores on the Mattis Dementia Rating Scale, the Montgomery−Asberg Depression Rating Scale, or the Brief Psychiatric Rating Scale between groups, but did not report comprehensive neuropsychological testing results. Schuepbach et al. (2013) randomized patients with earlier staged PD to best medical therapy or bilateral STN, and found that, within the stimulation group at 24-month follow-up, there was no significant change in scores on the Unified PD Rating Scale (UPDRS)—Part I or on the Montgomery−Asberg Depression Rating Scale. However, patients had significant improvement in scores on the Brief Psychiatric Rating Scale Score, the Beck Depression Inventory—II score, and the Starkstein Apathy Scale.

Yaguez et al. (2014) performed formal neuropsychological evaluations an average of 7.1 months before and 9.4 months after bilateral STN DBS in 30 patients with nondementia PD and found a significant decline in the immediate story recall portion of verbal memory with a large effect size. The best predictors of this decline were preoperative Full Scale Intelligence Quotient and list learning, which suggests that patients with mild baseline deficits in general intellectual function and list learning may be at higher risk of decline in other areas of verbal memory after bilateral STN DBS. Preoperative mild executive dysfunction was unchanged postoperatively.

Subthalamic Nucleus Versus Pallidal Stimulation

Four randomized trials of STN versus GPi DBS with a follow-up of 7−36 months have compared neurocognitive outcomes between the two targets (Table 11.1). The two smaller studies found no significant difference in cognitive outcomes (Boel et al., 2016; Okun et al., 2009). However, one of the two larger studies found a significantly greater decline in processing speed index in STN patients than in GPi patients (Follett et al., 2010), and one study demonstrated worse scores on the Mattis Dementia Rating Scale and on the Hopkins Verbal Learning Test in STN than in GPi patients (Weaver et al., 2012). Although three of the four studies found no significant difference in depression scores between the two targets at last follow-up (Boel et al., 2016; Okun et al., 2009; Weaver et al., 2012), the largest study, which followed 147 STN and 152 GPi patients over 24 months, found that the scores of the GPi group improved on the Beck Depression Inventory—II while the scores of the STN group worsened at a statistically significant level (Follett et al., 2010).

Unilateral Versus Bilateral Stimulation and Side of Surgery

Studies directly comparing the neurocognitive effects of unilateral and bilateral DBS are limited. Sjöberg et al. (2012) evaluated 10 PD patients who

TABLE 11.1 Randomized Controlled Trials Evaluating Cognition in STN Versus GPi DBS

Study	Year	Follow-Up	Unilateral or Bilateral	No. of STN	No. of GPi	Significant Cognition Findings	Mood Findings
Okun et al.	2009	7 months	Unilateral	22	23	1. No significant differences at the optimal clinical setting 2. STN subgroup with greater decline in letter verbal fluency than GPi ($P = .03$), but did not reach predefined $P < .025$ significance level	No difference in visual analogue mood, Beck Depression Inventory, or State-Trait Anxiety
Follet et al.	2010	24 months	Bilateral	147	152	Decline in processing speed index greater for STN than in GPi groups ($P = .03$)	Beck Depression Inventory—II improved in GPi group but worsened in STN group ($P = .02$)
Weaver et al.	2012	36 months	Bilateral	70	89	1. Mattis Dementia Scale: STN scores slightly worse at 6 months versus unchanged in GPi ($P = .03$), gradual decline over time, but more in STN than GPi ($P = .01$) 2. Hopkins Verbal Learning Test: GPi slightly worse at 6 months, then stable, versus STN stable at 6 months, then worse at 36 months ($P = .01$). Hopkins Verbal Learning Test Delayed Recall with similar trend over time ($P = .004$) 3. Decline in both groups over time: Wisconsin Card Sorting Test ($P = .03$), Brief Visuospatial Memory Test ($P = .001$), Brief Visuospatial Memory test delayed recall ($P = .001$)	No differences in Beck Depression Inventory—II over time
Boel et al.	2016	36 months	Bilateral	39	39	No significant difference in any test ($P = .17$ to $P = .87$)	No difference in Young mania rating Scale, Hospital Anxiety and Depression Scale (HADS) total score, or HADS anxiety and depression subscores

DBS, Deep brain stimulation; GPi, globus pallidus internus; STN, subthalamic nucleus.

underwent bilateral STN DBS for treatment of primarily bilateral motor symptoms and compared them to six PD patients who received left-sided STN DBS for treatment of primarily unilateral motor symptoms. They found a significantly greater decline in verbal fluency with bilateral STN stimulation than with unilateral stimulation in the speech-dominant hemisphere. Rothlind et al. (2007) evaluated the cognitive effects of patients undergoing staged bilateral DBS (one electrode placed in an initial operation followed by placement of a second electrode in a subsequent operation). A total of 42 patients were randomized to the STN ($n = 19$) or GPi ($n = 23$) target and underwent baseline neuropsychological evaluations. The decision regarding initial treatment side was determined on the basis of the laterality of symptom severity and hand dominance. Patients with significant motor symptom asymmetry underwent initial DBS lead implantation in the contralateral hemisphere, whereas those without symptom asymmetry had the first lead implanted in the hemisphere opposite the dominant hand.

A total of 29 of the original 42 participants proceeded to the second surgery for implantation of the contralateral electrode, which occurred an average of 7 months after the first procedure (Rothlind et al., 2007). Full neuropsychological evaluations were performed with patients on-medication and on-stimulation an average of 6 months after the first surgery, and an average of 15 months after the second operation. After unilateral DBS, a significant difference was noted between STN and GPi groups only for the Wechsler Adult Intelligence Scale—Revised (WAIS-R) Digit Symbol test, which evaluates fine manual dexterity and visuomotor coordination. No significant differences were noted between the two groups for any other test, and thus the STN and GPi were combined into one sample. Compared to 13% of 42 patients at baseline, 36% of these patients had scores in the mild impairment range ($T < 40$), and reductions were also decreased in phonemic fluency and performance on the word trial of the Stroop Color and Word Test. With bilateral DBS, patients in the STN group had a significant decrease in scores on the Digit Symbol Substitution Test, unlike patients in the GPi group. The GPi patients demonstrated a significant reduction in the WAIS-R Digit Span backward score, which evaluates auditory working memory, whereas the STN patients showed no change. When the GPi and STN groups were combined, the overall cohort demonstrated significant reductions in animal fluency, phonemic fluency, and WAIS-R arithmetic.

From the second neuropsychological evaluation after the first electrode placement to the third evaluation after the second electrode placement, there were no significant differences when the group was analyzed collectively (Rothlind et al., 2007). However, there was a larger effect size (0.51) for bilateral DBS compared to baseline, and the STN patients had a significant decline in scores on the Stroop Color and Word test. Within the unilateral DBS group, patients who had left-sided electrode placement demonstrated a larger effect size compared to those who had right-sided placement for

Animal Naming fluency (-0.79 vs -0.29), and the overall effect on Animal Naming fluency by unilateral DBS was attributed to reduced scores in the left-sided group alone. The patients who underwent right-sided electrode placement had a significant decline in Animal Naming fluency only after placement of the left-sided electrode, which suggests a potential to interpret this result as a negative effect of left-sided electrode placement. Overall, unilateral DBS (either GPi or STN) resulted in mild but statistically significant decreases in measures of function such as working memory and verbal fluency, and similar declines were observed after bilateral stimulation. Significant declines in verbal association fluency were found only after left-sided stimulation.

PRACTICAL APPLICATION AND CASE EXAMPLES

Case 1: Target Selection Based on Baseline Neurocognitive Risk

A 71-year-old woman with a 17-year history of PD had predominant symptoms of bilateral tremor, rigidity, balance difficulties, and significant daily motor fluctuations. Her medication regimen consisted of carbidopa−levodopa 25−250 mg five times per day and a 2-mg rotigotine transdermal patch daily. Her UPDRS-III scores improved from 50 off medication to 23 on medication. Neuropsychological evaluation demonstrated moderate impairments in learning and encoding, as well as deficits in verbal and visual retention. She also experienced difficulties on isolated executive functioning tasks. All other cognitive domains were within normal range. The scores of the patient on the Beck Depression Inventory indicated mild depression, and her scores on the Beck Anxiety Inventory indicated mild anxiety. Her daytime sleepiness was within the normal range, and concerns were minimal for impulsive behavior and apathy.

The results of her overall evaluation were consistent with mild cognitive impairment, amnestic in multiple domains, which might place her at higher risk for confusion and worsening of her baseline cognitive impairment after DBS implantation. Furthermore, her baseline degree of depression and anxiety indicated that she was at risk of for mood worsening after surgery. Despite these risk factors, the treatment team believe that she would benefit from DBS, given the substantial degree to which her bilateral motor symptoms were impairing her quality of life. In the context of the patient's age and neurocognitive concerns, a multidisciplinary team consisting of the patient's neurologist, neuropsychologist, and neurosurgeon agreed upon bilateral GPi DBS as the optimal surgical plan. The patient subsequently underwent successful bilateral DBS and internal pulse generator placement in one operation and began programming 2 weeks later (Fig. 11.1).

At 6-month follow-up, the patient demonstrated a good response to stimulation and significant improvement in her tremor, rigidity, bradykinesia,

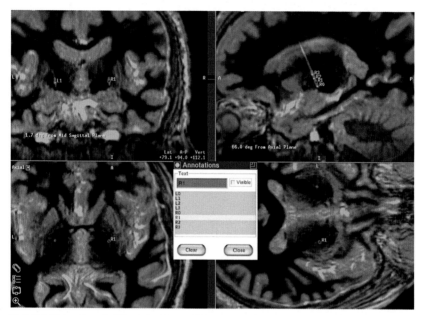

FIGURE 11.1   Case 1: Bilateral globus pallidus interna deep brain stimulation. *Used with permission from Barrow Neurological Institute, Phoenix, Arizona.*

and motor fluctuations. A repeat neuropsychological evaluation demonstrated relatively stable neurocognitive performance compared to her preoperative testing. Although she experienced a mild decline in delayed verbal recall, she had minor improvements in verbal fluency, visual memory, and basic attention. Her mood continued to be stable postoperatively (mildly anxious, with no worsening of preoperative depressive symptoms). This case illustrates how the GPi target might be selected instead of the STN because of concerns about potential worsening of mood and cognition in a patient with baseline mild cognitive impairment, including specific deficits in learning, encoding, executive functioning, verbal and visual retention, and mild depression and anxiety.

Case 2: Electrode Staging Due to Baseline Neurocognitive Risk

A 67-year-old woman with a 9-year history of PD presented primarily with symptoms of tremor, rigidity, and bradykinesia, and a score on the UPDRS-III that improved from 51 to 27 from the off-medication state to the on-medication state. Neuropsychological evaluation demonstrated mild cognitive impairment, amnestic subtype with deficits in perceptual reasoning, visuospatial learning and memory, and executive functioning and processing speed. Her medication regimen included 1.5 mg pramipexole three times

daily, 150 mg carbidopa−levodopa−entacapone every day, 100 mg amanta-
dine hydrochloride twice daily, and 25−100 mg carbidopa−levodopa three
times daily. The patient's deficits in multiple domains of neurocognition performance led to concern about greater risk of further neurocognitive decline
in one or more of these neurocognitive domains.

This potential postoperative risk was minimized by having a multidisciplinary team consisting of the patient's neurologist, neuropsychologist, and
neurosurgeon agree on a surgical plan to perform staged bilateral STN DBS,
with the left electrode to be placed first. The patient was discharged on postoperative day 2, and had no episodes of confusion or reported cognitive
changes. She recovered well after the surgery and 6 months later underwent
repeat neuropsychological testing. Her intelligence level was stable, although
she had selective declines in cognitive function (e.g., learning and memory
for unstructured verbal information). Other cognitive measures were
stable compared to baseline.

The patient subsequently underwent placement of the contralateral electrode 1 year after placement of the first electrode (Fig. 11.2). She recovered
well postoperatively, with no episodes of confusion, and her global cognitive
function was stable at her third formal neuropsychological evaluation 6
months after the second operation. Her memory improved on this final evaluation, and her diagnosis changed from mild cognitive impairment amnestic

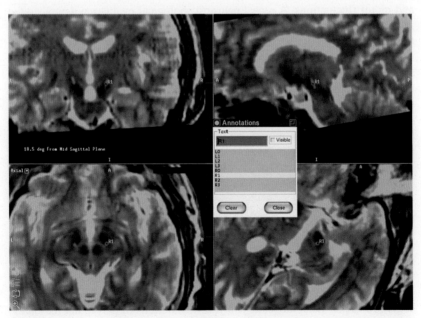

FIGURE 11.2   Case 2: Bilateral staged subthalamic nucleus deep brain stimulation. *Used with
permission from Barrow Neurological Institute, Phoenix, Arizona.*

subtype to the nonamnestic subtype. This improvement was believed to be less likely related to the stimulation and more likely related to her overall improved quality of life and reduction in medication.

FUTURE DIRECTION: DEEP BRAIN STIMULATION FOR COGNITIVE DISORDERS

The future direction of DBS in neurocognition lies in ongoing efforts to investigate the potential therapeutic and disease-modifying roles of electrical stimulation in Alzheimer's disease (AD), PD dementia (PDD), and mild cognitive impairment in PD. Because these pathologies are not only neurodegenerative disorders but also a result of neural circuitry dysfunction linked to cortical and subcortical sites regulating cognition and memory, interest has been growing in the potential role of modulating these malfunctioning circuits to restore or maximize their remaining function (Mirzadeh, Bari, & Lozano, 2016; Xu & Ponce, 2017).

Stimulation of the Nucleus Basalis of Meynert

Pathophysiology and Mechanisms of Action

Severe atrophy in the basal forebrain cholinergic system, particularly in the nucleus basalis of Meynert (NBM), is found both in PDD patients and in AD patients (Gratwicke et al., 2013). Numerous postmortem human studies demonstrate up to a 96% loss of NBM neurons in both PDD and AD patients compared with age-matched controls (Etienne et al., 1986; Gaspar & Gray, 1984). Functional neuroimaging with positron emission tomography (PET) using an acetylcholinesterase radioligand demonstrates cholinergic deficits in the cortex of both PDD and AD patients (Bohnen et al., 2003), and in both diseases, positive correlations have been demonstrated between the degree of NBM neuronal loss, consequent cholinergic deficits in the cortex, and degree of overall cognitive impairment. Volumetric magnetic resonance imaging in both PDD and AD patients also confirms the presence of significant NBM degeneration compared to that in age-matched controls, and the severity of degeneration significantly correlates with the degree of cognitive decline determined by objective neuropsychological tests (Choi et al., 2012; Hanyu et al., 2002).

Early animal studies have reinforced the role of the NBM in behavior and cognition and the potential benefit of electrical stimulation. The nucleus basalis network has demonstrated a critical role in neocortical arousal by direct neocortex activation. Cholinergic input deprivation in rats has been shown to hinder information transmission to the cortex (Buzsaki et al., 1988). Stimulation of the rodent NBM or optogenetic stimulation of NBM projection axons can cause a neocortical desynchronization evident on electroencephalography, with fast gamma oscillations indicating an alert awake state (Kalmbach, Hedrick, & Waters, 2012; Metherate, Cox, & Ashe, 1992). Lesioning of the rodent NBM has been demonstrated to prevent cortical desynchronization on electroencephalography and to produce slow synchronized delta waves corresponding with coma and unresponsiveness (Buzsaki et al., 1988; Fuller et al., 2011). Because NBM degeneration may serve a dominant role in the cognitive deficits of dementia, stimulating any remaining output networks in and from this structure has been postulated to enhance abilities to increase cholinergic output to the cortex.

In contrast to the neural signal inhibition caused by the high-frequency stimulation (100–180 Hz) used to treat patients with movement disorders, low-frequency stimulation (20–50 Hz) of the NBM in rats appears to incite stimulation and increase cholinergic release from cortical acetylcholine terminals (Kurosawa, Sato, & Sato, 1989). The next step in evaluating the specific effects of NBM stimulation on cognition in animal models is challenging because of the lack of good models available for specific NBM cholinergic deficits in PDD and AD. However, evidence does exist that cognition is improved after low-frequency stimulation of the NBM in healthy animals. When Montero-Pastor et al. (2001, 2004) stimulated unilateral rat NBMs with DBS at low frequencies (60–100 mA, 1 Hz, and 500 ms), they found significant improvement on a retention test reflective of memory consolidation. McLin, Miasnikov, and Weinberger (2002) found improvement in associative memory with unilateral brief NBM stimulation (50–150 mA, 0.2 ms, and single pulse). Another theory suggests that cholinergic neurons play an indirect role in modulating the cerebral cortex concentration of β-amyloid (Aβ) (Ovsepian & Herms, 2013). Because cholinergic muscarinic M1 receptor activation can decrease Aβ production and reduce the concentration of cerebrospinal fluid, increasing cholinergic output to the cortex via NBM stimulation could potentially slow the neurotoxic Aβ plaque production observed in patients with AD (Nitsch et al., 2000).

Current Evidence and Trials

In 1985, Turnbull et al. (1985) reported on implanting a left NBM DBS electrode in a 74-year-old man diagnosed with probable AD. During the procedure, the diagnosis was confirmed with a biopsy, which revealed neurofibrillary tangles, amyloid angiopathy, and senile plaques. Stimulation cycled 15 seconds on and 12 minutes off, with stimulation parameters of 3 V, 50 Hz, and 210 μs. After 8 months of stimulation, the patient had no beneficial clinical response, although formal neuropsychological assessments were not performed. Some effect on cortical glucose metabolism was seen on fludeoxyglucose PET (FDG-PET) studies 2 months after stimulation, with a less significant reduction in glucose on the right, unstimulated hemisphere than on the left, stimulated hemisphere.

Subsequently, in 2009, to address both the motor and cognitive symptoms of a patient with PDD, Freund et al. (2009) implanted bilateral STN and NBM leads using the coordinates $X = 12.5$ mm lateral to the third ventricular wall, $Y = 4$ mm posterior to the anterior commissure, and $Z = 5$ mm below the intercommissural plane. The ventral two contacts were in the STN, and although motor symptoms improved with STN stimulation, cognition remained unchanged. After initiation of NBM stimulation, the patient experienced rapid cognitive improvement as demonstrated in several tasks, including the Rey Auditory Verbal Learning Test, the clock drawing task, and the Trail Making Test, part A.

This success led Kuhn et al. (2015a) to conduct a phase 1 study of bilateral low-frequency NBM DBS in six patients with a 1-month double-blind sham-controlled phase, followed by an 11-month open-label phase. The primary outcome, difference in postoperative scores on the AD Assessment Scale—cognitive subscale (ADAS-Cog) compared to baseline, declined by a 3-point average over 1 year, which was not significant. Four patients were deemed to be therapeutic responders on the basis of an improved or stable score at 1 year. The secondary outcome, score on the Mini-Mental State Examination (MMSE), demonstrated no average change at 1 year. In three of four patients who underwent FDG-PET examination, studies demonstrated that cortical glucose metabolism decreased by 2%−5%, particularly in temporal and amygdalo-hippocampal areas. This decrease in the rate of metabolism is lower than that reported by Lo et al. (2011), who previously demonstrated a 5.2% per year average decline in FDG uptake in untreated AD patients. Further subanalysis suggested that patients with milder AD were more likely to remain stable or to improve throughout the course of stimulation (Kuhn et al., 2015a).

To evaluate the hypothesis that younger patients with less advanced AD benefit more from NBM DBS, Kuhn et al. (2015b) performed bilateral NBM DBS in two younger patients (ages 61 and 67 years) who had higher level cognitive functioning and who underwent the same surgical and stimulation protocols that were used in the phase 1 pilot study, with the exception of the crossover design. The first patient had an ADAS-Cog score that initially improved by 1 point at 12 months, but then decreased by 7 points at 26 months. His ADAS-Cog memory score remained stable, and his MMSE score declined by 1 point at 26 months. The second patient had an ADAS-Cog that improved by 4 points at 12 months and then returned to baseline at 24 months. His ADAS-Cog memory score remained stable, and his MMSE improved by 2 points at 26 months.

The same group of researchers subsequently reported long-term 24-month follow-up on all eight patients in 2016 (Hardenacke et al., 2016). The four patients who initially had lower baseline ADAS-Cog scores (fewer mistakes, less severe AD) remained relatively stable at 24-month follow-up, whereas three of the four patients who initially had higher baseline ADAS-Cog scores (more mistakes, more severe AD) declined in ADAS-Cog performance over time. On the basis of these results, the authors concluded that DBS may have a more beneficial impact on disease progression and cognitive function when it is performed earlier in the disease course when patients are younger. This effect is most likely due to modulatory effects of DBS on cholinergic networks. Because previous experimental data have demonstrated that cholinergic upregulation inhibits the formation of neurotoxic proteins, NBM DBS may not only have symptomatic relevance but may also have pathologic and disease-modifying capabilities.

Two ongoing randomized controlled trials of NBM DBS are being conducted to evaluate the treatment of neurocognitive disorders. One study at Barrow Neurological Institute, St. Joseph's Hospital and Medical Center, Phoenix, Arizona, is evaluating the safety and efficacy of NBM DBS for cognitively impaired patients with mild PD (clinicaltrials.gov; NCT:02924194). Patients receive simultaneous implants of bilateral GPi electrodes through the same bur hole to treat the motor symptoms of PD. They then undergo an initial randomized phase of 3 months "on" stimulation and 3 months "off" stimulation before proceeding to an open-label stimulation phase if their neurocognitive status is stable or improved. The second trial at the University Hospital of Rouen, France, is evaluating the impact of NBM stimulation on cognitive behavioral performance, particularly memory tasks, in patients with Lewy body dementia (ClinicalTrials.gov; NCT:01340001). In addition to long-term evaluation of the safety and efficacy of NBM DBS, the study will examine stimulation parameters, programming protocols, and delayed side effects of stimulation.

Stimulation of the Fornix

Pathophysiology and Mechanism

In addition to the NBM, other targets that have been investigated in healthy rodents were found to have potential cognitive enhancement abilities. One such target is the fornix, which is a large fiber bundle that connects the hippocampus and subiculum to the mammillary bodies and septal area via the circuit of Papez (Tsivilis et al., 2008). Both the fornix and the perforant pathway (originating from the entorhinal cortex) are major input pathways to the hippocampus, and thus they both play a significant role in memory formation (Bliss & Collingridge, 1993). In 2003, while recording bilateral hippocampal electroencephalography in a rat model, Williams and Givens (2003) found that administering low-frequency stimulation to the fornix or perforant pathway (i.e., stimulation of either anatomical structure) reset the theta rhythm, which occurs during the presence of a novel stimulus. This stimulation is associated with the initiation of long-term potentiation that, in turn, correlates with new memory formation (Buzsaki, 2002; Vertes, 2005).

Lee et al. (2013) later demonstrated that low-frequency stimulation of the medial septal nucleus (which projects via the fornix to the hippocampus) restored hippocampal theta rhythm in rats after epileptic seizures or brain injuries, subsequently improving spatial work memory task performance. Additional animal studies have demonstrated that neurogenesis induced by stimulation of the anterior nucleus of the thalamus (Hamani et al., 2011) or the entorhinal cortex (Stone et al., 2011) has potential memory-enhancing effects. Although these animal models may not accurately represent the neurodegenerative disease states of PDD and AD, they do suggest that delivering low-frequency stimulation to input structures of the hippocampus, such as the fornix, can modulate memory-encoding and can improve cognitive function (Laxton, Lipsman, & Lozano, 2013).

Current Evidence

While attempting hypothalamic stimulation to modulate appetite in an obese patient, Hamani et al. (2008) discovered that stimulation of specific electrode contacts and thresholds elicited vivid past autobiographical memories. Neuropsychological testing preoperatively and at 3-week postoperative follow-up also demonstrated improvement in memory as demonstrated by significant increases (>1.5 SD from baseline) in scores on the California Verbal Learning Test and the Spatial Associative Learning test. The FrameLink system (Medtronic plc, Dublin, Ireland) was used to evaluate active electrode contacts, and the Schaltenbrand–Wahren atlas (Shaltenbrand & Wahren, 1977) indicated whether the active contacts were adjacent to the forniceal columns.

Lesions in the fornix are known to result in severe memory impairment (Wilson et al., 2008), and axonal degeneration in the fornix has been associated with progression to AD. On the basis of these findings in a hypothalamic stimulation case and data from prior animal studies, Laxton et al. (2010) designed a phase 1 trial of fornix stimulation in six patients with mild to moderate AD. DBS electrodes were placed in a trajectory parallel to the columns of the fornix, approximately 5 mm lateral to the midline and 2 mm anterior to the fornix. The deepest contacts were 2 mm superior to the dorsal portion of the optic tract. The stimulation parameters were based on those used in the hypothalamic stimulation case, with a monopolar configuration of the most ventral contacts at 3–5 V, 130 Hz, and 90 μs.

After 12 months of stimulation, there was a mean increase of 4.2 points in ADAS-Cog scores, which is less than the historically reported mean increases in ADAS-Cog of 6–7 points per year (range, 3–10 points per year) (Ito et al., 2010; Mayeux & Sano, 1999). When the rate of decline in MMSE during the 11 months before and after surgery were examined, it was found to decrease by a mean of 0.8–2.9 points per year, or slightly less than the average expected rate of decline (3 points per year) in the general AD population (Ito et al., 2010; Mayeux & Sano, 1999). Although the results showed a slight benefit in decreasing the rate of disease progression in comparison to that in historical controls, the variability in the rate of AD progression within a population and among individuals precluded definitive conclusions about the efficacy of the procedure. There were no reports of seizures, hypothalamic dysfunctions, weight changes, sleep alterations, or endocrinologic disturbances, thus demonstrating the safety and tolerability of fornix DBS for up to 1 year.

Standardized low-resolution electromagnetic tomography imaging did demonstrate that the ipsilateral hippocampus and parahippocampal gyrus were activated by DBS. Preoperative FDG-PET scans also demonstrated decreased temporoparietal glucose metabolism compared with that in healthy controls, which is an expected AD finding; postoperative scans 2 weeks after stimulation initiation demonstrated increased temporoparietal glucose metabolism compared to baseline. This increased metabolism was sustained 1 year after surgery, which suggests that fornix DBS induced a sustained neurophysiological change in limbic and cognitive areas of the brain known to be negatively affected in AD (Laxton et al., 2010).

In a recent study, Fontaine et al. (2013) demonstrated unchanged memory scores (ADAS-Cog, MMSE, and Free and Cued Selective Reminding Test) 1 year after bilateral fornix DBS in one AD patient, who also showed increased bilateral temporal lobe metabolism. Koubeissi et al. (2013) performed low-frequency fornix stimulation with depth electrodes in 11 intractable epilepsy patients and found improved MMSE scores after a 4-hour period. However, neither study was controlled.

ADvance Trial

The results of Laxton et al.'s (2010) initial phase 1 study prompted a randomized, controlled, 12-month, double-blinded multicenter phase 2 study of fornix DBS in patients with mild, probable AD. The ADvance trial randomized patients between the ages of 45 and 85 years with an ADAS-Cog score of 12–24 in a double-blinded manner to be either on or off stimulation for 1 year. The primary endpoint was the safety of fornix DBS and the secondary endpoint was therapeutic efficacy.

The initial results, reported by Ponce et al. (2016), included 90-day adverse events, surgical technique, stereotactic accuracy, and hospital course. Patients underwent bilateral fornix DBS electrode placement, with the electrodes running parallel and 2 mm anterior to the columns of the fornix, through the hypothalamus, and terminating in a posteromedial position in relation to the optic tracts (Fig. 11.3). After the procedure, stereotactic magnetic resonance imaging was obtained to confirm stereotactic accuracy. Activation or no activation of the device occurred 2 weeks after the procedure in a double-blinded, randomized fashion. Initial 90-day postoperative

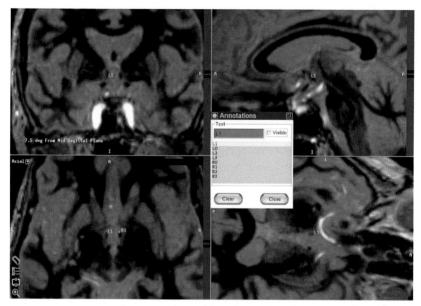

FIGURE 11.3 Bilateral fornix deep brain stimulation for Alzheimer's disease. *Used with permission from Barrow Neurological Institute, Phoenix, Arizona.*

safety results in 42 patients from seven participating sites demonstrated accurate targeting, a mean length of stay of 1.4 ± 0.8 days, and relative tolerability of the procedure. Seven serious adverse events occurred in five patients (11.9%). Return to surgery was necessary in four patients (9.5%): two for device explantation due to infection, one for lead repositioning, and one for evacuation of a chronic subdural hematoma (Ponce et al., 2016).

In the follow-up study evaluating 12-month safety and efficacy outcomes, Lozano et al. (2016) reported three long-term serious therapy-related adverse events in one patient in the off-stimulation cohort (depression, worsening confusion, and suicidal ideation) and no events in the on-stimulation cohort. The scores on the ADAS-Cog 13 and Clinical Dementia Rating Scale-Sum of Boxes that changed were comparable in the off ($n = 21$) and on ($n = 21$) stimulation groups, demonstrating similar degrees of decline from baseline to 12 months. On-stimulation patients had significantly increased glucose metabolism on FDG-PET in several brain areas (hippocampus, temporal association cortex, parietal association cortex, occipital cortex, cerebellar hemispheres, precentral gyrus, and postcentral gyrus) at 6 months compared to the off-stimulation group, but this difference was not sustained at 12-month follow-up.

In a subsequent post hoc multivariate regression analysis, the authors found a significantly positive association between age and cognitive outcome. Patients ≥ 65 years of age who were on stimulation had a trend toward greater benefit in clinical outcomes compared to patients <65 years of age, who trended toward worse clinical outcomes compared to those who were off stimulation. Within the FDG-PET data, the ≥65-year-old group demonstrated an increase in metabolism on stimulation of greater magnitude than that of the overall group at both 6 and 12 months, while the <65-year-old group generally demonstrated decreased metabolism in both stimulation states. Cognitive variables were similar for both groups, but post hoc analysis of preoperative PET scans showed decreased metabolism in the younger patients in both temporal and parietal regions.

The authors attributed these observed age-related clinical differences to a more malignant disease process in the younger AD patients that may have been associated with greater degree of cerebral atrophy and metabolic derangement than in their counterparts. Differences in clinical responsiveness are also potentially attributable to varying genetic and clinical phenotypes in younger AD patients who are less responsive to neuromodulation. Overall, the ADvance trial demonstrated that bilateral fornix DBS in AD patients was safe and well tolerated. Stimulation was associated with increased glucose metabolism at 6 months, but not at 12 months; cognitive outcomes were similar for the on and off groups at 12 months; and the subcategory of patients ≥65 years old was more likely to benefit from stimulation that may result in worsening in patients <65 years (Lozano et al., 2016).

Future directions in fornix DBS include exploring stimulation protocols and formulating more systematic methods for programming. Authors in the ADvance trial noted that stimulation parameters were selected somewhat arbitrarily on the basis of those most frequently used in DBS for other targets, in particular for PD and ET. However, it is uncertain whether selected parameters were optimal enough for AD, given the lack of immediately observable clinical outcomes (i.e., tremor or rigidity in PD), to base further adjustments for longer term benefits. In addition, in animal models, delivering current beyond the optimal therapeutic range can become counterproductive and interfere with memory function (Hamani et al., 2010), so the absence of benefit in this trial may be attributable to suboptimal stimulation dosing or lead location. Factors that require further exploration include the modification of specific parameters (voltage, frequency, and pulse width), stimulated contact, and intermittent stimulation paradigms.

Other Potential Targets

Another potential therapeutic strategy for improving neurocognition is to directly modulate medial temporal structures. Although some studies have demonstrated that direct low-frequency stimulation of the hippocampus has a contradictory effect and impairs recognition memory in healthy persons (Coleshill et al., 2004), Suthana et al. (2012) have demonstrated that acute low-frequency stimulation to adjacent entorhinal cortex enhances spatial memory in healthy persons during the learning phase. When low-frequency bipolar stimulation (3.0 V, 50 Hz, 300 μs) was applied unilaterally to the entorhinal cortex in seven epilepsy patients, spatial navigation task performance improved by approximately 64%. Although stimulation of the entorhinal cortex may improve memory function in AD, its clinical effect is likely to be limited by the lack of effect on other cognitive deficits caused by the disease, such as attention, perception, and executive function.

Basal ganglia stimulation in animal models, specifically within the caudate, has been previously explored as a potential method to enhance learning (Williams & Eskandar, 2006); however, applications to humans have not yet been evaluated and require future exploration. In patients with depression, subcallosal gyrus DBS has resulted in cognition improvements that are independent of improvements in depression, suggesting that there may be a direct correlation between stimulation and cognition enhancement. The underlying mechanism is unclear, but increased blood flow to specific areas in the frontal cortex suggest that activation in these areas may be induced by stimulation (McNeely et al., 2008).

CONCLUSION

Neuropsychology, cognition, and behavior have always played an important role in DBS for the treatment of movement disorders. DBS involves a thorough neuropsychological evaluation to assess and minimize any potential neurocognitive side effects of surgery and chronic stimulation. As the therapeutic efficacy has improved for the control of movement disorders over the past two decades, efforts are now focusing on the neurocognitive symptoms of PD, AD, and other neurodegenerative disorders. The two primary targets of investigation currently are the NBM and the fornix. Phase 1 trials for NBM DBS are under way, and a phase 3 trial for fornix DBS is being planned. Results from these studies will help elucidate the efficacy and future role of DBS in neurocognition.

ACKNOWLEDGMENTS

We thank the staff of Neuroscience Publications at Barrow Neurological Institute for assistance with manuscript preparation.

ABBREVIATIONS

Aβ	β-amyloid
AD	Alzheimer's disease
ADAS-Cog	Alzheimer's Disease Assessment Scale—cognitive subscale
DBS	deep brain stimulation
ET	essential tremor
FDA	US Food and Drug Administration

FDG-PET	fludeoxyglucose positron emission tomography
GPi	globus pallidus internus
MMSE	Mini-Mental State Examination
NBM	nucleus basalis of Meynert
PD	Parkinson's disease
PDD	PD dementia
PET	positron emission tomography
STN	subthalamic nucleus
UPDRS	Unified PD Rating Scale
VIM	ventral intermediate nucleus
WAIS-R	Wechsler Adult Intelligence Scale—Revised

REFERENCES

Baron, M. S., Vitek, J. L., Bakay, R. A., et al. (1996). Treatment of advanced Parkinson's disease by posterior GPi pallidotomy: 1-year results of a pilot study. *Annals of Neurology, 40,* 355–366.

Bartholow, R. (1874). Experimental investigations into the functions of the human brain. *The American Journal of the Medical Sciences, 67,* 305–313.

Benabid, A. L., Pollak, P., Gervason, C., et al. (1991). Long-term suppression of tremor by chronic stimulation of the ventral intermediate thalamic nucleus. *Lancet, 337,* 403–406.

Benabid, A. L., Pollak, P., Gross, C., et al. (1994). Acute and long-term effects of subthalamic nucleus stimulation in Parkinson's disease. *Stereotactic and Functional Neurosurgery, 62,* 76–84.

Benabid, A. L., Pollak, P., Louveau, A., et al. (1987). Combined (thalamotomy and stimulation) stereotactic surgery of the VIM thalamic nucleus for bilateral Parkinson disease. *Applied Neurophysiology, 50,* 344–346.

Benazzouz, A., Gao, D. M., Ni, Z. G., et al. (2000). Effect of high-frequency stimulation of the subthalamic nucleus on the neuronal activities of the substantia nigra pars reticulata and ventrolateral nucleus of the thalamus in the rat. *Neuroscience, 99,* 289–295.

Bliss, T. V., & Collingridge, G. L. (1993). A synaptic model of memory: Long-term potentiation in the hippocampus. *Nature, 361,* 31–39.

Boel, J. A., Odekerken, V. J., Schmand, B. A., et al. (2016). Cognitive and psychiatric outcome 3 years after globus pallidus pars interna or subthalamic nucleus deep brain stimulation for Parkinson's disease. *Parkinsonism & Related Disorders, 33,* 90–95.

Bohnen, N. I., Kaufer, D. I., Ivanco, L. S., et al. (2003). Cortical cholinergic function is more severely affected in parkinsonian dementia than in Alzheimer disease: An in vivo positron emission tomographic study. *Archives of Neurology, 60,* 1745–1748.

Brice, J., & McLellan, L. (1980). Suppression of intention tremor by contingent deep-brain stimulation. *Lancet, 1,* 1221–1222.

Buzsaki, G. (2002). Theta oscillations in the hippocampus. *Neuron, 33,* 325–340.

Buzsaki, G., Bickford, R. G., Ponomareff, G., et al. (1988). Nucleus basalis and thalamic control of neocortical activity in the freely moving rat. *Journal of Neuroscience, 8,* 4007–4026.

Caparros-Lefebvre, D., Blond, S., Pecheux, N., et al. (1992). Neuropsychological evaluation before and after thalamic stimulation in 9 patients with Parkinson disease. *Revue Neurologique (Paris), 148,* 117–122.

Choi, S. H., Jung, T. M., Lee, J. E., et al. (2012). Volumetric analysis of the substantia innominata in patients with Parkinson's disease according to cognitive status. *Neurobiology of Aging, 33,* 1265–1272.

Coleshill, S. G., Binnie, C. D., Morris, R. G., et al. (2004). Material-specific recognition memory deficits elicited by unilateral hippocampal electrical stimulation. *Journal of Neuroscience*, *24*, 1612−1616.

Delgado, J. M. (1952). Permanent implantation of multilead electrodes in the brain. *Yale Journal of Biology and Medicine*, *24*, 351−358.

DeLong, M. R. (1990). Primate models of movement disorders of basal ganglia origin. *Trends in Neurosciences*, *13*, 281−285.

Deuschl, G., Schade-Brittinger, C., Krack, P., et al. (2006). A randomized trial of deep-brain stimulation for Parkinson's disease. *New England Journal of Medicine*, *355*, 896−908.

Eidelberg, D., Moeller, J. R., Kazumata, K., et al. (1997). Metabolic correlates of pallidal neuronal activity in Parkinson's disease. *Brain*, *120*(Pt 8), 1315−1324.

Etienne, P., Robitaille, Y., Wood, P., et al. (1986). Nucleus basalis neuronal loss, neuritic plaques and choline acetyltransferase activity in advanced Alzheimer's disease. *Neuroscience*, *19*, 1279−1291.

Fields, J. A., Troster, A. I., Wilkinson, S. B., et al. (1999). Cognitive outcome following staged bilateral pallidal stimulation for the treatment of Parkinson's disease. *Clinical Neurology and Neurosurgery*, *101*, 182−188.

Follett, K. A., Weaver, F. M., Stern, M., et al. (2010). Pallidal versus subthalamic deep-brain stimulation for Parkinson's disease. *New England Journal of Medicine*, *362*, 2077−2091.

Fontaine, D., Deudon, A., Lemaire, J. J., et al. (2013). Symptomatic treatment of memory decline in Alzheimer's disease by deep brain stimulation: A feasibility study. *Journal of Alzheimers Disease*, *34*, 315−323.

Freund, H. J., Kuhn, J., Lenartz, D., et al. (2009). Cognitive functions in a patient with Parkinson-dementia syndrome undergoing deep brain stimulation. *Archives of Neurology*, *66*, 781−785.

Fuller, P. M., Sherman, D., Pedersen, N. P., et al. (2011). Reassessment of the structural basis of the ascending arousal system. *The Journal of Comparative Neurology*, *519*, 933−956.

Gaspar, P., & Gray, F. (1984). Dementia in idiopathic Parkinson's disease: A neuropathological study of 32 cases. *Acta Neuropathologica*, *64*, 43−52.

Ghika, J., Villemure, J. G., Fankhauser, H., et al. (1998). Efficiency and safety of bilateral contemporaneous pallidal stimulation (deep brain stimulation) in levodopa-responsive patients with Parkinson's disease with severe motor fluctuations: A 2-year follow-up review. *Journal of Neurosurgery*, *89*, 713−718.

Gironell, A., Kulisevsky, J., Rami, L., et al. (2003). Effects of pallidotomy and bilateral subthalamic stimulation on cognitive function in Parkinson disease: A controlled comparative study. *Journal of Neurology*, *250*, 917−923.

Gratwicke, J., Kahan, J., Zrinzo, L., et al. (2013). The nucleus basalis of Meynert: A new target for deep brain stimulation in dementia? *Neuroscience and Biobehavioral Reviews*, *37*, 2676−2688.

Hamani, C., Dubiela, F. P., Soares, J. C., et al. (2010). Anterior thalamus deep brain stimulation at high current impairs memory in rats. *Experimental Neurology*, *225*, 154−162.

Hamani, C., & Lozano, A. M. (2003). Physiology and pathophysiology of Parkinson's disease. *Annals of the New York Academy of Sciences*, *991*, 15−21.

Hamani, C., McAndrews, M. P., Cohn, M., et al. (2008). Memory enhancement induced by hypothalamic/fornix deep brain stimulation. *Annals of Neurology*, *63*, 119−123.

Hamani, C., Stone, S. S., Garten, A., et al. (2011). Memory rescue and enhanced neurogenesis following electrical stimulation of the anterior thalamus in rats treated with corticosterone. *Experimental Neurology*, *232*, 100−104.

Hanyu, H., Asano, T., Sakurai, H., et al. (2002). MR analysis of the substantia innominata in normal aging, Alzheimer disease, and other types of dementia. *American Journal of Neuroradiology, 23,* 27−32.

Hardenacke, K., Hashemiyoon, R., Visser-Vandewalle, V., et al. (2016). Deep brain stimulation of the nucleus basalis of Meynert in Alzheimer's dementia: Potential predictors of cognitive change and results of a long-term follow-up in eight patients. *Brain Stimulation, 9,* 799−800.

Heath, R. G. (1954). *Studies in schizophrenia: A multidisciplinary approach to mind-brain relationships.* Cambridge, MA: Harvard University Press.

Higginson, C. I., Fields, J. A., & Troster, A. I. (2001). Which symptoms of anxiety diminish after surgical interventions for Parkinson disease? *Neuropsychiatry, Neuropsychology and Behavioral Neurology, 14,* 117−121.

Ito, K., Ahadieh, S., Corrigan, B., et al. (2010). Disease progression meta-analysis model in Alzheimer's disease. *Alzheimers & Dementia, 6,* 39−53.

Kalmbach, A., Hedrick, T., & Waters, J. (2012). Selective optogenetic stimulation of cholinergic axons in neocortex. *Journal of Neurophysiology, 107,* 2008−2019.

Koubeissi, M. Z., Kahriman, E., Syed, T. U., et al. (2013). Low-frequency electrical stimulation of a fiber tract in temporal lobe epilepsy. *Annals of Neurology, 74,* 223−231.

Krause, M., Fogel, W., Heck, A., et al. (2001). Deep brain stimulation for the treatment of Parkinson's disease: Subthalamic nucleus versus globus pallidus internus. *Journal of Neurology, Neurosurgery, and Psychiatry, 70,* 464−470.

Kuhn, J., Hardenacke, K., Lenartz, D., et al. (2015a). Deep brain stimulation of the nucleus basalis of Meynert in Alzheimer's dementia. *Molecular Psychiatry, 20,* 353−360.

Kuhn, J., Hardenacke, K., Shubina, E., et al. (2015b). Deep brain stimulation of the nucleus basalis of Meynert in early stage of Alzheimer's dementia. *Brain Stimulation, 8,* 838−839.

Kurosawa, M., Sato, A., & Sato, Y. (1989). Stimulation of the nucleus basalis of Meynert increases acetylcholine release in the cerebral cortex in rats. *Neuroscience Letters, 98,* 45−50.

Laitinen, L. V. (2000). Leksell's unpublished pallidotomies of 1958−1962. *Stereotactic and Functional Neurosurgery, 74,* 1−10.

Laxton, A. W., Lipsman, N., & Lozano, A. M. (2013). Deep brain stimulation for cognitive disorders. *Handbook of Clinical Neurology, 116,* 307−311.

Laxton, A. W., Tang-Wai, D. F., McAndrews, M. P., et al. (2010). A phase I trial of deep brain stimulation of memory circuits in Alzheimer's disease. *Annals of Neurology, 68,* 521−534.

Lee, D. J., Gurkoff, G. G., Izadi, A., et al. (2013). Medial septal nucleus theta frequency deep brain stimulation improves spatial working memory after traumatic brain injury. *Journal of Neurotrauma, 30,* 131−139.

Lo, R. Y., Hubbard, A. E., Shaw, L. M., et al. (2011). Longitudinal change of biomarkers in cognitive decline. *Archives of Neurology, 68,* 1257−1266.

Loher, T. J., Gutbrod, K., Fravi, N. L., et al. (2003). Thalamic stimulation for tremor: Subtle changes in episodic memory are related to stimulation per se and not to a microthalamotomy effect. *Journal of Neurology, 250,* 707−713.

Lombardi, W. J., Gross, R. E., Trepanier, L. L., et al. (2000). Relationship of lesion location to cognitive outcome following microelectrode-guided pallidotomy for Parkinson's disease: Support for the existence of cognitive circuits in the human pallidum. *Brain, 123*(Pt 4), 746−758.

Lozano, A. M., Fosdick, L., Chakravarty, M. M., et al. (2016). A phase II study of fornix deep brain stimulation in mild Alzheimer's disease. *Journal of Alzheimers Disease, 54,* 777−787.

Mayeux, R., & Sano, M. (1999). Treatment of Alzheimer's disease. *New England Journal of Medicine, 341,* 1670−1679.

McLin, D. E., 3rd, Miasnikov, A. A., & Weinberger, N. M. (2002). Induction of behavioral associative memory by stimulation of the nucleus basalis. *Proceedings of the National Academy of Sciences of the United States of America, 99,* 4002−4007.

McNeely, H. E., Mayberg, H. S., Lozano, A. M., et al. (2008). Neuropsychological impact of Cg25 deep brain stimulation for treatment-resistant depression: preliminary results over 12 months. *The Journal of Nervous and Mental Disease, 196,* 405−410.

Medtronic. (2013). Medtronic DBS therapy for movement disorders: Indication-specific information for implantable neurostimulator.

Metherate, R., Cox, C. L., & Ashe, J. H. (1992). Cellular bases of neocortical activation: Modulation of neural oscillations by the nucleus basalis and endogenous acetylcholine. *Journal of Neuroscience, 12,* 4701−4711.

Mirzadeh, Z., Bari, A., & Lozano, A. M. (2016). The rationale for deep brain stimulation in Alzheimer's disease. *Journal of Neural Transmission (Vienna), 123,* 775−783.

Montero-Pastor, A., Vale-Martinez, A., Guillazo-Blanch, G., et al. (2001). Nucleus basalis magnocellularis electrical stimulation facilitates two-way active avoidance retention, in rats. *Brain Research, 900,* 337−341.

Montero-Pastor, A., Vale-Martinez, A., Guillazo-Blanch, G., et al. (2004). Effects of electrical stimulation of the nucleus basalis on two-way active avoidance acquisition, retention, and retrieval. *Behavioural Brain Research, 154,* 41−54.

Moretti, R., Torre, P., Antonello, R. M., et al. (2003). Neuropsychological changes after subthalamic nucleus stimulation: A 12 month follow-up in nine patients with Parkinson's disease. *Parkinsonism & Related Disorders, 10,* 73−79.

Nitsch, R. M., Deng, M., Tennis, M., et al. (2000). The selective muscarinic M1 agonist AF102B decreases levels of total Abeta in cerebrospinal fluid of patients with Alzheimer's disease. *Annals of Neurology, 48,* 913−918.

Okun, M. S., Fernandez, H. H., Wu, S. S., et al. (2009). Cognition and mood in Parkinson's disease in subthalamic nucleus versus globus pallidus interna deep brain stimulation: The COMPARE trial. *Annals of Neurology, 65,* 586−595.

Ovsepian, S. V., & Herms, J. (2013). Cholinergic neurons-keeping check on amyloid beta in the cerebral cortex. *Frontiers in Cellular Neuroscience, 7,* 252.

Pillon, B., Ardouin, C., Damier, P., et al. (2000). Neuropsychological changes between "off" and "on" STN or GPi stimulation in Parkinson's disease. *Neurology, 55,* 411−418.

Ponce, F. A., Asaad, W. F., Foote, K. D., et al. (2016). Bilateral deep brain stimulation of the fornix for Alzheimer's disease: Surgical safety in the ADvance trial. *Journal of Neurosurgery, 125,* 75−84.

Pool, J. L. (1954). Psychosurgery in older people. *Journal of the American Geriatrics Society, 2,* 456−466.

Roane, D. M., Yu, M., Feinberg, T. E., et al. (2002). Hypersexuality after pallidal surgery in Parkinson disease. *Neuropsychiatry, Neuropsychology, and Behavioral Neurology, 15,* 247−251.

Rodriguez-Oroz, M. C., Obeso, J. A., Lang, A. E., et al. (2005). Bilateral deep brain stimulation in Parkinson's disease: A multicentre study with 4 years follow-up. *Brain, 128,* 2240−2249.

Rothlind, J. C., Cockshott, R. W., Starr, P. A., et al. (2007). Neuropsychological performance following staged bilateral pallidal or subthalamic nucleus deep brain stimulation for Parkinson's disease. *Journal of the International Neuropsychological Society, 13,* 68−79.

Schuepbach, W. M., Rau, J., Knudsen, K., et al. (2013). Neurostimulation for Parkinson's disease with early motor complications. *New England Journal of Medicine, 368*, 610–622.

Sem-Jacobsen, C. W. (1965). Depth electrographic stimulation and treatment of patients with Parkinson's disease including neurosurgical technique. *Acta Neurologica Scandinavica. Supplementum, 13*(Pt 1), 365–377.

Shaltenbrand, G., & Wahren, W. (1977). *Atlas for stereotaxy of the human brain* (2nd Edition). Chicago, IL: Thieme.

Sjöberg, R. L., Lidman, E., Häggström, B., et al. (2012). Verbal fluency in patients receiving bilateral versus left-sided deep brain stimulation of the subthalamic nucleus for Parkinson's disease. *Journal of the International Neuropsychological Society, 18*, 606–611.

Smeding, H. M., Speelman, J. D., Koning-Haanstra, M., et al. (2006). Neuropsychological effects of bilateral STN stimulation in Parkinson disease: A controlled study. *Neurology, 66*, 1830–1836.

Stone, S. S., Teixeira, C. M., Devito, L. M., et al. (2011). Stimulation of entorhinal cortex promotes adult neurogenesis and facilitates spatial memory. *Journal of Neuroscience, 31*, 13469–13484.

Suthana, N., Haneef, Z., Stern, J., et al. (2012). Memory enhancement and deep-brain stimulation of the entorhinal area. *New England Journal of Medicine, 366*, 502–510.

Temel, Y., Kessels, A., Tan, S., et al. (2006). Behavioural changes after bilateral subthalamic stimulation in advanced Parkinson disease: A systematic review. *Parkinsonism & Related Disorders, 12*, 265–272.

Thobois, S., Guillouet, S., & Broussolle, E. (2001). Contributions of PET and SPECT to the understanding of the pathophysiology of Parkinson's disease. *Neurophysiologie Clinique, 31*, 321–340.

Trepanier, L. L., Kumar, R., Lozano, A. M., et al. (2000). Neuropsychological outcome of GPi pallidotomy and GPi or STN deep brain stimulation in Parkinson's disease. *Brain and Cognition, 42*, 324–347.

Troster, A. I. (2009). Neuropsychology of deep brain stimulation in neurology and psychiatry. *Frontiers in Bioscience (Landmark Ed), 14*, 1857–1879.

Troster, A. I., Fields, J. A., Wilkinson, S. B., et al. (1997). Unilateral pallidal stimulation for Parkinson's disease: Neurobehavioral functioning before and 3 months after electrode implantation. *Neurology, 49*, 1078–1083.

Troster, A. I., Wilkinson, S. B., Fields, J. A., et al. (1998). Chronic electrical stimulation of the left ventrointermediate (Vim) thalamic nucleus for the treatment of pharmacotherapy-resistant Parkinson's disease: A differential impact on access to semantic and episodic memory? *Brain and Cognition, 38*, 125–149.

Tsivilis, D., Vann, S. D., Denby, C., et al. (2008). A disproportionate role for the fornix and mammillary bodies in recall versus recognition memory. *Nature Neuroscience, 11*, 834–842.

Turnbull, I. M., McGeer, P. L., Beattie, L., et al. (1985). Stimulation of the basal nucleus of Meynert in senile dementia of Alzheimer's type: A preliminary report. *Applied Neurophysiology, 48*, 216–221.

Vertes, R. P. (2005). Hippocampal theta rhythm: A tag for short-term memory. *Hippocampus, 15*, 923–935.

Vingerhoets, G., van der Linden, C., Lannoo, E., et al. (1999). Cognitive outcome after unilateral pallidal stimulation in Parkinson's disease. *Journal of Neurology, Neurosurgery, and Psychiatry, 66*, 297–304.

Weaver, F. M., Follett, K., Stern, M., et al. (2009). Bilateral deep brain stimulation vs best medical therapy for patients with advanced Parkinson disease: A randomized controlled trial. *The Journal of American Medical Association, 301,* 63−73.

Weaver, F. M., Follett, K. A., Stern, M., et al. (2012). Randomized trial of deep brain stimulation for Parkinson disease: Thirty-six-month outcomes. *Neurology, 79,* 55−65.

Williams, A., Gill, S., Varma, T., et al. (2010). Deep brain stimulation plus best medical therapy versus best medical therapy alone for advanced Parkinson's disease (PD SURG trial): A randomised, open-label trial. *The Lancet Neurology, 9,* 581−591.

Williams, J. M., & Givens, B. (2003). Stimulation-induced reset of hippocampal theta in the freely performing rat. *Hippocampus, 13,* 109−116.

Williams, Z. M., & Eskandar, E. N. (2006). Selective enhancement of associative learning by microstimulation of the anterior caudate. *Nature Neuroscience, 9,* 562−568.

Wilson, C. R., Baxter, M. G., Easton, A., et al. (2008). Addition of fornix transection to frontal−temporal disconnection increases the impairment in object-in-place memory in macaque monkeys. *The Europe Journal of Neuroscience, 27,* 1814−1822.

Woods, S. P., Fields, J. A., Lyons, K. E., et al. (2001). Neuropsychological and quality of life changes following unilateral thalamic deep brain stimulation in Parkinson's disease: A one-year follow-up. *Acta Neurochirurgica (Wien), 143,* 1273−1277. (discussion1278).

Xu, D. S., & Ponce, F. A. (2017). Deep brain stimulation for Alzheimer's disease. *Current Alzheimer Research, 14,* 356−361.

Yaguez, L., Costello, A., Moriarty, J., et al. (2014). Cognitive predictors of cognitive change following bilateral subthalamic nucleus deep brain stimulation in Parkinson's disease. *Journal of Clinical Neuroscience, 21,* 445−450.

Chapter 12

Neuropsychology in the Outcome of Severe Traumatic Brain Injury

Erin D. Bigler[1,2]

[1]*Department of Psychology, Brigham Young University, Provo, UT, United States,*
[2]*Neuroscience Center, Brigham Young University, Provo, UT, United States*

The development of neuropsychological testing has been intimately connected with neurosurgery for straightforward reasons. As Masel and DeWitt (2010) point out, prior to the 20th Century improvements in the medical management and neurosurgical treatment of traumatic brain injury (TBI), moderate-to-severe TBI often was not survivable. Without surviving a brain injury and given middle-age life expectancy of just a century ago (De Flora, Quaglia, Bennicelli, & Vercelli, 2005), there was little need or reason for assessing the cognitive and neurobehavioral sequelae of a head injury. This changed dramatically with World War II, Korean and Vietnam Wars, and the development of field hospitals and mobile neurosurgical units (Stone, Patel, & Bailes, 2016). Surviving a moderate to severe TBI became more common, directly related to improved neurosurgical management of the initial brain injury, which not only saved lives but lessened neurologic morbidity. With experience learned from these past conflicts, survival following neurosurgical care in military conflicts post 9/11 Iraq and Afghanistan wars has been impressive (Risdall & Menon, 2011). What has been learned in the battlefields mid to late 20th century coincided with the movement in the 1960s to improve community-based emergent care and more uniform treatment of the civilian TBI patient (Mullins, 1999), which was directly linked to improved neurosurgical management of the TBI patient (Kolias, Guilfoyle, Helmy, Allanson, & Hutchinson, 2013). Similarly, in the early 1970s, computed tomography (CT) was introduced (Ambrose & Hounsfield, 1973), which quickly revolutionized acute assessment of the head injured patient (Dublin, French, & Rennick, 1977; Jennett, Galbraith, Teasdale, & Steven, 1976). With CT protocols for head injury being widely adopted in combined with the universal implementation of the Glasgow Coma Scale

Neurosurgical Neuropsychology. DOI: https://doi.org/10.1016/B978-0-12-809961-2.00013-8

(GCS) (Teasdale & Jennett, 1974) not only improved the clinical characterization of TBI and its neurosurgical treatment, but research applied to TBI. All of these developments improved the initial triage of the TBI patient, resulting in improved neurosurgical decision making, medical management and monitoring, and better outcome. What this meant for the patient was a dramatically improved chance of survival, but with severe brain injury comes a lengthy hospitalization and the likelihood of major structural and functional brain damage that adversely influences cognition, emotion, and behavior. How best to assess such changes? Can deficits be rehabilitated, if so which ones and with what techniques? How should cognitive, emotional, and behavioral recovery or adaption be quantified and monitored? The severe TBI patient who had been in neurosurgical care was in need of additional assessment to answer these types of clinical questions about cognition and behavior. Neuropsychology emerged in this identical, parallel timeframe with those tools to provide such services, the description of which will be the focus of this chapter.

HISTORICAL PERSPECTIVE

The basic assumption in neuropsychology is that behavior, emotion, and cognition can be measured, where performance on standardized tests permit inferences about the health or dysfunction of the brain. Standardized psychometric testing began in the service of education and identifying intellectual disability, with well-established tests for these purposes in place by mid-20th century (Lezak, Howieson, Bigler, & Tranel, 2012). As implied in the introduction, applying psychological measures to the brain injured patient did not begin in any systematic way until patients began to survive the injury. With the increased survival from head wounds that accompanied the World War II, Oliver Zangwill was one of the first psychologists to use standardized psychological measures extracted from what was being used in public education and mental health institutions to assess intellectual and memory functioning to those with TBI (see Collins, 2006; Zangwill, 1946). In the United States, Hans Lukas Teuber was the first to examine standardized psychological test results in soldiers who had survived penetrating brain injuries and neurosurgical treatment during World War II and the Korean War (Bender, Teuber, & Battersby, 1950; Weinstein & Teuber, 1957). The importance of these seminal contributions was that the neurosurgeon could verify, in a general sense, where the focal pathology was and the neurosurgical intervention to treat, which provided a rough independent description of where brain pathology resided (i.e., anterior versus posterior, left versus right). This facilitated the field of psychological assessment and neuropsychology to verify that impaired cognitive and behavioral functioning related to the brain being injured, and at times, to the influence of damage to specific brain areas. Remember, all of neuropsychological assessment in this era (1940s to mid-1970s) occurred prior to any type of contemporary neuroimaging, so establishing that cognitive changes could be related to specific brain pathology required the neurosurgeon to document where the brain "damage" was located.

The neurosurgical "proof" that brain pathology had occurred was critical for the evolution of neuropsychological assessment. With test performance on standardized measures in someone with documented, objective brain pathology exhibiting deficits on psychological/neuropsychological tests, this could now be extended to those who did not require neurosurgical intervention but met criteria for having sustained a TBI (Joseph et al., 2014). With that information in hand, similar test score patterns could be identified in those with head injury but not operated on. The so-called closed-head injury or CHI case, as it was originally referred to (Denny-Brown, 1943), could now be evaluated with neuropsychological measures and inferences about the "organicity" of the injury made by the level of psychometric impairment. Again, prior to contemporary neuroimaging, a patient with CHI may not have met any criteria for being operated on other than receiving an intracranial pressure (ICP) monitoring device (Johnston, Johnston, & Jennett, 1970). Other than routine skull films, prior to the advent of CT imaging there were basically three procedures available to neurosurgery to make imaging inferences about the brain: (1) cerebral angiography (Webster, Dawson, & Gurdjian, 1951) (2) pneumoencephalography, which was rarely done in the CHI patient, because of the complications with such an invasive procedure (Davies & Falconer, 1943), or (3) a radioisotope absorption scan, which when performed was typically done only during the subacute to chronic phase (Villani et al., 1975). So prior to modern neuroimaging techniques, for the severe CHI patient on a neurosurgical unit who recovered from coma but were never operated on except for ICP, but displayed cognitive and emotional sequelae, how could brain pathology be inferred? It was through neuropsychological test findings and clinical inference.

With this objective in mind to define a psychometric indicator of brain pathology, neuropsychological researchers, and clinicians began to map out how cognitive, behavioral, and emotional deficits could be demonstrated by standardized neuropsychological tests in those with TBI (Levin, Grossman, & Kelly, 1976; Lezak et al., 2012; Reitan, 1959; Spreen & Benton, 1965). Because of the low resolution of CT and its inability to detect more subtle structural brain pathology prior to the widespread implementation of magnetic resonance imaging (MRI) in the1990s, often the confirmation that "brain damage" was present in the postacute CHI patient with severe injury was via neuropsychological test findings. Accordingly, much of the early work in neuropsychology was in the development of psychometric methods to detect and infer the presence of "organic brain" pathology in those with CHI (Dencker, 1960).

With neuroimaging advancements at the end of the 20th Century, however, neuropsychology's role in detecting the so-called brain damage was diminished and gradually replaced by advanced neuroimaging methods. Nonetheless, neuropsychological assessment remained a critical clinical discipline in the evaluation of the TBI patient at a descriptive level, at any stage postinjury. Once the severe TBI patient was out of coma, neuropsychological test findings provided objective methods to document the cognitive and neurobehavioral sequelae of TBI, as well as assisting in plotting recovery or adaptation to brain injury and guide therapeutic treatments. As such by the late 1980s and early 1990s neuropsychological testing had become established in the care and management of the neurosurgical patient with severe TBI, and continues to this day, as a routine part of assessment in the TBI patient who has been on a neurosurgical unit (Clifton, Hayes, Levin, Michel, & Choi, 1992). However, with the emphasis shifting more to psychometric descriptions and documentation of the degree and type of cognitive impairment, the localization role of inferring brain pathology became less important for neuropsychology and therefore, not a major reason for performing neuropsychological evaluations in the TBI patient seen by a neurosurgeon.

TRADITIONAL NEUROPSYCHOLOGICAL STUDIES OF THE TRAUMATIC BRAIN INJURY PATIENT WITH SEVERE TRAUMATIC BRAIN INJURY

As outlined in Lezak et al. (2012) and characterized by Cipolotti and Warrington (1995), the standard neuropsychological approach ... makes the assumption that impairments in cognitive function can best be studied and understood by (a) assuming that there is a high degree of functional specialization in the cerebral cortex; (b) by undertaking a modularity approach to the analysis of complex cognitive skills; and (c) by assuming that brain damage can selectively disrupt some components of a cognitive system (p. 655). What emerged out of this historical period of neuropsychological test development (1950s through the 1980s) was a view that neuropsychological tests could assess performance within certain domains characterized by the following categories: Motor and sensory processing, language, memory, visual—spatial and perceptual—motor abilities, and executive function, academic and intellectual ability, along with personality/emotional functioning. Within these general domains just listed, there are a variety of neuropsychological tests that can be used to assess cognition, behavior and emotional functioning, some of which are presented in Table 12.1 that overviews some commonly used neuropsychological measures. This is not an exhaustive list by any means, but merely for the purposes of review, identifies some of the common metrics used in neuropsychological assessment of the TBI patient. As Lezak et al. (2012) review

TABLE 12.1 Representative Sample of Available Neuropsychological Tests and the Domain They Assess in Evaluating the Severe TBI Patient

Cognitive Domain	Name of Test	Description of Test	Time to Administer
General functioning/IQ	NIH Toolbox		30–45 min
	Neuropsychological Assessment Battery	The NAB is a comprehensive, modular battery covering five cognitive domains (attention, language, spatial, memory, and executive functioning)	Complete battery less than 4 h. Individual batteries between 30–60 min
	Wechsler Abbreviated Scale of Intelligence (WASI)	Screening instrument for cognitive functioning when full assessment of intellectual functioning is either not needed or not possible	15 min for two subtest form and 30 min for the four subtest form
	Repeatable Battery for the Assessment of Neuropsychological Status (RBANS)	12 subtest brief test to look at cognitive status, particularly in low functioning adults needing shorter testing sessions. The five cognitive domains are immediate memory, visuospatial, attention, language, and delayed memory	20–30 min
Motor	Finger Tapping	Measures self-directed manual motor speed	10 min
	Grip Strength	Measure the strength or intensity of grip across both hands	5 min
	Grooved Pegboard	Requires individuals to place grooved pegs into a pegboard as quickly as possible. Measures manual precision and dexterity	5 min
Sensory	Informal Sensory Perceptual Exam	Need info from Bigler	
	Smell Identification Test	Measures impairments in olfaction	Brief versions, 5 min; full version 10–15 min

(Continued)

TABLE 12.1 (Continued)

Cognitive Domain	Name of Test	Description of Test	Time to Administer
Attention	Symbol Digit Modalities Test	Simple substitution task that can be administered either on paper or orally and evaluates	2 min
	PASAT	Uses mental math to evaluate working memory, divided attention, and processing speed	10 min
	Trail Making Test	Paper test that measures attention, speed, and mental flexibility	5 min
Learning/ Memory	California Verbal Learning Test—II (CVLT)	Measures verbal learning and memory using multiple trial list learning	50 min including delay
	Rey-Osterrieth Auditory Verbal Learning Test (RAVLT)	Assesses verbal learning	10–15 min
	Rey Complex Figure Test	Simultaneously assesses visuospatial construction and visual memory	10–15 min excluding delay
	Wechsler Memory Scale (WMS-IV)	Evaluates visual memory, narrative memory, and list learning	45–60 min for primary subtests
Executive Functioning	Delis–Kaplan Executive Functioning System	Assesses nine areas of executive functioning. Individual subtests can be used	90 min for the full battery
	Wisconsin Card Sorting Task	Card game requiring patients to learn concepts without rules and shift responses based on feedback. Can be administered and scored on the computer	15–30 min
	Stroop Test	Presents visual stimulus (color/names) simultaneous and requires the patient to suppress habitual responses	5 min

Category	Test	Description	Time
Language	Boston Naming Test	Assess visual naming abilities using black and white drawings	10–20 min
Emotional Conditions/ Personality	Beck Depression Inventory and Beck Anxiety Inventory	Self-report measure of common symptoms of depression and anxiety	5–10 min for each
	Brief Rating of Executive Functioning	Self and observer report of common symptoms of executive functioning difficulties	10–15 min
	Personality Inventory of the DSM-V (PID-5)	Emerging personality inventory that asks common symptoms of psychiatric and personality disorders in the DSM-5. Public domain and can be downloaded and reproduced for free	45–60 min
Concussion	Immediate Post-Concussion Assessment and Cognitive Testing (ImPACT)	Repeatable online battery that measures attention span, working memory, sustained and selective attention time, nonverbal problem solving, and reaction time	25 min
	Standardized Assessment of Concussion (SAC)	Assesses common symptoms of concussion including loss of consciousness, amnesia, strength, concentration, and orientation	5–10 min
	Glasglow Coma Scale	Scale to assess consciousness after injury, looking at eye opening, verbal response, and motor response	5 min
	Galveston Orientation and Amnesia Test	Evaluates orientation to person, place, and time following head injury	5 min

there are over 300 standardized neuropsychological measures that have been used in neuropsychological assessment, so what is reflected in Table 12.1, is but a brief synopsis.

Later in this chapter discussion will be centered on viewing neuropsychological functioning related to underlying neural systems that drive the function being assessed, but there is also another general principle that needs to be mention. There is some degree of common hemispheric specialization, where the left hemisphere is more organized for language-based functions and the right hemisphere, more organized toward nonverbal and visual–spatial abilities. As a result of these general relations, when verbal-based functions are lower than nonverbal, visuospatial ability, the assumption is often made that left hemisphere deficits dominate, with the opposite pattern true for greater right hemisphere impairment (see Lezak et al., 2012).

A standard assumption in neuropsychology is that healthy, typical developing individuals possess cognitive abilities that are positively interrelated and function at least within the broad range of average, but may also be above average. Average is defined statistically as the range of scores that is ± 1 standard deviation (SD) around the mean. By definition then, scores that fall less than 1 SD below the mean are suspect of reflecting impairment. If in healthy, typical developing or developed individuals, psychometric test scores tend to congregate in the average range and low scores are not expected. In a practical sense, this means that in healthy, typical developing or developed individuals, neuropsychological test scores in one domain should generally relate to level of function in other domains and, at a minimum, exceed the lower bounds of average. For example, someone with a high school education and average intellectual ability, likely performs at similar levels in all other domains, such as memory and language, although there may also be areas of relative strength and weaknesses. To quantitatively address these assumptions, the standard bell curve as presented in Fig. 12.1 from Zasler and Bigler (2017), is key to understanding and interpreting neuropsychological test findings. In reference to Fig. 12.1, note that the vertical lines demarcate the average range denoted by ± 1.0 SD bars. For greater details on neuropsychological inference, please refer to Lezak et al. (2012), but what this boundary of average also means is that for a score to be "different"—altered in some way by a TBI—the obtained score needs to be less than would be expected for either normal variation between domains or what was the presumed native ability of the individual prior to injury. Sticking with the average individual with expected average scores across domains, until a specific score drops below one SD (e.g., 16th percentile) or more from the mean (50th percentile), it may not be indicative of a loss. Continuing with this logic, and knowing that typical variation in neuropsychological test performance in any individual tends to center around one-half to three-quarters of an SD of their "true" premorbid ability, then neuropsychologists have often set a 1.5–2.0 SD difference of a particular score

"Bell curve"and the interpretation of neuropsychological test results

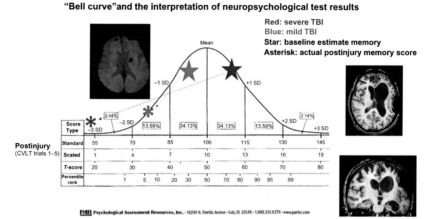

FIGURE 12.1   Bell curve and the interpretation of neuropsychological test results. Red, severe TBI; blue, mild TBI; asterisk, actual postinjury memory score; star, baseline estimate memory; SD, standard deviation. The axial and coronal T1-weighted MR images of the patient with severe (red) TBI are presented on the right, with image orientation reflecting right side on the viewer's right. The susceptibility weighted axial MR image in the upper left is from the patient (blue) which reflects multiple areas of hemosiderin deposition, likely reflective of shear injury. CVLT, California Verbal Learning Test, Second Edition. Dashed lines depict the discrepancy from assumed baseline level to actual neuropsychological performance level. *From Zasler and Bigler (2017), used with permission from Taylor & Francis.*

deviating from presumed premorbid ability or the midpoint of average from the hypothetical "normal" individual, before it is referred to as "abnormal."

Note under "Score Type" in Fig. 12.1 the four most common statistical metrics used by neuropsychologists are listed—standard scores, scaled scores, T-scores, and percentiles. All major neuropsychological tests in use today have been standardized in comparison to either healthy controls, or in reference to some type of patient group with or without the diagnosis as part of a reference study and results compared statistically. Based on appropriate reference samples, a distribution of scores is obtained on each neuropsychological test where the test findings from the normative sample is then fitted to a "smoothed" bell curve, as depicted in Fig. 12.1. Description of tests like those assessing intelligence, memory, or academic functioning are typically stated in terms of "standard scores" and either referred to as a quotient, like an intelligence quotient (IQ) or as some kind of an index score (i.e., memory index). As seen in Fig. 12.1, standard scores have a mean of 100 and an SD of 15. As such the IQ scores are most commonly stated in a neuropsychological report as a standard score. However, the individual subtests that go into creating the standard score are typically expressed as "scaled scores," since they represent the individualized scales used to compute the index scores, like IQ. Scaled scores have a mean of 10 and an SD of 3. Typically in a neuropsychological report, below index scores there will be a listing of individualized subtest scores, where the values are given in scaled score statistics. T-scores represent another method for reporting mean and SD, where the mean is 50 with an SD of 10. A number of neuropsychological measures are reported with T-score values, in particular measures of personality and emotional functioning.

Finally, note that all of the aforementioned metrics can also be referenced to a percentile score. Sometimes it is more straightforward to understand the level of impairment in terms of percentile scores. This may be especially true when neuropsychological test score feedback is given to the patient and/or family member. Standard, scaled, and T-scores may not have much meaning to someone not schooled in statistics, but as shown in Fig. 12.1, these scales can be quickly viewed as percentiles. As can be seen in the patient with a severe TBI (red star) there were profound short-term memory impairments where this individual performed less than what would be the case of 1% of the population. Indisputably, this reflects short-term memory impairment in this patient who had sustained a severe TBI.

In most TBI cases, there is no preinjury neuropsychological examination, so inferences have to be made about premorbid ability based on education, vocational abilities, and family history. Returning to Fig. 12.1 the patient characterized in blue was assumed to premorbidly be of average ability. This individual had completed high school, obtained grades in the C and B range and graduated without ever being classified as learning disabled or placed in special education or requiring any kind of assistance while in school. At the time of injury this individual was involved in routine office work and had no special training or certificates, other than the high school diploma. Given this information, as indicated by the blue star, preinjury estimates of cognitive functioning were in the average range, but probably below the 50th percentile. On a standardized memory test, the California Verbal Learning Test (Second Edition) this individual did perform below expectation (see the blue asterisk). This was a case of mild TBI (GCS = 14 with positive, brief loss of consciousness) but as visualized on the susceptibility-weight MRI in the upper left there were multiple hemorrhagic foci, indicative of shear injury. Given that impairments of memory are some of the most common in those with any severity level of TBI (Cristofori & Levin, 2015), this drop in memory performance from a low average ability, estimated to be at about the 30−35th percentile down to the ∼10th percentile likely represents a clinically significant reduction, although only an SD different.

Later in the chapter, more will be said about how premorbid ability levels are ascertained. But before that discussion, some commentary about the other patient's memory scores. The patient depicted in red sustained a severe TBI (GCS = 3) and required neurosurgical interventions including shunting to manage cerebral edema and to treat skull and facial fractures. This individual had a college degree and had worked in a professional capacity prior to injury. Accordingly, the assumption was that preinjury ability was above average. Actuarial studies demonstrate some of the limitations of this assumption (Baade & Schoenberg, 2004), but as a general rule average to above levels of preinjury IQ can be assumed in those with a college education, especially in those who have professional employment with commensurate level of cognitive demands to perform the job. Since this was the case with this patient, it was assumed that the preinjury level of intellectual function was ∼110. This level of premorbid ability also was consistent with college admission test percentiles taken in high school, reflective of above average academic ability. When administered the same memory test as the previously described patient with mild TBI, this patient with severe brain injury scored below the first percentile. In comparison to the normative standard this was at least a 3 SD drop but from premorbid ability level, closer to a 4 SD decline when compared to preinjury estimates.

In the nonaphasic patient, well-learned academic skills tend to be resistant to much change following acquired brain injury and therefore reflect premorbid ability level as well. Despite the profound memory deficits in this severe TBI patient, the patient's composite academic reading standard score was 101, reflective of at a minimum of average premorbid cognitive ability. This individual's average ability to read and intact language skills, including conversational abilities belied the fact that none of the information being read could be effectively retained. Being able to demonstrate the profound nature of memory impairments in comparison to some more intact cognitive abilities showed reduced capacity to function independently secondary to the severe TBI. In this case, for the neurosurgeon transferring care of this patient, the neuropsychological findings were critical in helping to establish what level of care and monitoring were necessary for this patient.

TIME POSTINJURY AND NEUROPSYCHOLOGICAL STATUS

Based now on half a century of neuropsychological studies (see Cristofori & Levin, 2015), the expected areas of neurocognitive functioning most likely affected by severe TBI include deficits in memory, processing speed, and executive functioning. In part, this is related to the likelihood for frontal and temporal lobe damage associated with TBI (Bigler, 2007).

Bendlin et al. (2008) and Farbota et al. (2012) examined neuropsychological test performance within ∼3 months postinjury (Visit 1) and up to 4 years postinjury in patients who had sustained moderate-to-severe TBI. Fig. 12.2 depicts initial deficits using another statistic—the z-score. In z-score terminology, the mean is set at 0.0 and SDs are in units of 1.0. As such, negative z-scores reflect below the mean, whereas positive values are above the mean, represented in SD units. Fig. 12.2 shows the initial major impairments, in particular affecting verbal fluency, processing speed, and short-term memory. Note that in this sample, when assessed more than a year postinjury substantial improvement in cognitive functioning occurred,

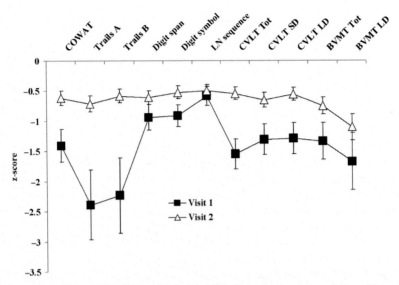

FIGURE 12.2 At ~2–3 months postinjury and then 1 year later, a group of 17 moderate-to-severe TBI patients were assessed with a range of neuropsychological tests. Initially almost all areas of assessment were below 1 SD, with the most profound deficits noted in processing speed. All areas showed some improvement in time although none returned to the mean. *From Bendlin et al. (2008) and Farbota et al. (2012), reprinted with permission from Elsevier and Cambridge University Press.*

but no area actually returned to the mean (0.0 z-score). As such, residual cognitive impairment of some degree is expected in the neuropsychological outcome from severe TBI, noting that none of the test values return to the mean, with some memory measures reflecting performance decrements that are about an SD below the midpoint of average. Konigs, Engenhorst, and Oosterlaan (2016) conducted a meta-analysis of IQ outcome in severe TBI, demonstrating that the in comparison to an expected average IQ (i.e., ~100) reflective of the normal sample mean, chronic IQ effects of severe TBI result in overall scores about three-quarters of an SD below the mean or a full scale I.Q. score of ~88. As such, it can be expected that the severe TBI patient will exhibit some detrimental cognitive effects of the TBI.

Examining the group findings of neuropsychological test outcome reveals another clinical caveat in establishing what a given patient's neuropsychological baseline or premorbid ability was and how to identify which cognitive domain has been most adversely influenced. The logic goes as follows. If brain injury alters a cognitive function it does so by lowering ability. While different abilities may be affected differentially by taking the maximum test score postinjury, and after a period of stability of several months postinjury, whatever scores are the highest, may also provide insight into what premorbid ability was. For example, reviewing Fig. 12.2, the fact that most of the measures reflecting different domains show improvement over time and fall within the "average" range, albeit the low average; the long delay on the Brief Visuospatial Memory Test was the lowest and remained ∼1.0 SD below the mean. One conclusion from this pattern of test scores is that memory performance in this cohort of TBI patients that involved retention of information at about 25 minutes postinitial test administration was not retained and exhibited the greatest degree of impairment.

Along these same lines of clinical logic, while a cognitive domain may not be affected by the TBI, it is unlikely that TBI would improve cognitive processes, only impair them. So postinjury, whatever cognitive domain exhibits the best neuropsychological test scores, likely provides some insight into the range of premorbid cognitive ability. As such, neuropsychologists when attempting to infer what the patient was like preinjury will use maximum score test performance to infer premorbid ability and the contrast between test scores with the highest and lowest scores to reflect where the greatest impairments are.

As another example, the case presented in Fig. 12.3 is from a patient with a severe, penetrating brain injury associated with a motor vehicle versus train collision, with a 5-month hospital stay. The patient required multiple

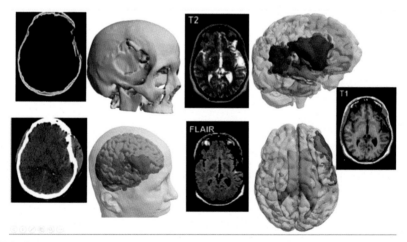

FIGURE 12.3   This patient sustained a severe TBI as a consequence of a train-motor vehicle collision with penetrating injury and depressed skull fracture. The bone window CT is shown in the upper left, outlining the depressed skull fracture. The bottom left image is the day-of-injury brain window. The colorized three-dimensional (3D) views are from the MRI obtained 2 years postinjury, showing a large area of encephalomalacia (dark red, in both lateral views) and in the dorsal 3D views the yellow depicts extensive underlying white matter signal abnormality detected on the FLAIR image. Given that the major areas of focal brain pathology where lateralized to the right hemisphere, it would be expected that neuropsychological impairments would likely be more reflective of nonverbal impairments, which was the case. Likewise, overall processing speed was reduced.

neurosurgical interventions for managing contusion, debridement, cerebral edema, and skull repair. By 2 years postinjury on the Wechsler Adult Intelligence Scale—IV Edition this college educated individual had a Full Scale IQ score of 92 with other Index Scores as follow: Verbal Comprehension = 105 (63rd percentile), Perceptual Reasoning = 96 (39th percentile), Working Memory Index = 92 (30th percentile) and Processing Speed = 74 (4th). On a generalized measure of memory function (Test of Memory and Learning, 2nd edition) the patient obtained the following: Verbal Memory Index = 100, NonVerbal Memory Index = 85, Composite Memory Index = 91; and a Learning Index of 89. Inspecting the MRI performed ~ 2 years postinjury, shows extensive focal encephalomalacia involving the posterior frontal and temporal lobe, in this right handed individual. Given the previous stated assumptions about laterality of cognitive abilities, absence of aphasia, right frontal, and temporal lobe focal pathology would more likely adversely influence nonverbal abilities, especially in the domain of memory. In that there would also be some nonspecific pathology, one would expect slowed processing speed, which is what was observed. Overall, one would also expect a lessening of general intellectual abilities, lowered from preinjury status. This individual had graduated from college and had held responsible professional employment prior to the severe brain injury. Accordingly, overall cognitive abilities reflected ~ 1 SD decline from previous ability level, but with the greatest impairment in processing speed and nonverbal memory, consistent with the more lateralized right hemisphere damage.

EFFORT AND ECOLOGICAL VALIDITY OF NEUROPSYCHOLOGICAL TESTS

There is another very important point about what is portrayed in Fig. 12.2 and that is neuropsychological testing changes over time. Neuropsychological testing is only appropriate in those severe TBI patients who have recovered sufficiently to cooperate with testing. Since severe TBI is associated with severe disorders of consciousness (Nakase-Richardson et al., 2013) as well as potentially severe motor and language impairments (Sbordone, Liter, & Pettler-Jennings, 1995), making some severe TBI cases inappropriate for neuropsychological referral because of lack of testability. Testing the patient too early in recovery may yield noninterpretable results given the level of patient compliance, which can invalidate any neuropsychological test result.

Also, a major limitation of neuropsychological testing comes from the following statements by Millis (2009) and Millis and Volinsky (2001): All cognitive tests require that patients give their *best effort* (italics added) when completing them. Furthermore, cognitive tests do not directly measure cognition: they measure behavior from which we make inferences about cognition. People are able to consciously alter or modify their behavior, including their behavior when performing cognitive tests. Ostensibly poor or "impaired" test scores will be obtained if an examinee withholds effort (e.g., reacting slowly to reaction time tests). There are many reasons why people may fail to give best effort on cognitive testing: financial compensation for personal injury; disability payments; avoiding or escaping formal duty or responsibilities (e.g., prison, military, or public service, or family support payments or other financial obligations); or psychosocial reinforcement for assuming the sick role (Slick, Sherman, & Iverson, 1999) and goes on to state that ".... Clinical observation alone cannot reliably differentiate examinees giving best effort from those who are not. (Millis, 2009, p. 2409)." As a means to assess test engagement a variety of measures referred to as "performance validity tests" or PVTs have been developed (see Lezak et al., 2012). These measures typically follow a format of appearing like a genuine cognitive measure, but in fact are relatively simple and assumed to require minimal amounts of cognitive effort. They are typically given in a binary, forced choice format so the patient either gets the item correct or not. PVT measures are often split into another dichotomy of the most easy and then some items that remain relatively easy, but a bit more challenging. For example, showing the patient the following string of numbers: 1-2-3 and then asking did I just show you 1-2-3 or was it 7-8-9? A more difficult but still easily doable discrimination would be something like showing the same string of numbers, 1-2-3 but this time the foil would be something like 1-2-4. Even in the presence of severe TBI, most TBI patients can readily pass PVT measures. All patients depicted in Figs. 12.1, 12.3, and 12.4 passed all PVT measures, some with 100% accuracy.

However, genuine caution needs to be used when interpreting PVT "failure," defined by below a certain cut point for the PVT measure being used, because severe TBI may affect attention and working memory, motivation and drive to engage in the testing process, where pain mediated problems may also affect PVT scores (Bigler, 2012, 2014, 2015). In most cases of severe TBI it is assumed that the patient is cooperating with the testing and test results reflect an accurate assessment, even when aspects of PVT measures are not necessarily "passed."

Despite that last statement, there is the issue of what is referred to as the ecological validity of the neuropsychological test findings. For neuropsychological test administration to be valid, it needs to follow a prescribed format. This means that the testing needs to be administered in a lab or clinic office setting, with minimal noise and distraction where all testing is done one-on-one. Furthermore, before each task is administered the examiner typically asks the examinee if they are OK, ready to proceed or if they need a break and likewise are encouraged at all times to always "try their best." So these are not "real-world" circumstances where distractions and outside pressures may affect cognitive performance. This is referred to as the ecological

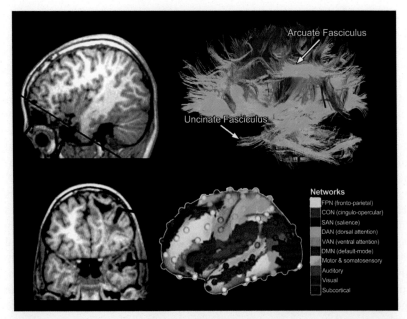

FIGURE 12.4 This child sustained a severe TBI with residual frontotemporal encephalomalacia. The colorized lateral view of the surface of the brain in the lower right depicts activation and resting-state derived brain networks. By plotting where focal damage has occurred it assists in understanding which brain networks may be affected. The upper right image is a whole brain DTI tractography plot highlighting the arcuate and uncinate fasciculi. The focal damage likely disrupted interconnectivity between the frontal lobe and other networks connected by these fasciculi, resulting in impaired processing speed and social-emotional functioning, as demonstrated by the neuropsychological test findings. *From Bigler (2016), used with permission from Frontiers in Systems Neuroscience, reproduced under the terms of the Creative Commons Attribution License (CC BY) at: http://journal.frontiersin.org/article/10.3389/fnsys.2016.00055/full.*

validity of neuropsychological tests which can be very limited, especially in the premorbidly high functioning individual with reasonably good recovery (Sbordone, 2014). Such patients may test well in the structured clinical setting, only to fail in a real-world home, social, or work setting. Accordingly, neuropsychological test findings are important, but there is also considerable clinical judgment that goes into making inferences about what test findings mean in the nonlaboratory setting.

A movement is afoot in neuropsychological test development that will hopefully improve ecological validity (Parsons, 2015). This involves using a virtual reality platform that makes the test items and tasks more like simulated daily cognitive tasks, with the assumption that the test abilities more closely linked to real-world challenges (Negut, Matu, Sava, & David, 2016).

As already mentioned neuroimaging supplanted part of the function of neuropsychological testing to infer "brain damage" but the integration of neuropsychological assessment as a cognitive probe using neuroimaging technology is beginning to emerge (Kane & Parsons, 2017; Oztekin, Long, & Badre, 2010). These developments will most likely be aided by progress with new virtual cognitive assessment techniques that could be adapted to the scan environment. None is ready for current clinical application for the TBI patient with severe brain injury, but will likely be part of standard clinical practice within another decade. In this scenario psychometric test findings will be derived concomitantly with brain activation patterns obtained as part of neuroimaging studies.

NEUROPSYCHOLOGICAL INTERPRETATION, BRAIN NETWORK THEORY, AND TRAUMATIC BRAIN INJURY

It is important to view neuropsychological status from the standpoint of brain networks that are affected (Bigler, 2016). Fig. 12.4 shows derived brain networks from a variety of neuroimaging techniques that permit the ability to not only identify regions that activate in response to a particular cognitive demand, but also areas with prominent white matter interconnectiveness, along with which networks are likely affected in a case of severe TBI with documented parenchymal damage. Note that these networks are multifaceted. There was an era in neuropsychology where it was assumed that "lesion localization" could be done merely from a cognitive and neurobehavioral perspective, but it is now well understood that the brain is a series of multilevel networks that are all integrated. So TBI should be viewed more as a disorder of white matter connectivity where focal damage demonstrated neurosurgically and/or via neuroimaging should be viewed from the perspective of which networks are disrupted and damaged and not so much as where the lesion or abnormality may be located or how large the lesion might be. Neuropsychological impairments that occur from TBI are, in part, most dependent upon the where the network damage in the brain occurs and how many networks are affected.

The point of network damage can be demonstrated by the case presented in Fig. 12.4. This child sustained a severe TBI (GCS = 3) where day-of-injury CT demonstrated multiple hemorrhagic contusions, especially in the frontal and temporal lobes. Cerebral edema and increased ICP were treated with intraventricular shunt, but no additional neurosurgical intervention was required. Frontotemporal encephalomalacia developed as a result of the focal frontotemporal hemorrhagic contusions as demonstrated in Fig. 12.4.

Despite the severe TBI, this child exhibited a relatively good outcome and was able to return to a regular classroom in a public school setting. However, there were deficits in social awareness, attention and processing speed. Although the child's intellectual level of function was measured with a full scale IQ score of 109, the standard score for processing speed was 95, nearly an SD lower. Although not reaching the level of significance previously described for 1.5−2.0 SDs difference, knowing that processing speed is one of the most common cognitive functions affected and that MRI studies show extensive white matter pathology, findings are consistent with disrupted white matter integrity from the TBI and the resultant slower processing speed. Furthermore, by examining the location of the focal pathology and the networks affected the frontoparietal network that participates in working memory, executive function and emotional control would certainly be affected as would the emotional control networks associated with the connections between the orbitofrontal and medial temporal and temporal polar areas.

UTILITY OF NEUROPSYCHOLOGICAL ASSESSMENT FINDINGS RELATED TO ADVERSE EVENTS IN THE LIFE OF THE SEVERE TRAUMATIC BRAIN INJURY PATIENT

Another important reason for TBI patients to undergo neuropsychological assessment is to provide baseline information should adverse future events occur subsequent to the brain injury. For example, having sustained a severe TBI increases the risk for posttraumatic epilepsy (PTE, see Annegers & Coan, 2000). Presence of PTE may be associated with its own alterations in neuropsychological status and functioning over time and require careful neuropsychological follow-up (Raymont et al., 2010). TBI is associated with a lifetime increased risk of developing new onset psychiatric disorder or worsening of a preexisting psychiatric disorder (Eapen, Allred, O'Rourke, & Cifu, 2015; Haarbauer-Krupa et al., 2017; Jones, Acion, & Jorge, 2017; Zaninotto et al., 2016). Again, follow-up neuropsychological testing especially when a prior baseline had been established is invaluable in following these patients. Also, there is increased risk for later-in-life dementia (Mendez, 2017; Washington, Villapol, & Burns, 2016). Knowing where a patient is functioning neuropsychologically at one stage is a critical clinical reference point for monitoring over time.

Of particular importance for baseline testing is the issue for later-in-life dementia. Since the severe TBI has likely compromised some aspect of cognitive functioning it is important to be able to track the stability of the deficit (Tomaszczyk et al., 2014). Serial neuropsychological assessments provide an ideal method for monitoring cognitive, emotional, and behavioral health over the long-term for the patient who has experienced a severe TBI.

PRACTICAL APPLICATIONS OF NEUROPSYCHOLOGICAL ASSESSMENT FINDINGS

Neuropsychological test findings can be very helpful in making clinical decisions about cognitive capacity, competency, ability to live independently or to drive an automobile, and return to work (Cattelani, Tanzi, Lombardi, &

Mazzucchi, 2002; Martin et al., 2012; Triebel et al., 2012; Wolfe & Lehockey, 2016). Likewise, in helping families understand their loved one who has been injured and who has changed cognitively and behaviorally, sharing neuropsychological test results with patient and family can be a therapeutic process to help all better understand the consequences of the brain injury (Dausch & Saliman, 2009; Eapen et al., 2015). When someone has sustained a severe TBI as the consequence of an accident or assault, there may be legal and forensic consequences, where neuropsychological test findings are well-suited to address issues related to how the injury affected the individual (Zasler & Bigler, 2017).

Given the improved success rate for surviving TBI, outcome research also needs to establish cognitive end points or objectives for improved recovery (Sharma et al., 2017; Turgeon et al., 2017). As such, emphasis has been placed on best defining how neuropsychological testing should be used in treatment outcome studies (Silverberg et al., 2017).

CONCLUSIONS

Neuropsychological assessment and consultation has become an integral part in the evaluation of patients with severe TBI. Neuropsychological tests provide a standardized assessment approach that informs the clinician about various domains of cognitive, behavioral, and emotional functioning and which areas may be preserved and where impairments may be. The basics of neuropsychological assessment are covered in this chapter.

ACKNOWLEDGMENTS

The assistance of Tara Austin, Elizabeth Passey, and Hilary Smith is gratefully acknowledged.

REFERENCES

Ambrose, J., & Hounsfield, G. (1973). Computerized transverse axial tomography. *The British Journal of Radiology*, *46*(542), 148–149. Retrieved from <http://www.ncbi.nlm.nih.gov/pubmed/4686818>.

Annegers, J. F., & Coan, S. P. (2000). The risks of epilepsy after traumatic brain injury. *Seizure*, *9*(7), 453–457. Available from https://doi.org/10.1053/seiz.2000.0458.

Baade, L. E., & Schoenberg, M. R. (2004). A proposed method to estimate premorbid intelligence utilizing group achievement measures from school records. *Archives of Clinical Neuropsychology*, *19*(2), 227–243. Available from https://doi.org/10.1016/S0887-6177(03)00092-1.

Bender, M. B., Teuber, H. L., & Battersby, W. S. (1950). Discrimination of weights by men with penetrating lesions of parietal lobes. *Transactions of the American Neurological Association*, *51*, 252–255. Retrieved from <http://www.ncbi.nlm.nih.gov/pubmed/14788126>.

Bendlin, B. B., Ries, M. L., Lazar, M., Alexander, A. L., Dempsey, R. J., Rowley, H. A., . . . Johnson, S. C. (2008). Longitudinal changes in patients with traumatic brain injury assessed with diffusion-tensor and volumetric imaging. *NeuroImage, 42*(2), 503−514. Available from https://doi.org/10.1016/j.neuroimage.2008.04.254.

Bigler, E. D. (2007). Anterior and middle cranial fossa in traumatic brain injury: Relevant neuro-anatomy and neuropathology in the study of neuropsychological outcome. *Neuropsychology, 21*(5), 515−531. Available from https://doi.org/10.1037/0894-4105.21.5.515.

Bigler, E. D. (2012). Symptom validity testing, effort, and neuropsychological assessment. *Journal of the International Neuropsychological Society, 18*(4), 632−640. Retrieved from <http://www.ncbi.nlm.nih.gov/pubmed/23057080>.

Bigler, E. D. (2014). Effort, symptom validity testing, performance validity testing and traumatic brain injury. *Brain Injury, 28*(13−14), 1623−1638. Available from https://doi.org/10.3109/02699052.2014.947627.

Bigler, E. D. (2015). Neuroimaging as a biomarker in symptom validity and performance valid-ity testing. *Brain Imaging and Behavior, 9*(3), 421−444. Available from https://doi.org/10.1007/s11682-015-9409-1.

Bigler, E. D. (2016). Systems biology, neuroimaging, neuropsychology, neuroconnectivity and traumatic brain injury. *Frontiers in Systems Neuroscience, 10*, 55. Available from https://doi.org/10.3389/fnsys.2016.00055.

Cattelani, R., Tanzi, F., Lombardi, F., & Mazzucchi, A. (2002). Competitive re-employment after severe traumatic brain injury: Clinical, cognitive and behavioural predictive variables. *Brain Injury, 16*(1), 51−64. Available from https://doi.org/10.1080/02699050110088821.

Cipolotti, L., & Warrington, E. K. (1995). Neuropsychological assessment. *Journal of Neurology, Neurosurgery, and Psychiatry, 58*(6), 655−664. Retrieved from <http://www.ncbi.nlm.nih.gov/pubmed/7608660>.

Clifton, G. L., Hayes, R. L., Levin, H. S., Michel, M. E., & Choi, S. C. (1992). Outcome mea-sures for clinical trials involving traumatically brain-injured patients: Report of a confer-ence. *Neurosurgery, 31*(5), 975−978. Retrieved from <http://www.ncbi.nlm.nih.gov/pubmed/1436429>.

Collins, A. F. (2006). An intimate connection: Oliver Zangwill and the emergence of neuropsy-chology in Britain. *History of Psychology, 9*(2), 89−112. Retrieved from <http://www.ncbi.nlm.nih.gov/pubmed/17152603>.

Cristofori, I., & Levin, H. S. (2015). Traumatic brain injury and cognition. *Handbook of Clinical Neurology, 128*, 579−611. Available from https://doi.org/10.1016/B978-0-444-63521-1.00037-6.

Dausch, B. M., & Saliman, S. (2009). Use of family focused therapy in rehabilitation for veter-ans with traumatic brain injury. *Rehabilitation Psychology, 54*(3), 279−287. Available from https://doi.org/10.1037/a0016809.

Davies, H., & Falconer, M. A. (1943). Ventricular changes after closed head injury. *Journal of Neurology and Psychiatry, 6*(1−2), 52−68. Retrieved from <http://www.ncbi.nlm.nih.gov/pubmed/21611417>.

De Flora, S., Quaglia, A., Bennicelli, C., & Vercelli, M. (2005). The epidemiological revolution of the 20th century. *The FASEB Journal: Official Publication of the Federation of American Societies for Experimental Biology, 19*(8), 892−897. Available from https://doi.org/10.1096/fj.04-3541rev.

Dencker, S. J. (1960). Closed head injury in twins. Neurologic, psychometric, and psychiatric follow-up study of consecutive cases, using co-twins as controls. *AMA Archives of General Psychiatry, 2*, 569−575. Retrieved from <http://www.ncbi.nlm.nih.gov/pubmed/13815890>.

Denny-Brown, D. (1943). The principles of treatment of closed head injury. *Bulletin of the New York Academy of Medicine, 19*(1), 3−16. Retrieved from <http://www.ncbi.nlm.nih.gov/pubmed/19312299>.

Dublin, A. B., French, B. N., & Rennick, J. M. (1977). Computed tomography in head trauma. *Radiology, 122*(2), 365−369. Available from https://doi.org/10.1148/122.2.365.

Eapen, B. C., Allred, D. B., O'Rourke, J., & Cifu, D. X. (2015). Rehabilitation of moderate-to-severe traumatic brain injury. *Seminars in Neurology, 35*(1), e1−e3. Available from https://doi.org/10.1055/s-0035-1549094.

Farbota, K. D., Bendlin, B. B., Alexander, A. L., Rowley, H. A., Dempsey, R. J., & Johnson, S. C. (2012). Longitudinal diffusion tensor imaging and neuropsychological correlates in traumatic brain injury patients. *Frontiers in Human Neuroscience, 6*, 160. Available from https://doi.org/10.3389/fnhum.2012.00160.

Haarbauer-Krupa, J., Taylor, C. A., Yue, J. K., Winkler, E. A., Pirracchio, R., Cooper, S. R., ... Manley, G. T. (2017). Screening for post-traumatic stress disorder in a civilian emergency department population with traumatic brain injury. *Journal of Neurotrauma, 34*(1), 50−58. Available from https://doi.org/10.1089/neu.2015.4158.

Jennett, B., Galbraith, S., Teasdale, G. M., & Steven, J. L. (1976). Letter: E.M.I. scan and head injuries. *Lancet, 1*(7967), 1026. Retrieved from <http://www.ncbi.nlm.nih.gov/pubmed/57432>.

Johnston, I. H., Johnston, J. A., & Jennett, B. (1970). Intracranial-pressure changes following head injury. *Lancet, 2*(7670), 433−436. Retrieved from <http://www.ncbi.nlm.nih.gov/pubmed/4195116>.

Jones, M., Acion, L., & Jorge, R. E. (2017). What are the complications and emerging strategies for preventing depression following traumatic brain injury? *Expert Review of Neurotherapeutics, 17*(6), 631−640. Available from https://doi.org/10.1080/14737175.2017.1311788.

Joseph, B., Aziz, H., Pandit, V., Kulvatunyou, N., Sadoun, M., Tang, A., ... Rhee, P. (2014). Prospective validation of the brain injury guidelines: Managing traumatic brain injury without neurosurgical consultation. *Journal of Trauma and Acute Care Surgery, 77*(6), 984−988. Available from https://doi.org/10.1097/TA.0000000000000428.

Kane, R. L., & Parsons, T. D. (2017). *The role of technology in clinical neuropsychology.* New York: Oxford University Press.

Kolias, A. G., Guilfoyle, M. R., Helmy, A., Allanson, J., & Hutchinson, P. J. (2013). Traumatic brain injury in adults. *Practical Neurology, 13*(4), 228−235. Available from https://doi.org/10.1136/practneurol-2012-000268.

Konigs, M., Engenhorst, P. J., & Oosterlaan, J. (2016). Intelligence after traumatic brain injury: Meta-analysis of outcomes and prognosis. *European Journal of Neurology, 23*(1), 21−29. Available from https://doi.org/10.1111/ene.12719.

Levin, H. S., Grossman, R. G., & Kelly, P. J. (1976). Short-term recognition memory in relation to severity of head injury. *Cortex; A Journal Devoted to the Study of the Nervous System and Behavior, 12*(2), 175−182. Retrieved from <http://www.ncbi.nlm.nih.gov/pubmed/954452>.

Lezak, M. D., Howieson, D. B., Bigler, E. D., & Tranel, D. (2012). *Neuropsychological assessment* (5th ed.). Oxford; New York: Oxford University Press.

Martin, R. C., Triebel, K., Dreer, L. E., Novack, T. A., Turner, C., & Marson, D. C. (2012). Neurocognitive predictors of financial capacity in traumatic brain injury. *The Journal of Head Trauma Rehabilitation, 27*(6), E81−E90. Available from https://doi.org/10.1097/HTR.0b013e318273de49.

Masel, B. E., & DeWitt, D. S. (2010). Traumatic brain injury: A disease process, not an event. *Journal of Neurotrauma*, *27*(8), 1529–1540. Available from https://doi.org/10.1089/ neu.2010.1358.

Mendez, M. F. (2017). What is the relationship of traumatic brain injury to dementia? *Journal of Alzheimer's Disease: JAD*, *57*(3), 667–681. Available from https://doi.org/10.3233/JAD-161002.

Millis, S. R. (2009). Methodological challenges in assessment of cognition following mild head injury: Response to Malojcic et al. 2008. *Journal of Neurotrauma*, *26*(12), 2409–2410. Available from https://doi.org/10.1089/neu.2008.0530.

Millis, S. R., & Volinsky, C. T. (2001). Assessment of response bias in mild head injury: Beyond malingering tests. *Journal of Clinical and Experimental Neuropsychology*, *23*(6), 809–828. Available from https://doi.org/10.1076/jcen.23.6.809.1017.

Mullins, R. J. (1999). A historical perspective of trauma system development in the United States. *The Journal of Trauma*, *47*(3 Suppl.), S8–S14. Retrieved from <http://www.ncbi. nlm.nih.gov/pubmed/10496604>.

Nakase-Richardson, R., Tran, J., Cifu, D., Barnett, S. D., Horn, L. J., Greenwald, B. D., ... Giacino, J. T. (2013). Do rehospitalization rates differ among injury severity levels in the NIDRR traumatic brain injury model systems program? *Archives of Physical Medicine and Rehabilitation*, *94*(10), 1884–1890. Available from https://doi.org/10.1016/j. apmr.2012.11.054.

Negut, A., Matu, S. A., Sava, F. A., & David, D. (2016). Virtual reality measures in neuropsychological assessment: A meta-analytic review. *The Clinical Neuropsychologist*, *30*(2), 165–184. Available from https://doi.org/10.1080/13854046.2016.1144793.

Oztekin, I., Long, N. M., & Badre, D. (2010). Optimizing design efficiency of free recall events for FMRI. *Journal of Cognitive Neuroscience*, *22*(10), 2238–2250. Available from https:// doi.org/10.1162/jocn.2009.21350.

Parsons, T. D. (2015). Virtual reality for enhanced ecological validity and experimental control in the clinical, affective and social neurosciences. *Frontiers in Human Neuroscience*, *9*, 660. Available from https://doi.org/10.3389/fnhum.2015.00660.

Raymont, V., Salazar, A. M., Lipsky, R., Goldman, D., Tasick, G., & Grafman, J. (2010). Correlates of posttraumatic epilepsy 35 years following combat brain injury. *Neurology*, *75* (3), 224–229. Available from https://doi.org/10.1212/WNL.0b013e3181e8e6d0.

Reitan, R. M. (1959). The comparative effects of brain damage on the Halstead impairment index and the Wechsler–Bellevue scale. *Journal of Clinical Psychology*, *15*(3), 281–285. Retrieved from http://www.ncbi.nlm.nih.gov/pubmed/13664813.

Risdall, J. E., & Menon, D. K. (2011). Traumatic brain injury. *Philosophical Transactions of the Royal Society of London. Series B, Biological Sciences*, *366*(1562), 241–250. Available from https://doi.org/10.1098/rstb.2010.0230.

Sbordone, R. J. (2014). The hazards of strict reliance on neuropsychological tests. *Applied Neuropsychology Adult*, *21*(2), 98–107. Available from https://doi.org/10.1080/ 09084282.2012.762630.

Sbordone, R. J., Liter, J. C., & Pettler-Jennings, P. (1995). Recovery of function following severe traumatic brain injury: a retrospective 10-year follow-up. *Brain Injury*, *9*(3), 285–299. Retrieved from <http://www.ncbi.nlm.nih.gov/pubmed/7541681>.

Sharma, B., Tomaszczyk, J. C., Dawson, D., Turner, G. R., Colella, B., & Green, R. E. A. (2017). Feasibility of online self-administered cognitive training in moderate–severe brain injury. *Disability and Rehabilitation*, *39*(14), 1380–1390. Available from https://doi.org/ 10.1080/09638288.2016.1195453.

Silverberg, N. D., Crane, P. K., Dams-O'Connor, K., Holdnack, J., Ivins, B. J., Lange, R. T., . . . Iverson, G. L. (2017). Developing a cognition endpoint for traumatic brain injury clinical trials. *Journal of Neurotrauma, 34*(2), 363−371. Available from https://doi.org/10.1089/neu.2016.4443.

Slick, D. J., Sherman, E. M., & Iverson, G. L. (1999). Diagnostic criteria for malingered neuro-cognitive dysfunction: Proposed standards for clinical practice and research. *The Clinical Neuropsychologist, 13*(4), 545−561. Available from https://doi.org/10.1076/1385-4046 (199911)13:04;1-Y;FT545.

Spreen, O., & Benton, A. L. (1965). Comparative studies of some psychological tests for cere-bral damage. *The Journal of Nervous and Mental Disease, 140*, 323−333. Retrieved from <http://www.ncbi.nlm.nih.gov/pubmed/14334254>.

Stone, J. L., Patel, V., & Bailes, J. E. (2016). Sir Hugh Cairns and World War II British advances in head injury management, diffuse brain injury, and concussion: An Oxford tale. *Journal of Neurosurgery, 125*(5), 1301−1314. Available from https://doi.org/10.3171/2015.8.JNS142613.

Teasdale, G., & Jennett, B. (1974). Assessment of coma and impaired consciousness. A practical scale. *Lancet, 2*(7872), 81−84. Retrieved from <http://www.ncbi.nlm.nih.gov/pubmed/4136544>.

Tomaszczyk, J. C., Green, N. L., Frasca, D., Colella, B., Turner, G. R., Christensen, B. K., & Green, R. E. (2014). Negative neuroplasticity in chronic traumatic brain injury and implica-tions for neurorehabilitation. *Neuropsychology Review, 24*(4), 409−427. Available from https://doi.org/10.1007/s11065-014-9273-6.

Triebel, K. L., Martin, R. C., Novack, T. A., Dreer, L., Turner, C., Pritchard, P. R., . . . Marson, D. C. (2012). Treatment consent capacity in patients with traumatic brain injury across a range of injury severity. *Neurology, 78*(19), 1472−1478. Available from https://doi.org/10.1212/WNL.0b013e3182553c38.

Turgeon, A. F., Lauzier, F., Zarychanski, R., Fergusson, D. A., Leger, C., McIntyre, L. A., . . . the Canadian Critical Care Trials, G. (2017). Prognostication in critically ill patients with severe traumatic brain injury: The TBI-Prognosis multicentre feasibility study. *BMJ Open, 7*(4), e013779. Available from https://doi.org/10.1136/bmjopen-2016-013779.

Villani, R., Gaini, S. M., Paoletti, P., Brambilla, G., Caneschi, S., & Frigeni, G. (1975). Radioisotope cisternography in head-injured patients. *Acta Neurochirurgica (Wien), 32*(1−2), 25−33. Retrieved from <http://www.ncbi.nlm.nih.gov/pubmed/1163316>.

Washington, P. M., Villapol, S., & Burns, M. P. (2016). Polypathology and dementia after brain trauma: Does brain injury trigger distinct neurodegenerative diseases, or should they be clas-sified together as traumatic encephalopathy? *Experimental Neurology, 275*(Pt 3), 381−388. Available from https://doi.org/10.1016/j.expneurol.2015.06.015.

Webster, J. E., Dawson, R., & Gurdjian, E. S. (1951). The diagnosis of traumatic intracranial hemorrhage by angiography. *Journal of Neurosurgery, 8*(4), 368−376. Available from https://doi.org/10.3171/jns.1951.8.4.0368.

Weinstein, S., & Teuber, H. L. (1957). Effects of penetrating brain injury on intelligence test scores. *Science, 125*(3256), 1036−1037. Retrieved from <http://www.ncbi.nlm.nih.gov/pubmed/13432753>.

Wolfe, P. L., & Lehockey, K. A. (2016). Neuropsychological assessment of driving capacity. *Archives of Clinical Neuropsychology, 31*(6), 517−529. Available from https://doi.org/10.1093/arclin/acw050.

Zangwill, O. (1946). Psychological work in a brain injuries unit. *Nursing Times, 42*, 44. Retrieved from <http://www.ncbi.nlm.nih.gov/pubmed/21066457>.

Zaninotto, A. L., Vicentini, J. E., Fregni, F., Rodrigues, P. A., Botelho, C., de Lucia, M. C., & Paiva, W. S. (2016). Updates and current perspectives of psychiatric assessments after traumatic brain injury: A systematic review. *Frontiers in Psychiatry*, *7*, 95. Available from https://doi.org/10.3389/fpsyt.2016.00095.

Zasler, N. D., & Bigler, E. (2017). Medicolegal issues in traumatic brain injury. *Physical Medicine and Rehabilitation Clinics of North America*, *28*(2), 379−391. Available from https://doi.org/10.1016/j.pmr.2016.12.012.

Section IV

Treatment & Management

Chapter 13

Neurosurgical Rehabilitation

Mark Barisa[1], Kier Bison[2], Kelley Beck[2] and Caitlin Reese[3]
[1]*Performance Neuropsychology, Dallas/Frisco, TX, United States,* [2]*Baylor Rehabilitation, Frisco, TX, United States,* [3]*UT Southwestern, Dallas, TX, United States*

BRIEF HISTORY

Rehabilitation is the process of helping an individual achieve the highest level of function, independence/autonomy, and quality of life possible after sustaining deficits associated with an injury, illness, or other factors affecting daily life functions. Rehabilitation services do not reverse or undo the damage caused by disease or trauma, but are instead designed to restore an individual to maximal level of functioning after illness or injury. Rehabilitation interventions are employed at an individual level to meet the specific needs of each patient through an individualized plan of care. As a result, the rehabilitation process can vary across patients. Nonetheless, rehabilitation care is a process involving similar interventions and trajectories across patients and some general principles have been developed for patients suffering traumatic brain injury (TBI), stroke, and other acquired neurologic insults. More recently, some recommendations for postneurosurgical rehabilitation care applying these guidelines have been suggested as a call for increased rehabilitation following neurosurgery (Thompson, Majumdar, Sheldrick, & Morcos, 2013).

As outlined in the World Report on Disability (World Health Organization and the World Bank, 2011), rehabilitation involves a comprehensive model of recovery, focusing on both acute and chronic care, utilizing an uninterrupted, interdisciplinary approach. Some general treatment components for rehabilitation programs include managing the acute medical needs associated with condition; preventing complications from medical interventions; identifying and treating the physical, functional, cognitive, and emotional deficits associated with the condition; improving daily self-care and other functional abilities; remediating deficits where possible; developing adaptive/compensatory tools and strategies to manage residual deficits; assisting in alterations to living environments to address deficits; and providing education, training, and support to patients and families to assist in adjustment/adaptation to the residual deficits, compensatory tools, and associated lifestyle changes.

Neurosurgical Neuropsychology. DOI: https://doi.org/10.1016/B978-0-12-809961-2.00014-X

Outcomes in rehabilitation are not measured by the amelioration of the cause or remediation of the residual deficits, but instead focus on the return to functional independence, autonomy, and return to prior life activities. The success of rehabilitation is dependent on a number of variables, including, but not limited to, the nature and severity of the disease, disorder, or injury; the type and degree of any resulting impairments and disabilities; the overall health of the patient; the patient's ability to engage in and benefit from the interventions being provided; and family support. Comorbid and premorbid medical factors also play a significant role in recovery including postsurgical or other medical complications, effects of ongoing medical interventions and medications, premorbid medical conditions, and psychological factors and resiliency that can negatively impact recovery and rehabilitation.

As outlined in Table 13.1, a number of physical, cognitive, functional, and psychological deficits are often associated with neurologic conditions. As a result, an integrated, collaborative, interdisciplinary approach is warranted. Neurorehabilitation programs typically include a core team of disciplines including physical medicine and rehabilitation physician, neuropsychologist, physical therapy (PT), occupational therapy (OT), speech-language pathology (SLP), rehabilitation nursing, and case management. A variety of areas are covered in rehabilitation programs including, but not limited to the following:

TABLE 13.1 Four Categories of Deficits Addressed in Rehabilitation

Area of Deficit	Examples
Physical	Motor deficits including loss of power, persistence, and initiation; diminished sensory function including vision changes; vestibular problems including loss of balance, dizziness, and vertigo; disorders of controlled or sequenced motor movements; debility and decreased conditioning; etc.
Cognitive and communication	Disturbance of one or more cognitive functions including orientation, attention/concentration, speeded information processing, learning and memory, visual spatial discrimination and reasoning, expressive and receptive language, executive reasoning and problem solving, self-awareness and insight, etc.
Psychological, psychosocial, and behavioral	Depression, anxiety, mood lability, adjustment reactions, disruptions of family structure, loss of social roles and involvement, impulsivity, disruptive behaviors, agitation emotional lability, anosognosia, thought disorders, etc.
Functional	Disruptions in daily self-cares, financial management, cooking, driving, return to school/work, household responsibilities, usual daily activities, exercise, sports, etc.

- Self-care skills and basic activities of daily living—Feeding, grooming, bathing, dressing, toileting, and sexual function.
- Physical and medical care needs—Dietary needs, medication management, skin and wound care, bathing and toileting assistance.
- Pain management—Medications and alternative methods of managing pain.
- Respiratory care—Ventilator care, if needed; breathing treatments and exercises to promote lung function.
- Mobility skills—Walking, transfers, and self-propelling a wheelchair.
- Cognitive skills—Attention/concentration, learning and memory, communication, judgment, problem solving, and organizational skills.
- Language/communication−expressive and receptive speech, comprehension, reading, writing, and other methods of communication.
- Psychological counseling—Identifying problems and solutions with thinking, behavioral, and emotional issues.
- Socialization and community reintegration skills—Interacting with others at home and within the community.
- Family support—Assistance with adapting to lifestyle changes, financial concerns, and discharge planning.
- Education—Patient and family education and training about the condition, medical care, and adaptive techniques.
- Vocational training—Work-related skills.

The Rehabilitation Team

In rehabilitation settings, the rehabilitation physician is part of a team of therapists and other providers that provide the direct therapeutic interventions with patients. For the purposes of this text, we will start this section discussing the various roles of a neuropsychologist on a rehabilitation team. The role of psychology on neurorehabilitation teams preceding and following neurosurgical procedures is multifaceted and includes a blending of neuropsychology, rehabilitation psychology, and health psychology services. In neurosurgical settings, clinical skills from all of these areas are necessary to help patients understand the complexities of the conditions that resulted in the need for neurosurgery, as well as the implications, complications, and expectations moving forward.

A clinical neuropsychologist is defined as a "professional psychologist trained in the science of brain−behavior relationships. The clinical neuropsychologist specializes in the application of assessment and intervention principles based on the scientific study of human behavior across the lifespan as it relates to normal and abnormal functioning of the central nervous system (Hannay et al., 1998—Houston Conference on Specialty Education and Training in Clinical Neuropsychology)." While diagnosis and treatment recommendations are important facets of neuropsychology, neuropsychologists also look at the functional impact of neuropsychological strengths and weaknesses to make recommendations that are tangible, timely, and relevant to the referral source and the patient. This is accomplished in a collaborative manner with the referral source, the patient, and additional rehabilitation team members to maximize continuity of care and best outcomes.

Rehabilitation Psychology is "a specialty area within professional psychology which assists the individual with an injury or illness which may be chronic, traumatic and/or congenital, including the family, in achieving optimal physical, psychological, and interpersonal functioning. The focus of rehabilitation psychology is on the provision of services consistent with the level of impairment, disability and handicap relative to the personal preferences, needs and resources of the individual with a disability. The rehabilitation psychologist consistently involves interdisciplinary teamwork as a condition of practice and services within a network of biological, psychological, social, environmental, and political considerations in order to achieve optimal rehabilitation goals (American Board of rehabilitation Psychology Specialty Definition—www.abpp.org)." Rehabilitation psychologists assist individuals in coping with, and adjustment to, chronic, traumatic or congenital injuries or illnesses that may result in a wide variety of physical, sensory, neurocognitive, emotional, and/or developmental disabilities. They work with patients and other professionals, typically within interdisciplinary teams, to increase function, reduce disability, and increase participation in everyday roles and responsibilities.

Clinical health psychologists apply "scientific knowledge of the interrelationships among behavioral, emotional, cognitive, social, and biological components in health and disease to: the promotion and maintenance of health; the prevention, treatment, and rehabilitation of illness and disability; and the improvement of the health care system. The distinct focus of Clinical Health Psychology (also known as behavioral medicine, medical psychology, and psychosomatic medicine) is at the juncture of physical and emotional illness, understanding and treating the overlapping challenges (American Psychological Association Clinical Health Psychology—http://www.apa.org/ed/graduate/specialize/health)." Clinical health psychologists assist individuals with health promotion, disease prevention, adjustment to illness, and treatment adherence. They evaluate, educate, and treat individuals whose thoughts, emotions, and behaviors have an adverse impact on their health while considering the biological, cognitive, behavioral, emotional, social, spiritual, psychosomatic, and environmental factors as they relate to health, illness, and health care.

Broadly, psychologists (including neuropsychologists, rehabilitation psychologists, and health psychologists) with foundations in neuropsychological assessment, functional neuroanatomy, knowledge of recovery courses, and knowledge of injury or illness mechanisms and severity serve a variety of roles on the neurorehabilitation team. A foundational knowledge of neuropsychological and neuroanatomical principles is important, but specific knowledge related to the presenting condition, neurosurgical interventions, and postsurgical recovery expectations are imperative to provide the best level of care. Additionally, the neuropsychologist must also be well-versed in the rehabilitation psychology and health psychology activities to meet the broad-based needs of patients and their families. This further extends to rehabilitation team activities where the neuropsychologist helps to integrate the biological, psychological, and social aspects of the patient.

The rehabilitation team includes a number of providers with specific training and skill in the rehabilitation of physical, functional, and cognitive deficits to restore maximal levels of functioning:

- Physical therapists "help individuals maintain, restore, and improve movement, activity, and functioning, thereby enabling optimal performance and enhancing health, well-being, and quality of life. Their services prevent, minimize, or eliminate impairments of body functions and structures, activity limitations, and participation restrictions (American Physical Therapy Association—www.apta.org/ScopeOfPractice)." Physical therapists assist in physical strengthening, gait training, vestibular functioning, environment navigation, and general conditioning.
- Occupational therapists focus on helping people with a physical, sensory, or cognitive disability be as independent as possible in all areas of their lives (American Occupational Therapy Association—www.aota.org/Practice. aspx). This includes basic functional activities such as managing self-care, dressing, and grooming as well as higher-level activities of daily living including cooking, driving, and other instrumental activities of daily living.
- Speech-language pathologists "work to prevent, assess, diagnose, and treat speech, language, social communication, cognitive communication, and swallowing disorders in children and adults (American Speech–Language–Hearing Association— www.asha.org)." This includes comprehensive evaluation and treatment of various disorders of swallowing as well as rehabilitation of cognitive-communication deficits.
- Rehabilitation nurses are an integral part of the rehabilitation team. "The goal of rehabilitation nursing is to assist individuals with disability and/or chronic illness to attain and maintain maximum function. The rehabilitation staff nurse assists clients in adapting to an altered lifestyle, while providing a therapeutic environment for client's and their family's development. The rehabilitation staff nurse designs and implements treatment strategies that are based on scientific nursing theory related to self-care and that promote physical, psychosocial, and spiritual health." (Association of Rehabilitation Nurses—www.rehabnurse.org/pubs/role/Role-Rehab-Staff-Nurse.html)
- Case managers and social workers work with the rehabilitation team to assist in communicating information to patients and families and assisting in discharge/disposition planning. This includes education; care coordination;

compliance with guidelines, requirements, and regulations; transition of care management and coordination; and utilization management (American Case Management Association—www.acmaweb.org).

- Additional specialties including neurology, neuro-ophthalmology, internal medicine, rehabilitation engineering and prosthetics, and others are also be involved depending on the needs of the patient.

Early in the neurorehabilitation process, psychologist interventions may include: family support and education, screenings (cognitive, aphasia, and emotional), neurobehavioral management, and adjustment counseling. Neuropsychological assessment becomes valuable as patients progress in recovery beyond the initial hospitalization, and prepare to return to differing levels of independence. Related recovery may also include a number of adjustment interventions. Awareness of strengths and weaknesses within the neurorehabilitation team is crucial regardless of setting, as the neuropsychologist may need to be at times team moderator, team supporter, as well as patient and family advocate. See a list of specific adjustment and cognitive rehabilitation interventions in Tables 13.2 and 13.3.

TABLE 13.2 Adjustment Interventions in Rehabilitation

Type of Intervention	Brief Explanation
Basic problem solving and goal setting	Family, peers, and rehabilitation staff help patients with coping and change.
Mindfulness interventions	Psychologists help people to become more aware of thoughts, emotions, and physical sensations and to react more objectively and less automatically.
Cognitive behavior therapy interventions	Psychologists help people to identify the types of thoughts and behaviors that maintain mood and anxiety disorders and to test and develop more helpful thoughts and behaviors (e.g., supporting an increased sense of control over situations and decreased behavioral avoidance)
Behavioral activation interventions	Psychologists help people to become more active, to notice which activities they find most enjoyable/satisfying, and to schedule these activities regularly to improve mood by boosting the drive system.
Compassion focused therapy interventions	Psychologists help people struggling with shame and self-criticism by supporting the development of compassion (toward self and others) to reduce any sense of threat.
Positive psychology interventions	Psychologists help people to recognize opportunities to experience pleasure, make full use of personal strengths, and use strengths to pursue meaning.

Note: Information from Ford (2017).

TABLE 13.3 Evidence-Based Cognitive Rehabilitation Interventions

Cognitive Domain	Examples of Intervention Techniques
Attention	Attention Remediation—Attention Process Training (APT)
	Time Pressure Management
	N-Back/Working Memory
Memory	External Memory Compensation (e.g., memory notebook, errorless learning, spaced retrieval)
	Memory Strategy Training (association and organizational techniques)
Executive functioning	Metacognitive Strategy Training
	Problem-Solving Strategies (e.g., Goal–Plan–Do–Review)
	Predict–Perform Techniques (awareness deficits)
Neglect/visuospatial impairment	Visual Scanning Training
	Visual Imagery-Lighthouse Strategy
	Limb Activation Strategies
Social communication	Group-based Interventions
	Individual Goal Setting
	Emotion Perception Deficits Training

Note: Adapted from the ACRM Cognitive Rehabilitation Manual (2012).

The Rehabilitation Continuum

Neurorehabilitation occurs across a continuum beginning in the acute medical setting and continuing through outpatient therapy and medical clinics. The most common stages of neurorehabilitation include the following:

- *Acute medical care*—Goal is to increase the survival and prevent medical complications that may delay recovery. Typically includes ICU or acute inpatient hospital units.
- *Acute intensive inpatient rehabilitation*—Goal is to assist the individual in maximizing independence in self-care, communication, and mobility in preparation for returning home safely.
- *Subacute rehabilitation*—Goal is to maintain medical stability and allow for additional recovery while rehabilitation efforts are less intense. Sometimes used for patients who are having a slow recovery in acute/intensive setting or for management of a disorder of consciousness (vegetative state, minimally conscious state, etc.).
- *Long-term/postacute/transitional rehabilitation*—Goals are cognitive rehabilitation, compensatory techniques to enhance cognitive capacity,

deficits awareness, social pragmatics, behavioral modification, community re-entry, and family support.

- *Long-term postrehabilitation follow-up*—Ongoing assistance in adjustment to new life situations, further develop appropriate coping strategies, and re-evaluate rehabilitation outcomes and future rehabilitation plans. This can occur in single or multidiscipline outpatient clinics with decreased frequency in terms of office-based therapies.

Where rehabilitation takes place depends on individual patient needs and will change relative to the patient's recovery. Many people recovering from injuries can be treated as outpatients in an office setting and do not require intensive inpatient rehabilitation services. Patients with severe deficits, but with continued acute medical needs, may need to begin rehabilitation interventions while remaining in an acute medical unit or facility. Research has shown that early rehabilitation interventions (i.e., in the intensive care or acute medical setting) improve overall functional outcomes (World Health Organization, 2011). As the patient's status improves and medical needs resolve or stabilize, the intensity, duration, and scope of rehabilitation interventions can increase. This occurs in an acute inpatient rehabilitation setting where therapy can be 3 or more hours per day in an interdisciplinary fashion. As the patient continues to progress and daily medical needs resolve, a transition to a postacute/transitional neurorehabilitation setting might be warranted to allow for an increase in the scope, duration, and intensity of therapies up to 6 hours of therapy per day, several days per week. This allows for mastery of higher-level functional abilities, including return to financial management, driving, work, and other high-level activities. Other patients progress in a sufficient fashion to return home and participate in outpatient therapeutic interventions with lower frequency and duration, if needed.

Research has not yet determined a ceiling for treatment intensity, but it is generally felt that more therapy is better than less (World Health Organization, 2011). However, rehabilitation therapies need to take place in an environment where the treatment team has a good underlying knowledge of the condition being treated and the appropriate therapeutic setting, dosing, and duration. The right environment and intensity of treatment should be based on the status and needs of the individual patient. Placement in a setting that is too low can limit recovery, while placement in a setting that is too high can result in psychological/behavioral changes, frustration, and disengagement.

Care and rehabilitation interventions at home can also be appropriate for people who have significant transfer, transportation, or medical limitations that prohibit them from attending more structured clinics/facilities. This allows for direct living environment observation and therapies in "real-life" settings. Also, many nursing homes/skilled nursing facilities have less intensive rehabilitation programs for patients that are not yet able to participate in the more rigorous programs at more structured rehabilitation facilities.

CURRENT PRACTICE

Neurorehabilitation is commonly utilized in TBI and hemorrhagic stroke, but is less often considered for other neurosurgical procedures such as tumor resection, lesion or lobar resection in cases of epilepsy, or new neurosurgical procedures for movement disorders and mood-related issues. From a neurosurgical perspective, rehabilitation needs, outcomes, and trajectory can vary greatly depending on a number of factors including the condition leading to the need for neurosurgery; the type of neurosurgical procedure and approach used; the location of the brain involved; relative motor, sensory, cognitive, and functional deficits associated with the surgery; additional treatment contributions; comorbid medical issues; and the age and overall health of the patient. While some neurosurgical patients can be discharged to home in relatively short order with minimal additional treatment needs, some conditions can result in problems with physical and cognitive abilities consistent with those commonly associated with TBI and stroke. In such cases, interdisciplinary rehabilitation interventions are appropriate to identify and treat the various residual deficits. Single discipline therapy is appropriate when the noted difficulties are circumscribed to a particular area of function, but more often an interdisciplinary approach is warranted.

Recent studies in the United Kingdom have stressed that patients recovering from neurologic surgery should be treated in specialist neurorehabilitation units and that the placement of such patients in facilities such as acute surgical, orthopedic, or general wards was not acceptable (Thompson et al., 2013). They noted, however, that many patients in need of neurorehabilitation during the acute phase of recovery following surgery still had to depend on the general orthopedic rehabilitation services of general hospitals. They noted that most often, patients spend the early weeks of their recovery being treated on neurology or neurosurgical wards where rehabilitation did not begin until discharge to a dedicated unit, thus failing to optimize input in the crucial early period.

The traditional notion of providing acute medical postsurgical treatment and rehabilitation in series rather than having rehabilitation interventions as part of the early continuum of care was felt to be inadequate. Instead, it was suggested that acute medical treatment and rehabilitation be conducted in parallel as is the case for postsurgical TBI and stroke patients. Research suggests that early rehabilitation interventions for neurosurgical patients resulted in cost-of-treatment savings in time and money as well as improved clinical outcomes (Thompson et al., 2013). While these studies focused primarily on trauma-based acquired brain injuries, it has been demonstrated that lengths of stay for acute neurorehabilitation patients are usually equal to or shorter than patients receiving treatment as usual (Greenwood, Strens, Watkin, Losseff, & Brown, 2004; Sirois, Lavoie, & Dionne, 2004); that acute neurorehabilitation units could significantly improve bed availability on acute neurosurgical wards (Bradley et al., 2006); and that early access to neurorehabilitation reduces overall treatment costs and may improve outcomes (Cowen et al., 1995; Sirois et al., 2004).

POTENTIAL APPLICATIONS

In lieu of specific case examples, this section outlines rehabilitation needs/activities relative to some specific patient populations. Acquired brain injury (ABI), including TBI and stroke, begin this section as rehabilitation is commonly associated with this patient population. This is followed by another common area in neurosurgical rehabilitation, neuro-oncology. Epilepsy and deep brain stimulation (DBS) round out this section to highlight areas where rehabilitation application may be overlooked or not obviously necessary. It is important to note that information included in each of these sections, while presented in specific population groups, is meant to be considered in a general context of postneurosurgical care and rehabilitation.

Acquired Brain Injury—TBI/CVA

Rehabilitation following ABI, including injuries of a traumatic nature as well as ischemic/hemorrhagic cerebrovascular accident (CVA), has one of the longest histories of any neurosurgical diagnostic group in this chapter. Origins date to the post World War I era, with efforts to treat and reintegrate soldiers returning from war with brain injuries, and were continued post World War II. Pioneers such as Kurt Goldstein in Germany documented the common cognitive sequelae after ABI in these returning soldiers, and laid a foundation for the rehabilitation of these sequelae through his attempts to assist veterans in returning to product work and family/social roles (Ben-Yishay & Diller, 2011). A.R. Luria, through his experiences treating brain injured soldiers in World War II, theorized about the processes involved in recovery of higher cortical functions after brain injury, and as such related these processes to particular rehabilitation approaches (Boake, 1989). Another influential early neurorehabilitation innovator in the United States, psychiatrist John Aita, is credited with development of a postacute rehabilitation program involving a multidisciplinary approach to treatment of head-injured veterans returning from World War II (Boake, 1989).

Advances in neuroimaging and neurosurgical techniques in the 1960s and 1970s led to greater numbers of severely injured patients surviving injuries that previously would have resulted in death. A greater number of individuals surviving catastrophic injuries resulted in a greater need for improved rehabilitation programs, funding, and tracking of longer-term outcomes. Establishment of the National Head Injury Foundation in 1980, which later became the Brain Injury Association of America, and the creation of brain injury standards by the Commission on Accreditation of Rehabilitation

Facilities (CARF) was a reflection of the increasing interest in brain injury rehabilitation. It was also during this time period that the first community-based rehabilitation day treatment programs for brain injury were developed by innovators such as Yehuda Ben-Yishay and Leonard Diller (Boake, 1989; Teasdale & Zitnay, 2013). TBI Model Systems was established in 1987 as a means to advance research through collaboration of multiple acute trauma hospitals and inpatient rehabilitation centers across the United States. Another highly influential brain injury collaboration was founded within the, American Congress of Rehabilitation Medicine (ACRM) in 1988, known as the Brain Injury-Interdisciplinary Special Interest Group (BI-ISIG), which has produced a number of publications, reviews, and courses guiding the practice of rehabilitation of ABI in the United States (ACRM).

Neuropsychology and Cognitive Rehabilitation in Acquired Brain Injury

While a detailed review of the evidence supporting cognitive rehabilitation in ABI is beyond the scope of this chapter, interested readers are referred to systematic reviews conducted by Cicerone and colleagues as well as in Europe by Cappa and colleagues (Cappa et al., 2005; Cicerone et al., 2011), as well as a metaanalytic review of the cognitive rehabilitation literature completed by Rohling and colleagues (Rohling, Faust, Beverly, & Demakis, 2009). Briefly, despite ongoing challenges related to research methodology that are well-documented throughout the rehabilitation literature, cognitive rehabilitation approaches have been shown to have positive effects on a number of areas post injury. Dams-O'Connor and Gordon (2010) describe seven roles of cognitive rehabilitation, including restoration of cognitive functioning, training in compensatory cognitive strategies, increasing aware-ness of deficits, improving mood and emotional regulation, facilitating return to work, encourage community reintegration, and prevention of self-injurious behavior. Thus, cognitive rehabilitation has a broader scope than a simple focus on improving performance in specific cognitive domains. And the fre-quency of neurobehavioral difficulties post ABI such as impulsivity, disinhi-bition, agitation, aggression, apathy/abulia, and difficulties with emotional regulation require careful evaluation and must be considered within the over-all cognitive rehabilitation program (McAllister, 2013). Support for family members of ABI survivors has traditionally been a key element in cognitive rehabilitation programs as well given the complex and long-term difficulties associated with ABI (Kreutzer, Kolakowsky-Hayner, Demm, & Meade, 2002).

The neuropsychologist working in cognitive rehabilitation of ABI will serve a variety of roles on the neurorehabilitation team. Skill set will need to include a foundation in neuropsychological assessment, functional neuroanat-omy, knowledge of recovery course post ABI varied by mechanism of injury and severity of injury, understanding and experience in applying cognitive rehabilitation interventions, ability to apply principles of psychotherapy with an ABI population, experience in family intervention, as well as a solid understanding of team dynamics and management. Specific job responsibilities will vary as a function of the setting, with a greater focus on family education and intervention early during recovery in the acute and subacute hospitalization period. Cognitive screening, neurobehavioral management, and application of cognitive rehabilitation principles to early recovery become important during the inpatient rehabilitation hospitalization. Neuropsychological assessment becomes valuable as patients progress in recovery beyond the initial hospitalization, and are ready to resume work or school responsibilities, management of daily life tasks (i.e., driving, medication/financial management), or to guide rehabilitation interventions. And of course, awareness of strengths and weaknesses of the team of professionals working with the ABI patient is crucial regardless of setting, as the neuropsychologist may need to be at times team moderator, team supporter, as well as patient advocate.

Neuro-Oncology

As outlined in Chapter 9, Neuropsychology in the Neurosurgical Management of Primary Brain Tumors, of this text, there is a great deal of variability in the course, presentation, interventions, and outcomes of various brain tumors. It is known that malignant brain tumors have a high likelihood of producing disabling effects on an individual's life and daily functional abilities. This relates to the tumor itself, as well as the subsequent effects of various treatment modalities including radiation, chemotherapy, and neurosurgical interventions. The types and severity of deficits are related to the location, size, and type of the tumor as well as the subsequent medication/medical interventions. Cognitive deficits have been noted to the most common complications of brain tumors (80%), followed by motor deficits (78%), visual-perceptual deficits (53%), sensory loss (38%), bowel/bladder impairment (37%), cranial nerve palsy (29%), dysarthria (27%), dysphagia (26%), aphasia (24%), and ataxia (20%) (Mukand, Blackinton, Cirncoli, Lee, & Santos, 2001). These deficits can be noted pre- and postsurgically. While one area of deficit can be significant in terms of daily life functions, patients typically present with combinations of deficits that can result in dramatic disturbance of daily life functions and alterations in family and social roles. Mukand et al. described that 74.5% of patients had three or more concurrent neurologic deficits, and 39% of patients had five or more deficits. In some cases, a progressive pattern of decline can be associated directly by the tumor progression or the related chemotherapy, radiation, and surgical treatments.

While survival rates are improving with new diagnostic and treatment options, the need for rehabilitation interventions has grown to manage residual cognitive, physical, functional, and emotional deficits in an effort to maximize functional independence and overall quality of life. Rehabilitation interventions can be applied in all stages of cancer care—from initial diagnosis through the various treatment interventions and beyond. However, the rehabilitation needs of the patient will change over the course of care with goals of treatment transitioning according to the patient's presenting deficits. As noted by Bartolo, Zucchella, and Pace (2012), rehabilitation should start early to reach the established and changing goals, prevent potential complications, and achieve better functional outcome. Rehabilitation, especially during the acute phase and immediately postoperatively, was found to improve functional outcome. The need for early intervention was said to be even more pressing in brain tumor patients where deterioration can be faster than patients with other acquired injuries such as stroke (Bartolo et al., 2012).

The Continuum of Care Across the Disease Process

The neuropsychologist's role across the continuum includes serial assessment of neurocognitive functioning and eventual comprehensive neuropsychological evaluation after a period of recovery and stability. Additionally, the neuropsychologist, with a knowledge of brain–behavior relationships, is available for education, support, and expectations management for the patient and family members. Issues related to energy conservation, pain management, mood/personality changes, vocational and family role changes, and other postsurgical/post-treatment changes can also fall under the purview of the neuropsychologist for assessment and intervention.

In the acute stage of brain cancer, initial symptoms such as headaches, seizures, cognitive declines, motor/sensory deficits, and/or nausea/vomiting often lead to a further medical work-up leading to the work-up and eventual diagnosis. Each person will experience these symptoms differently, but these initial symptoms set the stage for further medical/neurologic work-up and ultimately the diagnosis of the tumor. This diagnosis, in and of itself, can be traumatic and frightening for the patient. The psychologist/neuropsychologist has an early role in helping the patient understand the diagnosis and develop appropriate coping and adjustment strategies while appropriately managing expectations related to the diagnosis. Additional early rehabilitation interventions may be warranted early to manage the associated physical, cognitive, functional, and emotional deficits that may present in the initial stages of the disease course. This early rehabilitation interventions can also set the stage for future rehabilitation efforts through education and preparation for potential needs moving forward through the various stages of treatment.

Following the diagnosis, treatment protocols vary in relation to the size and type of tumor, location, genetic factors, and relative progression thus far. Treatments typically include combinations of chemotherapy, neurosurgery, and radiation interventions, each of which come with their own set of complications resulting in additional rehabilitation needs. Specific rehabilitation needs should be assessed early in the course of treatment to allow for early intervention in the hopes of maximizing function and outcomes across the continuum of cancer care. A common course would be early transition to an acute rehabilitation setting followed by subsequent treatment through an interdisciplinary postacute transitional rehabilitation program or multidisciplinary outpatient interventions that taper over time as goals are met. In the acute phase of rehabilitation, flexibility and frequent reassessments are required (Kirshblum, O'Dell, Ho, & Barr, 2001).

As noted, a variety of cognitive, physical, functional, and emotional deficits are related to brain tumors and related treatments, some of which are of particular importance for the neuropsychologist following these patients. Fatigue is a very common symptom and severe fatigue has been associated with poor functional status and poorer quality of life due to impaired physical functioning and sleep disturbances (Kim, Chun, & Han, 2012). This, in and of itself, can result in declines in some basic life functions, especially those with a high attentional or executive cognitive demand. As a result, fatigue can be a significant factor in determining the best environment and level of intensity for rehabilitative interventions—subacute, acute, postacute, or outpatient (Kos, Kos, & Benedicic, 2016). Even when fatigue is not a primary issue, deficits in cognitive functions are common, particularly shortly after neurosurgical procedures or in the context of chemotherapy and radiation treatments. The deficits may be temporary or more permanent, depending on whether the cause relates to structural changes or more transient factors such as initial brain swelling or medications. It is important to determine the severity and permanence of the cognitive impairments to appropriately determine rehabilitation plans and goals (Cheville, 2011). Affective disorders must be considered immediately after surgery as they too can negatively impact participation in rehabilitation interventions. It is important to note that mood/anxiety disorders are more common in patients with a history of depression and those with coincidental physical disability (Mainio, Hakko, Niemela, Koivukanga, & Rasanen, 2005).

A number of physical deficits can emerge following neurosurgery for brain tumor that warrant further rehabilitative interventions from PT, OT, and SLP professionals. It is not uncommon for patients to complain of significant headaches in the initial days/weeks following the surgery. These can be mild or more severe, and, when severe, they can have negative impacts on the patient's ability to fully participate in rehabilitation efforts until improvement is noted. Additional neurologic deficits following surgery can include paralysis, weakness and balance disturbances, vision changes, communication deficits, and swallowing difficulties. If these persist beyond the early days of recovery, rehabilitation interventions through PT, OT, and/or SLP are warranted early in the recovery process to maximize functional outcomes and to prevent/manage secondary complications (Kos et al., 2016).

In more severe cases, ongoing medical care, skilled nursing needs, and potentially hospice care bring about a different set of rehabilitation needs. The neuropsychologist's role transitions as well with a change in focus of care on managing the active needs of the patient and family rather than long-term planning in terms of daily life functions. This may include patients that are unable to participate in traditional active rehabilitation interventions and instead require passive rehabilitation modalities at bedside. These may include interventions such as verticalization, passive range of motions, assistive devices to help them with bed mobility and sitting, transfer training, maintenance of self-cares and grooming, modified self-care utensils, and wheelchairs for mobility. It is important to increase mobility even with the more severely impaired patients to allow them to leave their room as this has a positive effect on the patient's well-being, by preventing social isolation which may occur when the patient is constantly confined to their room (Kos et al., 2016).

As in all rehabilitation populations, it is extremely important to include support, education, and care for the patient's family as part of the rehabilitation continuum. By making families an integral part of the treatment team, they will be better able to care for the patient after rehabilitative care is completed. They will also be better prepared to identify and manage possible changes that occur over time and help identify possible signs of recurrence early.

Epilepsy

A detailed review of neuropsychology/neurosurgical aspects of epilepsy is provided in Chapter 8, Neuropsychology in Adult Epilepsy Surgery, of this text. A brief overview is provided here to give context to the role of neurorehabiltation interventions in the management of these patients pre- and post-surgically. Epilepsy is a chronic condition marked by recurrent, unprovoked seizures, and a seizure is characterized by a transient occurrence of signs and/or symptoms due to abnormal excessive or synchronous neuronal activity in the brain. An epilepsy diagnosis suggests that there is an underlying mechanism, intrinsic to the brain, and unrelated to transient factors. While epilepsy is not associated with one single pathology, it is fundamentally marked by an imbalance between inhibitory and excitatory neuronal activity resulting in recurrent excitatory circuits (Westerveld, 2014).

Video encephalogram (EEG) monitoring is key for detecting seizure activity, and both EEG monitoring, T2 weighted magnetic resonance imaging (MRI), and fluid-attenuated (FLAIR) sequences have become integral components of the presurgical evaluations of patients with medically intractable epilepsy (McMullen, 2011; Westerveld, 2014). Hippocampal sclerosis (within the temporal lobe) is the most common pathology in adults; as many as one third of patients with temporal lobe epilepsy and hippocampal sclerosis (especially in CA1 and CA3) develop medically intractable seizures (Westerveld, 2014).

Medication resistant temporal lobe epilepsy can be treated by resection of epileptogenic temporal lobe and mesial temporal structures (Westerveld, 2014) and has been found to be an effective treatment for medically refractory focal epilepsy (Noachtar and Borggraefe, 2009; Wiebe, Blume, Girvin, & Eliasziw, 2001). However, once the underlying pathology is understood by the presurgical surgical and functional neuroimaging, a neuropsychological evaluation is an integral part of the presurgical workup. Feedback from the evaluation can educate patients and families regarding the patient's cognitive strengths and weaknesses, to formally document cognitive deficits interfering with activities of daily living, and to support the treatment team by detailing possible risks and benefits of surgical resection of the epileptogenic focus (McMullen, 2011). From the neuropsychologist's perspective, the primary risk is the likelihood of cognitive decline following surgery. Factors that have been associated with cognitive decline include the side of surgery (Clusmann et al., 2002), age at onset of seizures (Hermann, Seidenberg, Haltiner, & Wyler, 1995), preoperative neuropsychological performance (Clusmann et al., 2002), and Wada memory asymmetry score (Lee et al., 2003). The Wada memory asymmetry score (defined as the difference between the memory score achieved by each hemisphere) is favorable when the nonepileptogenic mesial temporal lobe performs better on Wada memory testing than the epileptogenic mesial temporal lobe (McMullen, 2011).

Typical findings have suggested that in left anterior temporal lobectomy patients, the better the preoperative verbal memory and confrontation naming, the greater the risk of postoperative verbal memory decline and decline in naming, respectively (McMullen, 2011). In fact, postsurgical declines in verbal memory have been observed in left temporal lobe resection at three months (Helmstaedter & Elger, 1996; Helmstaedter et al., 2008), and associations between postoperative decreases in verbal fluency and naming test scores have been observed 4.5 months past surgery (Yogarajah et al., 2010). Temporal lobectomy also conveys some risk of superior quadrantanopsia (Westerveld, 2014).

The pre- and postsurgical deficits associated with epilepsy are amenable to neurorehabilitation efforts to assist in remediation when possible, and the development of compensatory/adaptive strategies to manage the residual difficulties that remain. Areas of particular focus can include memory functions, language functions, social comportment/pragmatics, mood/personality changes, and motor/sensory changes. The extent and pattern of deficits in epilepsy is location specific, limiting specific rehabilitation interventions for this section.

As an example of an intervention program for patients with epilepsy, following a neuropsychological evaluation and observation of memory problems across 2 weeks, a prioritized list of 10 concrete goals is created to target in treatment. Six to eight sessions are scheduled every 2 weeks. During these sessions, patients learn to use compensatory strategies for their specific treatment goals. In the first session, education about epilepsy and its relation to memory problems, memory functions, memory deficits in everyday life, and the treatment program is introduced. During subsequent sessions, memory compensatory strategies and memory aids are explained and applied to patient-specific memory problems. Patients are encouraged to choose the right memory strategies for them. Every third session, significant others are invited to participate in group sessions. Patients summarize a memory technique, and their application of the technique, before the group discusses the application. Significant others also receive guidance regarding how to encourage the use of memory techniques and how to address psychosocial problems related to memory deficits (and epilepsy). Overall, the plan of action is based on a cognitive cycle (observation—planning—decision—action—checking—evaluation). Concrete goals are planned, memory strategies are chosen and monitored during treatment. Progress is evaluated, and the plan of action is adjusted as necessary. Three months post treatment, participants return to clinic for a booster session during which the patient's plan of action is evaluated, and patients undergo a neuropsychological evaluation to assess treatment effects (Hendriks, 2001).

Despite some proposed interventions, a review of many studies finding correlations between resection and postoperative cognitive changes do not detail when and if neuropsychological rehabilitation programs are introduced. Moreover, the effects of cognitive rehabilitation on the cerebral substrate are not discussed in most studies (Nordvik et al., 2014). Future research directions may focus on the benefits of early cognitive rehabilitation postoperatively, which, if combined with fMRI use, may identify changes in cerebral activity concurrently (Mosca et al., 2014).

Movement Disorders—Deep Brain Stimulation

DBS is primarily used to treat movement disorders, with FDA approval for beginning in 1997 for Parkinson's disorder (PD) tremor. Since then, additional approvals for advancing stages have been awarded (Gardner, 2013). Chapter 11, Future Directions in Deep Brain Stimulation, of this book discusses the future directions for use of DBS. The most well-known and researched role of the neuropsychology in DBS for movement disorders has been with presurgical evaluations (Kubu, 2018); Chapter 10, The Role of the Neuropsychologist in Deep Brain Stimulation, of this book outlines this role in detail.

Support for the effectiveness of DBS for PD-related movement disorder, up to 36 months, was shown in a recent metaanalysis (Mansouri, Taslimi, & Badhiwala, 2018). However, a more stringent umbrella review of multiple metaanalytic studies recently challenged this claim (Papageorgiou, Deschner,

& Papageorgiou, 2017). The differing results are likely related to the lack of guidelines for rehabilitation following DBS surgery along with variability in the neurorehabilitation settings. Stabilization of the DBS efficacy in PD usually requires several weeks to months. In general, patients tended to start rehabilitation 20 days after implantation and stayed in the facility 29 days (Allert, Cheeran, & Deuschl, 2018). The major focus of initial postsurgical intervention and rehabilitation is the adjustment of stimulation parameters with the consideration of a vast number of side effects such as dyskinesia, behavioral changes, cognitive problems, infection, paresthesia, muscle contractions, dysarthria or hypophonia, akinesia, vision changes, etc. (Allert et al., 2018). While the level of evidence to support PT, Lee Silverman Voice Treatment, and speech therapy is established with PD patients, there is less data to support the efficacy of these treatments with PD patients after DBS. However, in a recent review of research on DBS rehabilitation Allert et al. (2018) suggested that PT is "probably" effective (limited populations or one single study). Other promising treatments noted in their review, but with limited data for both PD and PD post DBS placement, include OT, talk therapy, behavior therapy, cognitive training, and nursing.

Given the cognitive decline associated with the majority of PD patients, this area should not be ignored in presurgical evaluation and rehabilitation following DBS placement for this disease. While there are no standardized recommendations for neuropsychological/cognitive rehabilitation following DBS placement, the benefits of cognitive training in patients with PD irrespective of DBS placement have been shown. In a 2015 systematic review, Leung et al., found evidence to suggest that cognitive training lead to measurable improvements in working memory, executive functioning, and processing speed with PD patients. Depression is also fairly common in PD, with estimates of 50%−70% having this comorbidity. Both antidepressant medication and cognitive-behavioral therapy (CBT) have been found to be effective treatments for depression with PD patients via systematic review of the research (Bomasang-Layno, Fadlon, Murray, & Himelhoch, 2015). Neuropsychologists specializing in rehabilitation and/or rehabilitation psychologists are uniquely qualified to provide both cognitive rehabilitation training and CBT to address both of these common comorbidities in PD patients as well as to involve family education and interventions as needed.

FUTURE DIRECTIONS

Research indicates that early and intense neurorehabilitation provided by a multidisciplinary rehabilitation team within an appropriate rehabilitation environment can lead to clinical improvements, across almost all functional areas for patients following neurosurgical procedures greater than that noted for natural recovery alone (Thompson et al., 2013). However, the heterogeneity of the type and severity of deficits relative to the variety of conditions resulting in the need for neurosurgical interventions make it difficult to identify a single rehabilitation process for all neurosurgical patients. More work is needed to identify specific pre- and postsurgical pathways and processes that target the specific needs for the various diagnostic groups. This would include more short-term rehabilitation plans for epilepsy, DBS, and uncomplicated neuro-oncology cases, with more comprehensive plans for moderate to severe ABI and extensive neuro-oncology cases. With proper early rehabilitation interventions, neurosurgical patients have a greater potential to become more independent, have fewer postoperative complications, and, as some research has shown, have shorter lengths of stay and decreased costs. It is important that any rehabilitation protocols measure outcomes beyond surgical success and disease management, but also include improved function and quality of life.

Neuropsychologists are uniquely equipped to play a vital role in determining the rehabilitation needs for these various neurosurgical populations and can be an integral part of rehabilitation process/pathway development and research regarding the effectiveness of these plans. To help increase the potential for good postsurgical outcomes, neuropsychologists can also assist in presurgical preparation including evaluation, education, and expectation management. Setting the stage for recovery and rehabilitation prior to the surgery can also help in the development of appropriate coping and adjustment strategies to manage potential problems before they occur and develop resiliency to help manage any difficulties that do occur postsurgically. Prerehabilitation education and planning along with structured pathways/processes set the stage for improved outcomes following neurosurgical interventions regardless of the necessitating condition.

REFERENCES

Allert, N., Cheeran, B., Deuschl, G., et al. (2018). Postoperative rehabilitation after deep brain stimulation surgery for movement disorders. *Clinical Neurophysiology, 129*(3), 592–601.

American Congress of Rehabilitation Medicine (n.d.). *Brain Injury—Interdisciplinary Special Interest Group (BI-ISIG)*. Retrieved August 14, 2018, from https://acrm.org/acrm-communities/brain-injury/.

Bartolo, M., Zucchella, C., Pace, A., et al. (2012). Early rehabilitation after surgery improves functional outcome in inpatients with brain tumours. *Journal of Neuro-oncology, 107,* 537–544.

Ben-Yishay, Y., & Diller, L. (2011). *Handbook of holistic neuropsychological rehabilitation: Outpatient rehabilitation of traumatic brain injury.* New York, NY: Oxford University Press.

Boake, C. (1989). A history of cognitive rehabilitation of head-injured patients, 1915 to 1980. *Journal of Head Trauma Rehabilitation, 4*(3), 1–8.

Bomasang-Layno, E., Fadlon, I., Murray, A. N., & Himelhoch, S. (2015). Antidepressive treatments for Parkinson's disease: A systematic review and meta-analysis. *Parkinson Related Disorder, 21*(8), 833–842.

Bradley, L. J., Kirker, S. G. B., Corteen, E., Seeley, H. M., Pickard, J. D., & Hutchinson, P. J. (2006). Inappropriate acute neurosurgical bed occupancy and short term falls in rehabilitation: Implications for the National Service Framework. *British Journal of Neurosurgery, 20*, 36–39.

Cappa, S. F., Benke, T., Clarke, S., Rossi, B., Stemmer, B., & van Heugten, C. M. (2005). EFNS guidelines on cognitive rehabilitation: Report of an EFNS task force. *European Journal of Neurology, 12*, 665–680.

Cheville, A. L. (2011). Cancer rehabilitation. In L. B. Braddon, L. Chan, M. A. Harrast, et al. (Eds.), *Physical Medicine and Rehabilitation* (4th ed, pp. 1371–1401). Philadelphia, PA: Saunders.

Cicerone, K. D., Langenbahn, D. M., Braden, C., Malec, J. F., Kalmar, K., Fraas, M., et al. (2011). Evidence-based cognitive rehabilitation: Updated review of the literature from 2003 through 2008. *Archives of Physical Medicine and Rehabilitation, 92*, 519–530.

Clusmann, H., Schramm, J., Kral, T., Helmstaedter, C., Ostertun, B., Fimmers, R., & Elger, C. E. (2002). Prognostic factors and outcome after different types of resection for temporal lobe epilepsy. *Journal of Neurosurgery, 97*(5), 1131–1141.

Cowen, T. D., Meythaler, J. M., DeVivo, M. J., Ivie, C. S., Lebow, J., & Novack, T. A. (1995). Influences of early variables in traumatic brain injury on functional independence measures scores and rehabilitation length of stay and charges. *Archives of Physical Medicine and Rehabilitation, 76*, 797–803.

Dams-O'Connor, K., & Gordon, W. A. (2010). Role and impact of cognitive rehabilitation. *Psychiatric Clinics of North America, 33*, 893–904.

Ford, C. L. (2017). Mood. In R. Winson, B. A. Wilson, & A. Bateman (Eds.), *The Brain Injury Rehabilitation Workbook* (pp. 204–234). The Guilford Press.

Gardner, J. (2013). A history of deep brain stimulation: Technological innovation and the role of clinical assessment tools. *Social Studies of Science, 43*(5), 707–728.

Greenwood, R. J., Strens, L. H. A., Watkin, J., Losseff, N., & Brown, M. M. (2004). A study of acute rehabilitation after head injury. *British Journal of Neurosurgery, 18*, 462–466.

Hannay, H. J., Bieliauskas, L. A., Crosson, B. A., Hammeke, T. A., Hamsher, KdeS., & Koffler, S. P. (1998). Proceedings: The Houston conference on specialty education and training in clinical neuropsychology. *Archives of Clinical Neuropsychology, 13*(2), 157–158.

Helmstaedter, C., & Elger, C. E. (1996). Cognitive consequences of two thirds anterior temporal lobectomy on verbal memory in 144 patients: A three-month follow-up study. *Epilepsia, 37* (2), 171–180.

Helmstaedter, C., Loer, B., Wohlfahrt, R., Hammen, A., Saar, J., & Steinhoff, B. J. (2008). The effects of cognitive rehabilitation on memory outcome after temporal lobe epilepsy surgery. *Epilepsy Behavior, 12*(3), 402–409.

Hendriks, M. P. (2001). Neuropsychological compensatory strategies for memory deficits in patients with epilepsy. *Current Problems in Epilepsy, 16*, 87–94.

Hermann, B. P., Seidenberg, M., Haltiner, A., & Wyler, A. R. (1995). Relationship of age at onset, chronologic age, and adequacy of preoperative performance to verbal memory change after anterior temporal lobectomy. *Epilepsia, 36*(2), 137–145.

Kim, R. B., Chun, M. H., & Han, E. Y. (2012). Fatigue assessment and rehabilitation outcomes in patients with brain tumors. *Support Care Cancer, 20*, 805–812.

Kirshblum, S., O'Dell, M. W., Ho, C., & Barr, K. (2001). Rehabilitation of persons with central nervous system tumors. *Cancer, 92*(4 Suppl), 1029–1038.

Kos, N., Kos, B., & Benedicic, M. (2016). Early medical rehabilitation after neurosurgical treatment of malignant brain tumours in Slovenia. *Radiology Oncology, 50*(2), 139–144.

Kreutzer, J. S., Kolakowsky-Hayner, S. A., Demm, S. R., & Meade, M. A. (2002). A structured approach to family intervention after brain injury. *Journal of Head Trauma Rehabilitation*, *17*(4), 349−367.

Kubu, C. S. (2018). The role of a neuropsychologist on a movement disorders deep brain stimulation team. *Archives of Clinical Neuropsychology*, *33*(3), 365−374.

Lee, G. P., Park, Y. D., Westerveld, M., Hempel, A., Blackburn, L. B., & Loring, D. W. (2003). Wada memory performance predicts seizure outcome after epilepsy surgery in children. *Epilepsia*, *44*(7), 936−943.

Mainio, A., Hakko, H., Niemela, A., Koivukanga, S. J., & Rasanen, P. (2005). Depression and functional outcome in patients with brain tumours: A population-based 1-year follow-up study. *Journal of Neurosurgery*, *103*, 841−847.

Mansouri, A., Taslimi, S., Badhiwala, J. H., et al. (2018). Deep brain stimulation for Parkinson's disease: Meta-analysis of results of randomized trials at varying lengths of follow-up. *Journal of Neurosurgery*, *128*(4), 1199−1213.

McAllister, T. W. (2013). Emotional and behavioral sequelae of traumatic brain injury. In N. D. Zasler, D. L. Katz, & R. D. Zafonte (Eds.), *Brain Injury Medicine* (2nd ed, pp. 1034−1052). New York, NY: Demos Medical Publishing, LLC.

McMullen, W. J. (2011). Language dominant mesial temporal lobe epilepsy. In J. E. Morgan, I. S. Baron, & J. H. Ricker (Eds.), *Casebook of Clinical Neuropsychology* (pp. 382−388). Oxford University Press.

Mosca, C., Zoubrinetzy, R., Baciu, M., Aguilar, L., Minotti, L., Kahane, P., & Perrone-Bertolotti, M. (2014). Rehabilitation of verbal memory by means of preserved nonverbal memory abilities after epilepsy surgery. *Epilepsy & Behavior Case Reports*, *2*, 167−173.

Mukand, J. A., Blackinton, D. D., Cirncoli, M. G., Lee, J. J., & Santos, B. B. (2001). Incidence of neurologic deficits and rehabilitation of patients with brain tumors. *American Journal of Physical Medicine and Rehabilitation*, *80*, 346−350.

Noachtar, S., & Borggraefe, I. (2009). Epilepsy surgery: A critical review. *Epilepsy Behavior*, *15*(1), 66−72.

Nordvik, J. E., Walle, K. M., Nyberg, C., Fjell, A. M., Walhovd, K. B., & Westlye, L. T. (2014). Bridging the gap between clinical neuroscience and cognitive rehabilitation: The role of cognitive training, models of neuroplasticity and advanced neuroimaging in future brain injury rehabilitation. *NeuroRehabilitation*, *31*(1), 1−85.

Papageorgiou, P. N., Deschner, J., & Papageorgiou, S. N. (2017). Effectiveness and adverse effects of deep brain stimulation: Umbrella review of meta-analyses. *Journal of Neurological Surgery. Part A: Central European Neurosurgery*, *78*(2), 180−190.

Rohling, M. L., Faust, M. E., Beverly, B., & Demakis, G. (2009). Effectiveness of cognitive rehabilitation following acquired brain injury: A meta-analytic re-examination of Cicerone et al.'s (2000, 2005) systematic reviews. *Neuropsychology*, *23*(1), 20−39.

Sirois, H. M., Lavoie, A., & Dionne, C. E. (2004). Impact of transfer delays to rehabilitation of patients with severe trauma. *Archives of Physical Medicine and Rehabilitation*, *85*, 184−191.

Teasdale, G., & Zitnay, G. (2013). History of acute care and rehabilitation of head injury. In N. D. Zasler, D. L. Katz, & R. D. Zafonte (Eds.), *Brain Injury Medicine* (2nd ed, pp. 13−25). New York, NY: Demos Medical Publishing, LLC.

Thompson, J. H., Majumdar, J., Sheldrick, R., & Morcos, F. (2013). Acute neurorehabilitation versus treatment as usual. *British Journal of Neurosurgery*, *27*(1), 24−29.

Westerveld, M. (2014). Epilepsy and seizure disorders. In K. J. Stucky, M. W. Kirkwood, & J. Donders (Eds.), *Clinical Neuropsychology Study Guide and Board Review* (pp. 331–349). Oxford University Press.

Wiebe, S., Blume, W. T., Girvin, J. P., & Eliasziw, M. (2001). A randomized, controlled trial of surgery for temporal-lobe epilepsy. *New England Journal of Medicine, 345*(5), 311–318.

World Health Organization and the World Bank. (2011). *Rehabilitation. Chapter in* World Report of Disability. Malta: World Health Organization. <http://www.who.int/disabilities/world_report/2011/report.pdf> Accessed 08.12.18.

Yogarajah, M., Focke, N. K., Bonelli, S. B., Thompson, P., Vollmar, C., McEvoy, A. W., ... Duncan, J. S. (2010). The structural plasticity of white matter networks following anterior temporal lobe resection. *Brain, 133*(8), 2348–2364.

Chapter 14

Assessment of Functional Status After Neurosurgical Intervention

Michael W. Parsons[1] and Sally J. Vogel[2]

[1]*Massachusetts General Hospital/Harvard Medical School, Pappas Center for Neuro-Oncology & Psychological Assessment Center, Boston, MA, United States*, [2]*Advanced Behavioral Medicine, Tacoma, WA, United States*

HISTORY AND CURRENT PRACTICE

Neuropsychology has its roots in evaluating the severity of cognitive injury and aiding in lesion localization. As imaging techniques have developed, neuropsychology's role in localization has decreased; however, neuropsychology remains vital in understanding the cognitive and functional manifestations of illness and injury, including the general understanding of and individual differences in the clinical impact of a lesion. For neurosurgical patients, this includes the evaluation of an individual's strengths and weaknesses at baseline, changes through treatment and disease progression, how these factors may inform treatment considerations, and the evaluation of an individual's ability to function independently; ranging from an individual's ability to provide self-care to an individual's ability to return to work or to drive. Standard neuropsychological measures that were originally developed to aid in lesion localization have been extended for use in answering these questions. Additionally, measures have been created that are more specifically designed to assess real-world performance and abilities. Within the framework of functional ability, functional capacity and outcome have been separately identified and examined. Functional *outcome* is assessed via self-report questionnaires or clinician rated forms. Functional outcome can be considered measures of real-world performance, or what a person is doing in his/her daily life and may not necessarily reflect what they are capable of (Harvey, Velligan, & Bellack, 2007). Functional *capacity* is considered what a person is capable of given optimal circumstances (Harvey et al., 2007). Functional capacity is typically assessed using performance-based measures. These measures require an individual to complete various daily life tasks, such as making a shopping list or planning for a trip, and patients are rated

Neurosurgical Neuropsychology. DOI: https://doi.org/10.1016/B978-0-12-809961-2.00015-1

on his/her accuracy. Measures of both functional outcome and capacity may assess a number of skills, such as activities of daily living, social, vocational, or a combination.

Research has been conducted to determine the utility and validity of standard neuropsychological measures and functional ability measures to assess the likelihood of success or failure in various functional skills. This research will be summarized in the following sections regarding instrumental activities of daily living (IADL), employment, and driving, with particular focus on instrumental activities of daily living (IADL), employment, and driving. There are innumerous daily functional tasks, the summary of which would be beyond the scope of this chapter. The ones chosen to be reviewed are those that are most commonly researched and are most commonly requested by referrals.

Activities of Daily Living

Activities of daily living are typically separated into two categories: basic and instrumental. Basic activities of daily living include activities such as toileting, feeding, and bathing. IADL include more complex activities such as medication, appointment, and financial management. Patients, their loved ones, and their healthcare providers are often interested in the impact an illness or injury may have on the patient's ability to function independently. Broadly, there is a question of whether there has been a change in functional independence, and if so, to what extent. When there is a question regarding functional ability in a neuropsychological referral, it will generally be related to IADLs rather than ADLs, as ADL performance is more readily apparent to caregivers and treatment providers. IADL ability has important implications for diagnosis, treatment, and required level of care that can impact autonomy, among other things, making the accurate assessment and prediction of these abilities important. Loewenstein and Acevedo (Loewenstein & Acevedo, 2010) identify three elements to consider when evaluating whether cognitive ability is impacting IADL performance; causation, change, and specificity. The element of causation relates to the IADL difficulty being secondary to cognitive ability as opposed to other factors, such as physical limitations. The element of change refers to the comparison between an individual's ability to perform tasks currently to his/her ability in the past. A direct comparison in performance on assessment measures is often not possible as there is rarely a prior evaluation to which current performance can be compared. The clinical interview with the patient and informants will provide valuable information regarding what an individual was once capable of and whether there has been a change in their ability to complete those tasks. This element highlights the idea that if an individual never performed or mastered a skill in the past, his/her inability to complete it now is not sufficient evidence of a functional decline. It is important to evaluate for change in those skills that an individual once engaged in and determine whether a change has occurred. The element of specificity is the final element and it relates to the consideration that each IADL has multiple components and there are many avenues from which an individual may have difficulty completing the task. There may be task- (i.e., skills required to complete), person- (i.e., memory ability, occupational history, familiarity with task, etc.), and environment-specific (i.e., completing a well learned task in an unfamiliar environment) variables that are influencing a person's ability to complete a task.

Standard neuropsychological measures of global and domain specific cognitive abilities are often used to predict an individual's ability to perform IADL. A moderate relationship has been found between many neuropsychological measures and everyday functioning (Chaytor & Schmitter-Edgecombe, 2003; Farias, Harrell, Neumann, & Houtz, 2003). Global cognitive ability scores are generally the most reliable results and are often used to predict everyday functioning (Harvey, 2012; Pereira, Yassuda, Oliveira, & Forlenza, 2008). With regard to cognitive domains, executive functioning has been most extensively studied and has been found to relate to everyday functioning ability (Cahn-Weiner, Malloy, Boyle, Marran, & Salloway, 2000; Chaytor & Schmitter-Edgecombe, 2003). Memory has also frequently been found to predict everyday functioning (Ashendorf et al., 2017; Chaytor & Schmitter-Edgecombe, 2003; Wong, Wong, & Poon, 2011). Many of these studies have been conducted in aging populations as the question of independent living is often encountered in this cohort. Research in this area is lacking in a number of patient populations, including neurosurgical populations.

In addition to standard neuropsychological measures, methods of assessing functional ability that are more ecologically oriented and often more face valid are used. These include self or collateral report via interview and questionnaires, clinician rated forms, and performance-based measures of functional ability.

Self or collateral report: Important information pertaining to the individual and often his/her loved ones' perspectives on symptoms, functioning, and any changes that have occurred can be acquired through interview and questionnaires. There are many self- and collateral-report questionnaires of varying breadth and depth to choose from depending on the intended purposes. A few of these questionnaires are described in the following paragraphs but there are many others as well.

The frontal systems behavior scale [FrSBe; (Grace & Malloy, 2001)] is a rating scale of behaviors associated with frontal dysfunction, specifically apathy, disinhibition, and executive dysfunction. There are patient and collateral informant versions. Severity ratings are assigned to the presence of behaviors both before and after the injury or illness to allow for evaluation of change. The FrSBE has been found to be predictive of neuropsychological performance (Basso et al., 2008) and functional outcome (Basso et al., 2008; Reid-Arndt, Nehl, & Hinkebein, 2007).

The IADL scale (Lawton & Brody, 1969) is an eight-item questionnaire evaluating eight domains of functioning. The measure is quick to administer and can be completed by the patient and/or an informant. The functional activities questionnaire [FAQ; (Pfeffer, Kurosaki, Harrah, Chance, & Filos, 1982)] consists of 10 questions regarding level of assistance required in completing IADL.

While patient and collateral report are informative and have been shown to relate to neuropsychological performance and functional ability, they each have limitations. Patients' and informants' estimations of cognitive and functional ability may be impacted by various factors, resulting in possible over- or underestimations of abilities. Some factors that may impact such reports include neurological (i.e., anosognosia), emotional (i.e., depression), secondary-gain (i.e., litigation), or other factors. In their review of the ecological validity of neuropsychological measures, Chaytor and Schmitter-Edgecombe (Chaytor & Schmitter-Edgecombe, 2003) found self-report measures to more poorly relate to everyday functioning than informant or clinician rated measures in neurologically impaired individuals.

Performance-based: Performance-based tests of functional capacity require the patient to complete a number of everyday tasks and are rated on his/her accuracy.

The Texas Functional Living Scale [TFLS; (Cullum, Weiner, & Saine, 2009)] was initially designed as a performance-based measure of functional ability to be used with patients with Alzheimer's disease. It provides an overall T-score composed of four subscales evaluating various areas of IADL: (1) time, (2) money and calculation, (3) communication, and (4) memory.

The executive function performance test [EFPT; (Baum et al., 2008)] evaluates a patient's initiation, execution, and completion of four tasks, cooking, telephone use, medication management, and bill payment. The manual and materials can be downloaded for free online through the following link: http://www.ot.wustl.edu/about/resources/executive-function-performance-test-efpt-308

The UCSD performance-based skills assessment [UPSA; (Patterson, Goldman, McKibbin, Hughs, & Jeste, 2001)] was initially developed for use with patients with severe mental illness. Patients are asked to complete a number of tasks in five areas: Planning recreational activities, finance, communication, transportation, and household chores.

See Moore et al. (Moore, Palmer, Patterson, & Jeste, 2007) for a review of additional performance-based measures of everyday functioning. As with standard neuropsychological measures various factors can negatively impact performance, including motivation, secondary gain such as litigation, premorbid functioning, and time since injury or surgery to assessment. These measures have also been criticized for their length, many being time consuming and difficult to incorporate into limited available testing time. The artificial testing environment adds structure and task initiation that may be less available in everyday life as well as removes distraction and in many ways reduces the level of decision making that would be necessary without the instruction or materials that are present during testing. Some of the tasks are also becoming outdated, such as writing a check as online bill pay is becoming increasingly common.

A combination of traditional and ecologically oriented measures may better evaluate everyday functioning ability (Loewenstein, Rubert, Arguelles, & Duara, 1995). Research has found that while neuropsychological measures are associated with functional abilities, further specificity and additional variance can be accounted for by utilizing more ecologically oriented measures related to the task in question (Marcotte et al., 2006; Marcotte et al., 2004). Scaled scores have been recommended as the metric of choice in predicting everyday functioning to ease comparison across measures and to allow for the creation of summary scores (Marcotte, Scott, Kamat, & Heaton, 2010).

Driving Ability

There were 37,461 traffic related fatalities nationally in 2016 with an estimated economic cost of $242 billion in crash related expenses in 2010 based on the most recent reported statistics by the National Highway Traffic Safety Administration ("Traffic safety facts 2016", 2018). Driving has become an integral part of many individual's daily lives. It is a complex task requiring numerous cognitive skills, including sustained and divided attention, processing speed, visuospatial skills, inhibition, and problem solving to name a few. A variety of acute and chronic illnesses, injuries, or functional states (i.e., level of arousal, substance use, mood) that negatively impact cognitive functioning can in turn have a negative impact on driving ability.

Findings regarding the relationship between performance on standard neuropsychological measures and on-road as well as driving simulation performance are mixed (Marcotte & Scott, 2009) resulting in a lack of consensus on the best methods for predicting driving ability. Overall, cognitive impairment has been associated with poorer driving ability and the risk of a crash increases as the level of cognitive impairment increases (Withaar, Brouwer, & van Zomeren, 2000).

The neuropsychological evaluation is an important component of the multidisciplinary assessment of driving. There are multiple cognitive abilities that contribute to an individual's ability to drive safely, including visuospatial perception, shifting attention, and information processing speed. When questions about driving safety are a concern for a patient undergoing neuropsychological evaluation, the evaluation should be tailored to the specific illness or injury potentially impacting driving ability. For example, in a patient who underwent craniotomy for resection of a tumor from the right parietal lobe, assessment of visual perceptual abilities, hemispatial neglect, and the ability to shift attention would be important components of the neuropsychological evaluation. Additional questions to consider as part of the clinical interview and assessment include insight, mental health, driving experience, and premorbid personality characteristics. If the neuropsychological evaluation demonstrates very obvious deficits in some of these core capabilities, it may be appropriate to medically restrict a patient from driving. However, it is more commonly the case that a multidisciplinary evaluation of driving to make a determination if deficits are subtle on neuropsychological testing.

Although it is an important component of the driving evaluation process, neuropsychological testing in isolation may not be sufficient to assess important medical, physical, sensory, or other difficulties that can negatively impact driving ability. A qualitative and quantitative multidisciplinary approach to driving assessment has been suggested, with evaluations by neuropsychologists, physicians, occupational therapists, and driving specialists (Lundqvist, Alinder, Modig-Arding, & Samuelsson, 2011). Lundqvist and colleagues (2011) discuss this approach as used in a rehabilitation setting with 43 patients with congenital and acquired brain injuries. They found that a step-wise procedure using both qualitative and quantitative information from multidisciplinary assessments was the most accurate assessment of driving ability. They assert that no single test was able to provide enough information to make a determination of an individual's ability to drive safely.

Return to Other Activities

There are a potentially unlimited number of other activities that one may be interested in an individual's ability to return to following a major surgery, illness, or injury. Many such individualized special activities are covered in the next section regarding return to work. However, there are numerous non-work activities regarding which an individual's capacity may be of interest, including returning to school, parenting, volunteer activities, etc. In general, research suggests that executive functioning performance may be the best and most consistent predictor of everyday functioning (Chaytor & Schmitter-Edgecombe, 2003). Specific cognitive and functional skills to be assessed will depend on the functional skill in question and evaluations will need to be tailored appropriately.

Limitations to Ecological Validity

Although neuropsychological tests and procedures provide useful methods for dissecting cognitive function into domains and associating those abilities with neural structures and systems, they are less well suited to evaluating function in the real world. The carefully controlled and standardized methods that allow the tests to reliably measure function (e.g., administering the task the same way to every individual) also remove many aspects of normal behavior from the performance of cognitive tests. For instance, tests of executive function remove much of the spontaneity and initiative from task performance because the tester initiates the activity, redirects the subject, and sequences the steps. In the real world, making such decisions is an inherent part of performing a complex task, such as making decisions in one's job or taking care of a family at home. Furthermore, testing environments are carefully controlled to prevent nuisance factors from interfering with test performance and reducing the applicability of normative data that were gathered without any such interference. These distractions are a normal part of everyday life (e.g., interruptions, unexpected changes, and competing demands on one's attention) and being able to perform daily tasks often depends on the ability to simultaneously manage them (Poncet, Swaine, Dutil, Chevignard, & Pradat-Diehl, 2017). Although many of the "performance based" measures described earlier, such as the EFPT (Baum, Morrison, Hahn, & Edwards, 2003) have been normed and standardized in neurologic samples (Baum et al., 2016), they tend to be time consuming to administer and are rarely used in neuropsychological assessments. Furthermore, these tools have typically been developed for and normed on neurologic injury populations that are commonly seen in rehabilitation settings (e.g., traumatic brain injury, stroke) or those populations for whom functional independence is the primary concern (dementia), and their applicability to neurosurgical populations is not known.

PRACTICAL APPLICATIONS

Providing assessment for functional outcomes in neurosurgical practice is most likely to occur if there is close interaction between neurosurgeons, neuropsychologists, and rehabilitation professionals. To optimize the likelihood of positive functional outcomes, planning should begin prior to the neurosurgical procedure, and follow up will need to continue after the physical recovery from surgery for many patients. A collaborative relationship between the neuropsychologist and neurosurgeon will promote realistic expectations for both surgical outcomes and functional outcomes on the part of the patient and family. The following sections provide general guidelines for an integrated practice model and an example of a multidisciplinary neurosurgery/neuropsychology clinic.

Guidelines for Integrated Neuropsychology and Neurosurgery

The value of baseline assessment: When attempting to evaluate an individual's ability to return to a specific functional role after neurosurgery, it is extremely helpful to set a preoperative baseline. Of course, neuropsychological function prior to surgery is often influenced by the underlying disease (e.g., brain tumor), sequelae related to the illness (e.g., seizures), and medications used to treat that condition (e.g., steroids, antiepileptic drugs). Nonetheless, a preoperative evaluation provides valuable quantification of cognitive abilities that may be used for comparison postoperatively. If the individual was functioning in their role effectively prior to surgery, a return to their neuropsychological baseline provides strong support for the hypothesis that the individual can also return to that role.

In some situations, preoperative evaluation is not possible due to the urgency of the neurosurgical condition or due to very significant cognitive impairment, such as in the case of traumatic brain injury with dangerous intracranial hemorrhage or high grade brain tumor and herniation. However, neuropsychological evaluation prior to and following surgery is known to have a role in the functional outcome of patients in less urgent situations, such as those with slow growing mass lesions (Mandonnet et al., 2015) or candidates for epilepsy surgery (Loring, Hermann, & Cohen, 2010) or deep brain stimulator surgery. We strongly recommend that standard of care include a baseline neuropsychological evaluation by a board-certified neuropsychologist with specialty knowledge regarding the neurosurgical condition.

Identify tests that evaluate cognitive capacities required by the patient's occupation or other activity: Postoperative neuropsychological evaluations are often useful in identifying deficits that are known to be predictive of difficulties in important daily functions. Studies of the relationship between neuropsychological outcome and functional outcome are most commonly focused on the ability to return to work. For example, in a single institution case series, Mandonnet and colleagues (Mandonnet et al., 2015) completed cognitive evaluations in 25 patients with glioma shortly after surgery. When possible, these individuals had also completed baseline assessments (some could not due to the urgency of their presentation, typically with glioblastoma). Of the 3 patients who showed clear deterioration in function after surgery, all had significant difficulties returning to work and 2 were unable to successfully do so at all. This study built on prior evidence that information gleaned from neurocognitive assessment performed during intraoperative mapping of language function predicted neuropsychological performance and the likelihood of returning to employment after surgery (Moritz-Gasser, Herbet, Maldonado, & Duffau, 2012).

An example of integrated neurosurgery and neuropsychology, the neuro-oncology model: The integration of neuropsychology into neurosurgical practice as routine provides the most efficient framework for evaluating functional status of neurosurgical patients. This integrated practice model has been in place for decades in multidisciplinary epilepsy surgery centers, and is becoming accepted as standard of care in brain tumor surgical programs as well (Fox, Mitchell, & Booth-Jones, 2006; Noll, Bradshaw, Rexer, & Wefel, 2018; Parsons, Das, & Recinos, In Press). Integrated practice models can allow for baseline assessment prior to surgery when practical, thus enabling comparison of postsurgical objective cognitive performance to the presurgical level of function. To keep pace with the rapid turnaround required by urgent brain tumor surgery cases, neuropsychologists typically use focused cognitive batteries at the early time points (e.g., presurgery or preradiation), followed by more comprehensive batteries after recovery (Lageman et al., 2010; Noll et al., 2018). The combination of pre- and post-treatment assessments provides a basis to assess change in function, while also allowing time in the second evaluation for the more comprehensive assessment necessary to examine higher level abilities that may be relevant to the patient's functional abilities.

Integration of this sort is challenging from a health-care delivery perspective. It requires neuropsychological practices that have open access for evaluations that must be scheduled in an urgent manner. The neuropsychological schedule must be flexible enough to accommodate varying assessment lengths, from 60−90 minutes brief evaluations up to 4−8 hours comprehensive assessments, depending on the needs of the patient. Furthermore, as detailed in the recent clinical description by Noll and colleagues (Noll et al., 2018), a specialized set of skills and experiences related to the neurosurgical condition is critical. Although few centers can boast the kind of integrated practice described at MD Anderson cancer center, a number of neuropsychologists are actively working through international organizations, such as the International Neuropsychological Society's Oncology Special Interest Group and the Children's Oncology Group, to enhance clinical program development, academic training, and research productivity related to cognitive outcomes in brain tumor patients.

CASE EXAMPLES

To clarify some of the issues that arise in predicting functional ability and managing postsurgical outcomes in brain tumor patients, we present the following cases. One has brief preoperative assessment available for comparison, while the other case has only a postsurgical evaluation and subsequent follow up with rehabilitation. These cases demonstrate some of the strengths and limitations of neuropsychological evaluation in predicting functional outcome.

Case 1: The patient is a 65-year-old structural engineer in good health with no prior cognitive issues. After returning from an international conference at which he had given the keynote address the week before, his assistant noticed facial asymmetry and left sided weakness. Upon presentation to ED, he was found to have an ~ 8 cm mass lesion centered in right temporal cortex with mass effect on the surrounding frontal and parietal cortices as well as midline shift. He was hospitalized and seen for brief neuropsychological evaluation in the ICU the day prior to undergoing surgery.

At the time of the baseline evaluation, the patient reported no cognitive symptoms. He had been working full time and had received exemplary feedback on his recent presentation, which included highly technical details related to his profession. A family member was also interviewed who agreed that there had been no obvious problems with the patient's memory, concentration, speech/language, or perceptual abilities. She volunteered that he had seemed more short tempered and in the past several months, and seemed disengaged or unemotional at times.

A brief evaluation was conducted in the ICU. It was frequently interrupted by visits from nursing staff, well-wishers stopping by to see the patient, and background noise related to equipment and other patients in the ICU. The patient performed in the normal range on tests of language and memory, but performed below expectations on tests of information processing speed, visuospatial function, and executive function. These deficits were particularly notable given that the expected premorbid level of functioning in these domains would be well above average for this patient, given his high level of occupational and educational attainment.

The patient underwent resection and the mass was found to be a WHO Grade III Anaplastic Oligodendroglioma. Postoperative magnetic resonance imaging (MRI) showed a gross total resection of contrast enhancing tumor, with residual areas of diffuse infiltrating tumor versus edema. He was treated with concomitant chemotherapy and partial field radiation to the area of resection (daily treatment for 6 weeks). During that time, he was off work, checking in periodically from home on his email and some ongoing projects. After completion of his treatment, he expressed interest in returning to work full time as a structural engineer, involved in design and execution of large scale projects. The patient continued to feel that he had no change in cognitive function from his normal level of ability. Neuropsychological re-evaluation was requested by the neurosurgeon to contribute to the return-to-work discussion prior to release to return to full-time work in this high demand and high risk job.

The postoperative evaluation included the tests that had been given at baseline (using alternate versions of the task where available) in addition to a more extensive battery of tests that examine attention, visuospatial processing, memory, and executive function. In order to illustrate the test findings, many of the individual test scores were grouped by domain and averaged to create a composite score across five domains (Fig. 14.1). Although it is necessarily an oversimplification of the actual neuropsychological data, the graph serves a useful purpose in communicating results to patients and caregivers.

Based on the persistent weaknesses in visuospatial skills, executive function, and processing speed, we recommended a cautious approach to reintegration into the workplace. Specifically, we met with the patient and family members to map out a several week period of observationally based work.

The patient identified a colleague in his office who agreed to review some of his preliminary work and give direct feedback to the patient. The patient agreed to work under his colleague's supervision until such time as his performance at work could be understood. Ultimately, the patient decided to reduce his role at work to consultative, rather than directive. He continued to serve as a consultant, teacher, and mentor (Fig. 14.2).

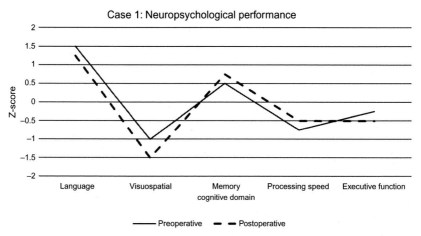

FIGURE 14.1 Neuropsychological performance across cognitive domains. The values plotted are average performance from multiple tests, displayed as Z-scores, both prior to and after surgical intervention. Negative values are below average compared with the population.

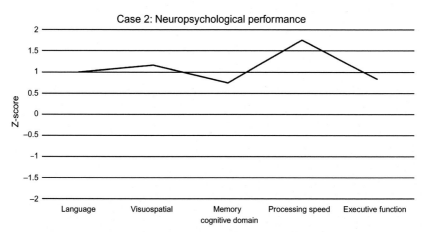

FIGURE 14.2 Neuropsychological performance across cognitive domains for case 2. The values plotted are average performance from multiple tests, displayed as Z-scores. Note that this individual is scoring above the population mean in all domains, as expected given her high baseline level of cognitive ability.

This case demonstrates several important principles regarding neuropsychological assessment and return to work:

- The patient was already showing deficits on neuropsychological testing at the time of initial presentation, likely due to extensive edema related to his large mass lesion. Nonetheless, these deficits had gone undetected in the workplace, other than some mild change in personality/emotion regulation.
- There is a correspondence between known aspects of neuroanatomy, cognitive performance, and the demands of the patient's occupation that are relevant to the case. Specifically, the right hemisphere lateralization of the tumor resulted in deficits in spatial skills, visual search, and spatial aspects of executive function. Given that his occupation required high level design capabilities, complex reasoning, and decision making, these specific deficits were relevant to his ability to perform crucial job tasks.
- There is no test of the capability to perform this patient's occupation. The neuropsychologist must use clinical judgement, knowledge of functional neuroanatomy, and clear interpersonal communication to provide input that is specific to this individual situation.
- In cases such as this one, a collaborative relationship between the patient, the workplace, and the neuropsychologist can help to navigate a situation in which there are unknowns (e.g., how the patient will perform) and to identify a workable solution that promotes the patient's well-being and removes risk of error or harm.

Case 2: The patient is a 28-year-old woman who had begun training to become an airline pilot when she was discovered to have a brain mass in the context of progressively worsening headaches. The lesion was in the anterior frontal pole on the right, measuring about 1 cm in diameter. It was felt to be an incidental finding, unrelated to her headaches, which were subsequently treated as typical migraines. The mass was resected (gross total resection) and found to be a low-grade astrocytoma. The neuro-oncology team recommended following the patient with serial imaging and no further treatment until progression. Postoperatively, the patient desired to return to her training program and work toward pilot licensure.

The Federal Aviation Administration (FAA) has developed a uniquely detailed set of guidelines for the management of neuropsychological problems in pilots and air traffic controllers ("Guide For Aviation Medical Examiners; Disease Considerations, Disease Protocols, Psychiatric, and Psychological Evaluations; Specification for Psychiatric and Psychological Evaluations", 2018). The guidelines were developed based on research that has sought to identify core cognitive underpinnings of the skills needed for flight (Kay, 2002). The FAA lists a specific battery of *required* neuropsychological tests. These include a computerized test that was developed specifically for this purpose and can only be administered by a certified neuropsychologist who has purchased the appropriate equipment (specific touch screen). The FAA also requires that the evaluation be administered and interpreted by a board certified or board eligible neuropsychologist. A complete list of test scores must be sent to the FAA for review by specialty neuropsychologists, who consult to the FAA and provide the opinion regarding fitness for duty.

The patient reported no cognitive symptoms either prior to or after the surgical intervention. Although no preoperative baseline was available, her neuropsychological test scores were all consistently above average, with strengths in processing speed and recent memory. The evaluation was conducted according to FAA guidelines, requiring about 4—5 hours of face-to-face testing by the neuropsychologist and psychometrist, in addition to interview time. All relevant information was submitted to the FAA for review and action related to her pilot training.

The case illustrates a unique approach to the problem of using laboratory based cognitive tests to predict performance in a specific real-world activity. The federal government contracted with specialists in the field of cognition and aviation to develop a method for evaluating pilot safety. The experts in this area have conducted normative studies demonstrating that airline pilots perform better than age-matched controls on specific measures of processing speed, attention, and visuospatial skill. Based on the assumption that any decrement in cognitive ability is associated with a high level of risk to the public, this professional group has mapped out a clear process by which cognitive data can be used to make a decision about fitness for duty.

FUTURE DIRECTIONS

Evaluation of functional status in real-world activities is a challenging problem for clinicians who operate in medical centers, separated from the activity they wish to measure. There are many discrepancies between the testing environment used by neuropsychologist and the world in which the patient needs to perform their daily functions. As we have illustrated in this chapter, there are multiple techniques to approach this challenge, all of which could be further developed:

- Ecologically valid assessment procedures that either mimic or take advantage of real-world activities.
- Integration of ongoing evaluation, rehabilitation, and re-evaluation related to an individual's ability to perform a specific task.
- The development of empirically validated cognitive tests that measure key capabilities involved in performing a specified task, and explicit guidelines for determining capacity to perform by experts in that area.

In addition to the challenges of cognitive assessment, there are additional layers of complexity that relate to interpersonal behavior, social cognition, and interpersonal skills. As difficult as it is to measure cognitive ability, these social and emotional skills are more challenging to quantify and more vulnerable to nuisance variables. However, creative cognitive neuroscientists have been developing new tools (e.g., using virtual reality technology) to improve the validity of such assessments (Canty, Neumann, Fleming, & Shum, 2017). These abilities, colloquially known as "emotional intelligence," are critical to successful functioning in many life activities and are frequently disrupted by neurosurgical problems such as brain tumor, epilepsy, and cerebrovascular disorders.

Given the extreme demand for their specialized skills, it is unrealistic for neurosurgeons to expect to care for the long term needs that are common in their patient populations. Furthermore, the need to evaluate functional abilities in a naturalistic setting and the complex psychometric issues that are inherent in such activities are not the forte of neurosurgeons. Building multidisciplinary treatment teams for neurosurgical patient populations is the best way to ensure a continuum of care that promotes optimal functional outcome after neurosurgery. We believe that frequent integration of neuropsychological evaluation, including preoperative baseline assessments whenever possible, is a good strategy for improving functional outcome. Continued collaboration using a multidisciplinary team model will be essential to improving our ability to predict, identify, and assist patients with difficulties functioning in their everyday activities.

REFERENCES

Ashendorf, L., Alosco, M. L., Bing-Canar, H., Chapman, K. R., Martin, B., Chaisson, C. E., et al. (2017). Clinical utility of select neuropsychological assessment battery tests in predicting functional abilities in dementia. *Archives of Clinical Neuropsychology*, 1–11.

Basso, M. R., Shields, I. S., Lowery, N., Ghormley, C., Combs, D., Arnett, P. A., et al. (2008). Self-reported executive dysfunction, neuropsychological impairment, and functional outcomes in multiple sclerosis. *Journal of Clinical and Experimental Neuropsychology*, *30*(8), 920–930.

Baum, C., Morrison, T., Hahn, M., & Edwards, D. (2003). *Test manual: Executive function performance test*. St. Louis, MO: Washington University.

Baum, C. M., Connor, L. T., Morrison, T., Hahn, M., Dromerick, A. W., & Edwards, D. F. (2008). Reliability, validity, and clinical utility of the Executive Function Performance Test: A measure of executive function in a sample of people with stroke. *The American Journal of Occupational Therapy*, *62*(4), 446–455.

Baum, C. M., Wolf, T. J., Wong, A. W. K., Chen, C. H., Walker, K., Young, A. C., et al. (2016). Validation and clinical utility of the executive function performance test in persons with traumatic brain injury. *Neuropsychological Rehabilitation*, *27*(5), 603–617.

Cahn-Weiner, D. A., Malloy, P. F., Boyle, P. A., Marran, M., & Salloway, S. (2000). Prediction of functional status from neuropsychological tests in community-dwelling elderly individuals. *Clinical Neuropsychology*, *14*(2), 187–195.

Canty, A. L., Neumann, D. L., Fleming, J., & Shum, D. H. K. (2017). Evaluation of a newly developed measure of theory of mind: The virtual assessment of mentalising ability. *Neuropsychological Rehabilitation, 27*(5), 834−870.

Chaytor, N., & Schmitter-Edgecombe, M. (2003). The ecological validity of neuropsychological tests: A review of the literature on everyday cognitive skills. *Neuropsychology Review, 13* (4), 181−197.

Cullum, C. M., Weiner, M. F., & Saine, K. (2009). *Texas functional living scale professional manual*. San Antonio, TX: Pearson Assessments.

Farias, S. T., Harrell, E., Neumann, C., & Houtz, A. (2003). The relationship between neuropsychological performance and daily functioning in individuals with Alzheimer's disease: Ecological validity of neuropsychological tests. *Archives of Clinical Neuropsychology, 18* (6), 655−672.

Fox, S. W., Mitchell, S. A., & Booth-Jones, M. (2006). Cognitive impairment in patients with brain tumors: Assessment and intervention in the clinic setting. *Clinical Journal of Oncology Nursing, 10*(2), 169−176.

Grace, J., & Malloy, P.F. (2001), *Frontal systems behavior scale professional manual*.

Guide for aviation medical examiners; disease considerations, disease protocols, psychiatric and psychological evaluations; specification for psychiatric and psychological evaluations. (2018). Retrieved July 6, 2018, from <https://www.faa.gov/about/office_org/headquarters_offices/avs/offices/aam/ame/guide/dec_cons/disease_prot/ppevals/>

Harvey, P. D. (2012). Clinical applications of neuropsychological assessment. *Dialogues in Clinical Neuroscience, 14*(1), 91−99.

Harvey, P. D., Velligan, D. I., & Bellack, A. S. (2007). Performance-based measures of functional skills: Usefulness in clinical treatment studies. *Schizophrenia Bulletin, 33*(5), 1138−1148.

Kay, G. G. (2002). Guidelines for the psychological evaluation of air crew personnel. *Occupatinal Medicine, 17*(2), 227−245, iv.

Lageman, S. K., Cerhan, J. H., Locke, D. E., Anderson, S. K., Wu, W., & Brown, P. D. (2010). Comparing neuropsychological tasks to optimize brief cognitive batteries for brain tumor clinical trials. *Journal of Neurooncology, 96*(2), 271−276.

Lawton, M. P., & Brody, E. M. (1969). Assessment of older people: Self-maintaining and instrumental activities of daily living. *Gerontologist, 9*(3), 179−186.

Loewenstein, D., & Acevedo, A. (2010). The relationship between instrumental activities of daily living and neuropsychological performance. In T. D. Marcotte, & I. Grant (Eds.), *The neuropsychology of everyday functioning* (pp. 93−112). New York: Guilford Press.

Loewenstein, D. A., Rubert, M. P., Arguelles, T., & Duara, R. (1995). Neuropsychological test performance and prediction of functional capacities among Spanish-speaking and English-speaking patients with dementia. *Archives of Clinical Neuropsychology, 10*(2), 75−88.

Loring, D. W., Hermann, B. P., & Cohen, M. J. (2010). Neuropsychological advocacy and epilepsy. *Clinical Neuropsychology, 24*(3), 417−428.

Lundqvist, A., Alinder, J., Modig-Arding, I., & Samuelsson, K. (2011). Driving after brain injury: A clinical model based on a quality improvement project. *Psychology, 2*, 615−623.

Mandonnet, E., De Witt Hamer, P., Poisson, I., Whittle, I., Bernat, A. L., Bresson, D., et al. (2015). Initial experience using awake surgery for glioma: Oncological, functional, and employment outcomes in a consecutive series of 25 cases. *Neurosurgery, 76*(4), 382−389, discussion 389.

Marcotte, T. D., Lazzaretto, D., Scott, J. C., Roberts, E., Woods, S. P., & Letendre, S. (2006). Visual attention deficits are associated with driving accidents in cognitively-impaired HIV-infected individuals. *Journal of Clinical and Experimental Neuropsychology, 28*(1), 13−28.

Marcotte, T. D., & Scott, J. C. (2009). Neuropsychological performance and the assessment of driving behavior. In I. Grant, & K. Adams (Eds.), *Neuropsychological assessment of neuropsychiatric and neuromedical disorders* (3rd ed, pp. 652−687). New York, New York: Oxford University Press.

Marcotte, T. D., Scott, J. C., Kamat, R., & Heaton, R. K. (2010). Neuropsychology and the prediction of everyday functioning. In T. D. Marcotte, & I. Grant (Eds.), *Neuropsychology of everyday functioning* (pp. 5−38). New York, New York: The Guilford Press.

Marcotte, T. D., Wolfson, T., Rosenthal, T. J., Heaton, R. K., Gonzalez, R., Ellis, R. J., et al. (2004). A multimodal assessment of driving performance in HIV infection. *Neurology, 63* (8), 1417−1422.

Moore, D. J., Palmer, B. W., Patterson, T. L., & Jeste, D. V. (2007). A review of performance-based measures of functional living skills. *Journal of Psychiatric Research, 41*(1-2), 97−118.

Moritz-Gasser, S., Herbet, G., Maldonado, I. L., & Duffau, H. (2012). Lexical access speed is significantly correlated with the return to professional activities after awake surgery for low-grade gliomas. *Journal of Neurooncology, 107*(3), 633−641.

Noll, K. R., Bradshaw, M. E., Rexer, J., & Wefel, J. S. (2018). Neuropsychological practice in the oncology setting. *Archives of Clinical Neuropsychology, 33*(3), 344−353.

Parsons, M.W., Das, P., & Recinos, P. (In Press). Neurosurgery for Meningiomas. In K. Sanders (Ed.), *Physician's Field Guide to Neuropsychology: Collaborative Instruction Through Clinical Case Example*. New York, NY, USA: Springer.

Patterson, T. L., Goldman, S., McKibbin, C. L., Hughs, T., & Jeste, D. V. (2001). UCSD Performance-Based Skills Assessment: Development of a new measure of everyday functioning for severely mentally ill adults. *Schizophrenia Bulletin, 27*(2), 235−245.

Pereira, F. S., Yassuda, M. S., Oliveira, A. M., & Forlenza, O. V. (2008). Executive dysfunction correlates with impaired functional status in older adults with varying degrees of cognitive impairment. *International Psychogeriatrics, 20*(6), 1104−1115.

Pfeffer, R. I., Kurosaki, T. T., Harrah, C. H., Jr., Chance, J. M., & Filos, S. (1982). Measurement of functional activities in older adults in the community. *Journal of Gerontology, 37*(3), 323−329.

Poncet, F., Swaine, B., Dutil, E., Chevignard, M., & Pradat-Diehl, P. (2017). How do assessments of activities of daily living address executive functions: A scoping review. *Neuropsychological Rehabilitation, 27*(5), 618−666.

Reid-Arndt, S. A., Nehl, C., & Hinkebein, J. (2007). The frontal systems behaviour scale (FrSBe) as a predictor of community integration following a traumatic brain injury. *Brain Injury, 21*(13-14), 1361−1369.

Traffic safety facts 2016, (2018). Retrieved July 6, 2018, 2018, from <https://crashstats.nhtsa. dot.gov/Api/Public/ViewPublication/812554>.

Withaar, F. K., Brouwer, W. H., & van Zomeren, A. H. (2000). Fitness to drive in older drivers with cognitive impairment. *Journal of the International Neuropsychological Society, 6*(4), 480−490.

Wong, G. K. C., Wong, R., & Poon, W. S. (2011). Cognitive outcomes and activity of daily living for neurosurgical patients with intrinsic brain lesions: A 1-year prevalence study. *Hong Kong Journal of Occupational Therapy, 21*(1), 27−32.

Index

Note: Page numbers followed by "*f*" and "*t*" refer to figures and tables, respectively.

Printed in the United States
By Bookmasters